ISOTOPES: ESSENTIAL CHEMISTRY
AND APPLICATIONS

Special Publication No. 35

Isotopes: Essential Chemistry and Applications

The Lectures delivered at a Review Symposium organised by The Education Division of The Chemical Society in conjunction with The University of Surrey

University of Surrey, July 23rd-25th 1979

Edited by
J. A. Elvidge and J. R. Jones,
University of Surrey

The Chemical Society
Burlington House, London W1V 0BN

British Library Cataloguing in Publication Data

Isotopes. - (Chemical Society. Special publications;
   No.35   ISSN 0577-618X).
   1. Isotopes - Congresses
   I. Elvidge, John A   II. Jones, John Richards
   III. Chemical Society. Education Division
   IV. University of Surrey   V. Series
   541'.388   QD466.5

   ISBN 0-85186-830-4

Chem
QD
466.5
.I7
1980

Printed in Great Britain by
Whitstable Litho Ltd., Whitstable, Kent

Introduction

For the chemist considering the use of isotopes there are usually three aspects to be considered. These are (a) the preparation and purification of the labelled compound, (b) the determination of the degree and specificity of labelling by means of a suitable analytical technique and (c) the application. In the proceedings of the Review Symposium an attempt has been made to reflect the importance of these aspects. In some of the applications a stable isotope has been prefered, in others the first choice is a radioisotope. Time and time again one is struck by the complementary nature of the two kinds of isotopes and left to regret the absence of radioactive isotopes of oxygen and nitrogen of sufficiently long half-lives. That most of the attention is focussed on the isotopes of hydrogen and carbon is, in view of their importance in organic chemistry, inevitable.

In a single volume of this kind one can not hope to provide a comprehensive coverage of the large number of topics involved. It is however possible to illustrate the most important aspects of the use of isotopes by means of selected examples and this is what each author has striven to do.

In view of the increasing availability of compounds labelled with stable and radioactive isotopes, together with the important advances that have been made in instrumental methods of analysis, it seems likely that the number of applications involving isotopes will continue to grow and provide a fertile area of research in both chemistry and allied subjects. It is our hope that this volume can play a small part in this development.

> J. A. Elvidge
>
> J. R. Jones

v

CONTRIBUTORS

J.A. ELVIDGE.   Chemistry Department, University of Surrey.

E.A. EVANS.   The Radiochemical Centre Ltd., Amersham,
Buckinghamshire.

D.R. HAWKINS.   Huntingdon Research Centre, Huntingdon,
Cambridgeshire.

F. HIBBERT.   Department of Chemistry, Birkbeck College, London.

J.R. JONES.   Chemistry Department, University of Surrey.

I.M. LOCKHART.   BOC Ltd., Prochem, London.

D.S. MILLINGTON.   VG-Organic Ltd., Altrincham, Cheshire.

D.M. RACKHAM.   Lilly Research Centre Ltd., Windlesham, Surrey.

W.D. UNSWORTH.   VG-Organic Ltd., Altrincham, Cheshire.

D.W. YOUNG.   School of Molecular Sciences, University of Sussex.

CONTENTS

1.   Stable Isotopes - Separation and Application   1

     By I.M. Lockhart

     Introduction
     Expression of Isotopic Content
     Separation of Stable Isotopes
     Separation by Low Temperature Distillation
      Carbon Isotopes
      Nitrogen and Oxygen Isotopes
     Separation of Oxygen Isotopes by Distillation of
                         Water
     Separation of Isotopes by Chemical Exchange
     Separation by Thermal Diffusion
     Other Methods of Isotope Separation
     Compounds Labelled with Stable Isotopes
       Chemical Synthesis
       Preparation of Labelled Compounds by Biosynthesis
     Analytical Techniques
       Mass Spectrometry
       Nuclear Magnetic Resonance Spectroscopy
       Optical Emission Spectroscopy
       Infrared Spectroscopy
       Miscellaneous Methods
     Applications of Stable Isotopes
       Chemical Studies
       Applications in Biochemistry
       Applications in Pharmacology
       Clinical Applications of Stable Isotopes
       Agricultural Studies
       Environmental Applications

2.   Isotopic Labelling with Carbon-14 and Tritium   36

     By E.A. Evans

     Introduction
     Properties of Carbon-14 and Tritium

Labelling Techniques
  Types of Labelling
  General Methods
Purification and Analysis of Labelled Compounds
  Purification Methods
  Quality Control
Handling and Health Physics
Summary and Concluding Remarks

3.   Purity and Stability of Radiochemicals   67

     By E.A. Evans

     Introduction
     Purity
       Types of Purity
       Measurement of Purity
     Self-decomposition of Radiochemicals
       Factors Affecting Decomposition
       Mechanisms of Decomposition
       Decomposition of Macromolecules and Polymers
       Isotope Exchange during Storage of Tritium
                   Labelled Compounds in Aqueous Solutions
       Decomposition of L-[^{35}S]Methionine
       Decomposition of ^{32}P-Labelled Nucleotides
       Summary of Decomposition Rates
     Effects of Impurities in Tracer Experiments

4. ^{13}C N.M.R. Spectroscopy in Medical Chemistry   97

     By D.M. Rackham

     Introduction
     Historical Background
     Continuous Wave (CW) and Fourier Transform (FT) N.M.R.
     Information Derived from the ^{13}C N.M.R. Experiment
       The Chemical Shift
       Spin-Spin Splitting
       Integrated Peak Areas
       Spin-Lattice (T_1) Relaxation Measurements
       Effects of Chemical Exchange and Rate Processes
       Isotopic and Chemical Derivatisation
     Recent Progress in ^{13}C Methodology

Lanthanide Shift Reagents
Quantitative Analysis by ^{13}C N.M.R.
Analysis of Small Quantities
Two Dimension (2-D) N.M.R. Spectroscopy
^{13}C N.M.R. Spectra of Solids
Study of Biomolecules by ^{13}C N.M.R.
The Use of ^{13}C Labelled Materials in Studies of
    Biosynthetic Pathways
Structural Analysis of Biomolecules
Interactions of Bioactive Materials with Small
    Molecules

5. Deuterium and Tritium Nuclear Magnetic Resonance   123
    Spectroscopy

    By J.A. Elvidge

Introduction
    Proton Magnetic Resonance Spectroscopy
    Characteristics of ^{1}H N.M.R. Spectroscopy and
        Notes on the Interpretation of Spectra
    Nuclear Properties of the Proton
    Main Methods of Producing N.M.R. Spectra
^{2}H N.M.R. Spectroscopy
    Deuterium
    Nuclear Properties of the Deuteron and Character-
        istics of ^{2}H N.M.R.
    Applications of ^{2}H N.M.R.
    Summary
    N.M.R. Applications of ^{2}H as a Quadrupolar Nucleus
^{3}H N.M.R. Spectroscopy
    Tritium
    Nuclear Magnetic Properties of Tritium
    Characteristics of ^{3}H N.M.R. Spectroscopy
    Applications of ^{3}H N.M.R.
    Summary
Conclusions

6. Mass Spectrometric Methods of Isotope Analysis   195

    By D.S. Millington and W.D. Unsworth

Introduction

Part I. Isotope Ratio Mass Spectrometry
   The Notation
   General Description of the Instruments
   Resolution and Abundance Sensitivity
      Resolution
      Abundance sensitivity
   Stable Isotope Enrichment Spectrometers
      Sample introduction
      Ionisation method
      Ion detection
   Applications of S.I.E. Spectrometers
      Pollution monitoring
      Monitoring of food substitutes
      Geochemistry
      Medical applications
      Analysis of nuclear fuel and waste
   Thermal Ionisation Mass Spectrometers
      The ionisation process
   Applications of T.I. Spectrometry
      Atomic weights and nuclear constants
      Age determination
      Pollution monitoring
Part II. Organic Mass Spectrometric Methods
   Determination of the Isotopic Composition of
        Molecules by Accurate Mass Measurement
   Precision of Measurement of Relative Isotopic
        Abundance
   Elemental composition by isotope ratio measurement
   Tracer Experiments with Isotope Labelling
   Other Applications of Isotope Labelling in Metabolic
        Studies
   Isotope Labelling in Organic Mass Spectrometry
   Quantitative Mass Spectrometry by Isotope Dilution
      Quantification from the direct insertion probe
      Quantification using the gas chromatograph
   Limit of Detection by G.C. - M.S. Assays using
        Isotope Dilution
      Methods of increasing selectivity
   Quantification using Field Desorption
   Scope of HPLC-MS in Quantification

7. Applications of Isotopes in Drug Metabolism   232

    By D.R. Hawkins

    Important Isotopes
      Radioisotopes
      Stable Isotopes
    Detection and Measurement of Isotopes
    Synthesis of Labelled Compounds
      Radioisotopes
      Stable Isotopes
    Radioisotope Studies in Animals
      Excretion Studies
      Tissue Distribution Studies
      Investigation of Metabolites
    Radioisotope Studies in Man
      Use and Problems Associated with Labile Radiolabels
      Non-metabolite Residues
    Stable Isotopes
      Detection and Identification of Metabolites
      Measurement of Drug and Metabolite Concentrations
      Mechanistic Studies
      Use of N.M.R. for Metabolite Identification

8. Applications of Isotopes in Biosynthesis   276

    By D.W. Young

    Introduction
      Radioactive Isotopes
      Stable Isotopes
    Isotopes of Carbon
    Isotopes of Hydrogen
    Isotopes of Oxygen
    Isotopes of Nitrogen
    Biosynthesis of Penicillins and Cephalosporin C

9. Isotopes and Organic Reaction Mechanisms   308

    By F. Hibbert

    Introduction
    Studies of Mechanism Using Isotope Labels

    Benzyne Intermediate
    Hydrolysis of Esters
    Rearrangement of Carbocations
    Enzyme Catalysed Phosphoryl Transfer
    Acid-Base Catalysed Proton Exchange in Carbon-
        Hydrogen Bonds
Kinetic Isotope Effects
    Introduction
    Hydrogen Isotope Effects on Proton Transfer
        Reactions
      Primary isotope effects
      Solvent isotope effects
    Isotope Effects on Elimination Reactions

10. Hydrogen Isotope Exchange Reactions   349

    By J.R. Jones

    Introduction
    Procedures
    Exchange Reactions
      Acid-catalysed
      Base-catalysed
      Metal-catalysed
      Photochemical and Radiation-induced
      Enzymic
      Miscellaneous
    Factors Affecting Exchange
    Conclusion

Stable Isotopes - Separation and Application

I.M. Lockhart

BOC Limited, Prochem, Deer Park Road, London SW19 3UF

Introduction

Following their discovery in the period 1927-1932, the heavy stable iso-
topes of hydrogen (^{2}H), carbon (^{13}C), nitrogen (^{15}N), oxygen (^{17}O and ^{18}O),
and sulphur (^{33}S and ^{34}S) rapidly assumed an important role in biochemical
studies.   Many examples of their use in the study of the metabolism of
proteins, carbohydrates, and lipids, were presented at a symposium held in
1947.[1]   The first supplies of ^{14}C had been made available in 1945 and this was
probably one of the earliest occasions on which work with radioisotopes was
also reported.   The subsequent widespread use of radioactive isotopes in a
multitude of applications resulted in a decline of interest in their stable
counterparts where cost and lack of instrumentation put them at a disadvantage.
However, since the mid-1960s, dramatic developments in quantitative detection
techniques, especially gas chromatography linked with mass spectrometry and of
Fourier transform nmr, have opened up new opportunities, which, coupled with
large reductions of the price of the isotopes, have led to a renaissance in
their use.   Much important work has been reported in the Proceedings of International Symposia[2-8] and review articles[9-12] which provide testimony to their
importance in the life sciences.

It is unlikely that stable isotopes will achieve the widespread utility
of their radioactive counterparts.   Nevertheless, in many circumstances, they
offer important advantages.   Using techniques such as ^{13}C-nmr, not only may
the label be detected, but its position can be determined in the intact
molecule.   In contrast, the non-specific detection methods used with radioisotopes require elaborate degradative investigations to define the location
of the tracer.   Furthermore, it has been claimed that the sensitivity of the
specific detection method can, in certain circumstances, be comparable with
that of the methods used with radioisotopes.[13]   With nitrogen and oxygen,
there are no competing radioisotopes with a half-life adequate for their widespread application.   There are no associated problems of radiation in the
synthesis and handling of labelled compounds, while in many clinical studies,
as well as some environmental applications, the lack of radioactivity is
particularly important.   The complementary use of stable and radioisotopes can
be a potent research tool, especially in biosynthetic studies.

In this review, methods used for the separation of stable isotopes will

be described.   The synthesis of labelled compounds, techniques for detection
and assay, and areas of application will also be discussed.   Particular
attention will be paid to the isotopes of carbon, nitrogen, and oxygen;   to
date, sulphur isotopes have only assumed a minor role.   The field of deuterium
chemistry is too extensive for adequate treatment;   it will therefore be
essentially excluded.

Expression of Isotopic Content

Isotope content is normally quoted as atom% (or mole%) which is defined
as the ratio of the number of atoms (or moles) of the isotope to the total
number of atoms (or moles) of that element expressed as a percentage.
Definition of the isotopic composition of a compound normally only refers to
the atoms in the labelled position.   For example, in sodium $[2-^{13}C]$ acetate
at 90 atom%, 90% of the methyl carbon atoms are ^{13}C while on the carbonyl
group ^{13}C only occurs to the extent of the natural abundance.   If both atoms
are labelled, as in sodium $[1,2-^{13}C]$ acetate, then, at 90 atom%, the statistical distribution of the various species is 81% doubly labelled, 8.5% of each
of the single labelled possibilities, and 2% of unlabelled material.

Isotopic enrichment refers to a change in isotopic composition and most
commonly represents the increase above the natural level;   it is expressed as
atom% excess.   Geochemists, oceanographers, and other environmentalists who
are involved in determining variations in the natural abundance of isotopic
species, express their results in the $\delta\permil$ notation.   This is the deviation
in parts per thousand of the isotopic ratio of the sample compared to a
reference substance;   in the case of ^{18}O, it can be represented by

$$\delta^{18}O = \frac{^{18}O/^{16}O \text{ (sample)} - {}^{18}O/^{16}O \text{ (standard)}}{^{18}O/^{16}O \text{ (standard)}} \times 1000$$

Care must be exercised in quantifying labelled compounds.   Whilst
reference to the total mass of a substance with a statement of its isotopic
content is common practice, various alternative forms are also used.   Definition of the mass of a compound in terms of grams of isotope will frequently
be encountered.   For example, 100g of $[^{15}N]$ ammonium sulphate at 10 atom%
^{15}N may alternatively be described as 2.3g ^{15}N as $[^{15}N]$ ammonium sulphate at
10 atom% ^{15}N.   Although the expressions are equivalent and each has its merits,
the indiscriminate use of such alternatives does lead to misunderstanding and
confusion.   Great care must therefore be exercised to ensure correct interpretation.

Separation of Stable Isotopes

The natural abundance of the heavy isotopes that form the subject of this review is as follows:

^{2}H 0.015, ^{13}C 1.1, ^{15}N 0.36, ^{17}O 0.04, ^{18}O 0.2, ^{34}S 4.3 atom%.

The various methods that have been successfully used for the separation of stable isotopes exploit the very small changes in the physical and chemical properties that are produced on substitution of a heavier isotope in a molecule. The basic principles have been well documented[9] and current commercial practices reviewed.[11] The lighter isotopes have also assumed some practical significance in recent years; in the course of this review, ^{12}C and ^{14}N denote carbon and nitrogen that have been depleted of ^{13}C and ^{15}N respectively while the natural material is referred to as carbon and nitrogen.

When two isotopes (A and A') are in equilibrium between two phases (I and II), the equilibrium constant of the reversible process in equation 1 will determine the feasibility of their fractionation.

$$AX + A'Y \rightleftharpoons A'X + AY$$

phase I phase II phase I phase II (1)

This constant, known as the separation factor, α, is represented by

$$\alpha = \frac{A'/A \text{ (phase I)}}{A'/A \text{ (phase II)}}$$

The principal methods that have been used commercially for the separation of the isotopes of carbon, nitrogen, oxygen, and sulphur are based on distillation, chemical exchange, and, to a lesser extent, on thermal diffusion. They are summarised in Table 1 and will be discussed in some detail. Reference will also be made to other methods that have been studied but have not as yet revealed any significant potential in this area.

Since there are only small differences in the properties of the isotopic species, the separation factors are close to unity and to obtain significant enrichments a large number of separation stages must be employed. The use of long fractionating columns or cascades of several units is therefore necessary, and the packing of the columns is of critical importance in the construction of a reliable unit. As continuous operation is essential, the plant must be fully automated with alarm systems incorporated. Finally, the construction and operation of isotope separation plants requires a high level of engineering and scientific expertise. These factors, together with the capital investment required, dictate that isotope separation is most economically achieved in an industrial-type unit; outputs are normally of the order of a few kg per annum.

Table 1.   Principal Methods for the Separation of Stable Isotopes

Isotope	Method of separation and material used		
	Distillation	Chemical exchange	Thermal diffusion
^{12}C	CO		
^{13}C	CO	CO_2/carbamate	CH_4
^{14}N	NO		
^{15}N	NO	NO_x/HNO_3	
^{16}O	NO		
^{17}O	NO, H_2O, D_2O		
^{18}O	NO, H_2O, D_2O, O_2		
^{34}S		SO_2/bisulphite	CS_2

Separation by Low Temperature Distillation

The separation of isotopes by distillation depends upon the small differences in the vapour pressure between the various labelled species of a substance.

Carbon Isotopes.   Most, if not all, of the world's ^{13}C is now produced by the cryogenic distillation of carbon monoxide.   The separation factor at the operating temperature is 1.008.   The separation plant basically consists of a boiler, fractionating columns, and a liquid nitrogen cooled reflux condenser as illustrated diagramatically in Figure 1.   In the system operated in the UK by Prochem,[12,14] carbon monoxide is freed from carbon dioxide and fed to two randomly packed, stepped columns which are 20m in length and vary in diameter from 6.5cm at the condenser to 2.5cm at the boiler end.   On a fixed time cycle, carbon monoxide at about 12 atom% ^{13}C is alternately transferred to a third column that has a diameter of 2.5cm throughout its length while high-boiling impurities such as methane remain in the boilers of the primary columns.   The high efficiency second stage unit lifts the enrichment to the range 90-93 atom% ^{13}C.   The three columns, mounted in a common vacuum jacket, are served by a single liquid nitrogen condenser.   The process is fully automated, and, since no bulky ancillaries are required at the boiler end, housing the columns in a 21m bore hole gives a compact, efficient unit.   An initial equilibrium time of 10-12 weeks is required on start-up which underlines the importance of

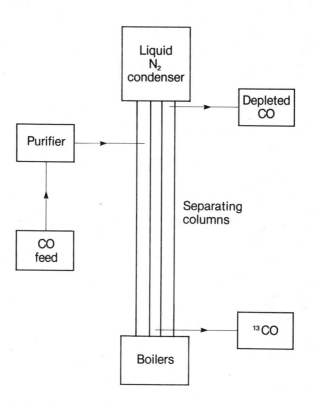

Figure 1.   Separation of ^{13}C by cryogenic distillation – schematic.

continuous running. The Prochem unit was maintained at working temperature for some six years of uninterrupted production on its initial run.

$[^{12}C]$ carbon monoxide at 99.95 atom% ^{12}C can be produced by rearranging the operation of the column or by feeding the waste gas to a similar unit for further fractionation.

Similar methods but employing a somewhat different column arrangement have been used for the separation of the carbon isotopes (^{12}C and ^{13}C) in the U.S.A. at the Los Alamos Scientific Laboratories.[15]

It is generally agreed that the maximum enrichment attainable in a process of this type is of the order of 93 atom% ^{13}C. The limiting factor is the difficulty of separating $^{13}C^{16}O$ from the $^{12}C^{18}O$ which also accumulates at the base of the column. This problem was partially overcome at Prochem by transferring the product to a second similar unit[16] when the isotopic content of the carbon monoxide was raised to 98 atom% ^{13}C. A recently commissioned plant at Los Alamos[17] employs a 250m column assembly for the initial fractionation. The product, which as previously contains some 5-10% of the $^{12}C^{18}O$ species, is passed through an exchanger consisting of a tungsten tube at $1000^{\circ}C$ when random exchange takes place and most of the ^{18}O is then present as $^{13}C^{18}O$. This material is finally fed to a 54m column suspended below the main column; the product from the bottom boiler reaches 99 atom% ^{13}C. Again, the whole unit is suspended in a hole in the ground.

It should be noted that since the feedstock for the ^{13}C separation processes derives from a geological source, there is no problem of contamination of the product with ^{14}C.

Nitrogen and Oxygen Isotopes. There are two important methods for the separation of the isotopes of nitrogen that are used commercially. The first, which utilises chemical exchange, is discussed later, and the second involves the cryogenic distillation of nitric oxide. The latter method is of particular interest as it enables the isotopes of oxygen to be separated simultaneously. Alternative separations of ^{17}O and ^{18}O by distillation of water or heavy water are described in the next section.

The potential of cryogenic distillation of nitric oxide for isotope separation was originally demonstrated by Clusius,[18,19] and has been developed at the Los Alamos Scientific Laboratories[20,21] to produce several kilograms of ^{15}N, ^{17}O, and ^{18}O; ^{14}N may be separated on the tonne scale. Fundamentally, the technique is similar to the method used for the low temperature rectification of carbon monoxide.   However, from the theoretical point of view, a six component system must be considered and from the practical point of view, operation is more complex for a number of reasons.   It is essential that an elaborate purification procedure of the feed gas is undertaken.   Nitric oxide disproportionates to give nitrous oxide and nitrogen dioxide; either of these compounds may form solid plugs in the column.   Disproportionation may also occur slowly within the column and precautions are required to ensure that any contamination is washed away.   Liquid nitrogen cannot be used as a direct coolant since the freezing point of nitric oxide is -163°C; somewhat complex arrangements are therefore required for the condenser cooling system.   Finally, the serious potential explosion hazard of liquid nitric oxide must be emphasised: in a separation plant of this type it poses special handling problems.

At Los Alamos, the separation plant consists of two stainless steel columns of 57m and 89m in length.   The separating sections, which are 2.5cm or 5cm in diameter, are filled with a stainless steel packing.[21]   As with the ^{13}C separation plants, the columns are suspended in a vacuum jacket and housed in a hole in the ground.   In operation, the shorter, high capacity primary column gives a well-stripped top stream of $^{14}N^{16}O$ while a mixture that is partially enriched in ^{15}N, ^{17}O and ^{18}O is obtained from the base of the column.   However, since the ^{15}N product is present essentially as $^{15}N^{16}O$ and the ^{18}O as $^{14}N^{18}O$, these species are more difficult to separate than $^{14}N^{16}O$ and $^{15}N^{18}O$, and randomisation is therefore necessary.   This may be achieved at room temperature and elevated pressure (10 atmos).   Gas from the base of the first column is passed through the exchanger and fed to the longer high enrichment column where $^{15}N^{18}O$ is removed from the base and $^{15}N^{17}O$ and $^{15}N^{16}O$ from progressively higher points as indicated diagramatically in Figure 2. Material from the 'waste' stream of the second column is recycled through the exchanger.

Because of the unsuitability of nitric oxide as a final product, and since both nitrogen and oxygen isotopes are simultaneously present and require separation, a number of conversion procedures have been adopted.   Nitric oxide can be oxidised to nitric acid or reduced to nitrogen, whilst reaction with hydrogen affords ammonia and water.[21]

^{18}O at low enrichment (25 atom%) can be very economically produced by cryogenic distillation of oxygen.[14]   The isotope in the feedstock is essentially present as $^{16}O^{18}O$ and therefore a suitable exchanger would be required in the system if the technique was adopted for the production of ^{18}O oxygen at

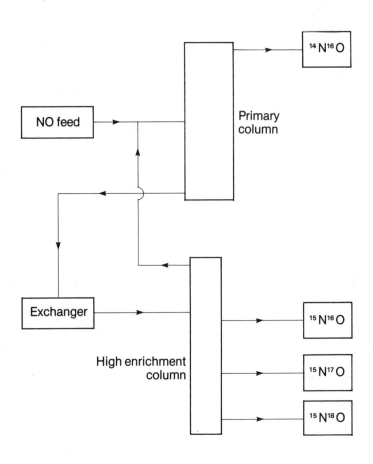

Figure 2.   Low temperature distillation of nitric oxide[21] - schematic.
             (The exchanger also receives a recycle stream from the
             high enrichment column)

high enrichment levels.

Separation of Oxygen Isotopes by Distillation of Water

The first commercially produced oxygen isotopes were obtained by distillation of water. Column packing is of crucial importance in the effective operation of an isotope separation plant and, in the case of water distillation, it was the invention of wettable wire mesh rings (Dixon rings) that led to the vital breakthrough on the industrial scale. It has been established that the optimum temperature range for water separation plants is 55 - 80°C at pressures of 120 - 360 torr.

The original separation plant at the Weizmann Institute[22] was subsequently modified to optimise ^{17}O production.[23] The extensively branched system comprises 27 columns of 17 to 100mm inside diameter with lengths of 7.6 to 14m. Separation factors of ^{17}O and ^{18}O in H_2O at 72.5°C are 1.0034 and 1.0063 respectively. Product is taken from the plant as $H_2^{17}O$ (20 atom% ^{17}O) and as $D_2^{18}O$ (98 atom% ^{18}O). Thermal diffusion of the oxygen obtained on electrolysis of the former enabled enrichments of 96 atom% ^{17}O to be attained.

Over the last decade, oxygen isotopes have been separated at the Karlsruhe Nuclear Research Centre by counter-current distillation of heavy water enriched to 0.12 atom% ^{17}O and 1.4 atom% ^{18}O.[24] The plant consists of a total of 44 steam heated, packed columns, normally 12m long and having diameters of 100, 34, or 12mm. The set-up is essentially in three sections. After passing through a preliminary stage, the D_2O feed enters the ^{18}O enrichment unit of 20 columns. The ^{18}O content of the emerging product can exceed 99.9 atom%; electrolysis and conversion of the $[^{18}O]$ oxygen to $[^{18}O]$ water removes the tritium that is present in the heavy water used. Product from an intermediate point is removed for ^{17}O enrichment. It is electrolysed and converted to H_2O in order to utilise the improved separation factor (1.0031 in H_2O compared with 1.0026 in D_2O) before admission to the third stage. This is a water distillation unit consisting of 18 narrow columns; the enrichment of the ^{17}O approaches 30 atom%. Using a single column and a batch process, $[^{16}O]$ water may be produced at 99.99 atom% ^{16}O.

Separation of Isotopes by Chemical Exchange

The small differences that exist at equilibrium between two isotopic species in a reversible reaction involving two chemical substances are utilised in chemical exchange processes. Of the many examples reported, the so-called Nitrox process for the separation of ^{15}N has been the most widely adopted. It utilises exchange between oxides of nitrogen and nitric acid (equation 2)[25-27] and is operated commercially at a number of centres in Europe.[14,28-30]

$$^{15}NO \; + \; H^{14}NO_3 \; \rightleftharpoons \; ^{14}NO \; + \; H^{15}NO_3 \qquad (2)$$

$\text{(gas)} \qquad \text{(liq)} \qquad\qquad \text{(gas)} \qquad \text{(liq)}$

The optimum separation factor is 1.055 and the highly efficient and reliable
system is capable of producing significant quantities of ^{15}N assaying up to
99.9 atom%.   The most serious disadvantage of the method is the highly
corrosive nature of the refluxing acids and gases.   Separation of the oxygen
isotopes is not apparently feasible, and so far as can be ascertained   it has
not been used for the isolation of ^{14}N.

The Nitrox process utilises groups of columns which are in three sections
as illustrated in Figure 3.   Oxides of nitrogen are produced in the bottom
refluxer by the action of sulphur dioxide on nitric acid;   sulphuric acid goes
to waste while the gas stream passes into the separating column where it meets
a downflow of nitric acid.   The rising gas is reconverted to the liquid phase
in the top refluxer by addition of water and oxygen.   Enriched nitric acid
is taken from the bottom of the separating column.   The plants in use at
Prochem consist of three-stage systems of packed columns 7m long and varying
in diameter from 7.5 to 2.2cm.   They are linked with suitable metering pumps
capable of handling the gas saturated acid in the system.   Product can be
obtained from the column in the liquid phase as nitric acid or in the gaseous
phase as mixed oxides of nitrogen.

Other chemical exchange processes that have been used for the separation
of nitrogen isotopes include ammonia and ammonium nitrate,[31] ammonia and
ammonium carbonate,[32] and ammonia and its complexes with aliphatic alcohols.[33]

Prior to the recent developments in the cryogenic distillation of carbon
monoxide, high enrichment ^{13}C was prepared by a chemical exchange process based
on the fact that salt-like carbamates are formed when carbon dioxide combines
with amines in non-aqueous solutions (equation 3).[34-36]

$$2R_2NH \; + \; CO_2 \; \rightleftharpoons \; (R_2NCO_2)^-(H_2NR_2)^+ \qquad (3)$$

Enrichment of ^{18}O occurred in the gaseous carbon dioxide and ^{13}C in the carba-
mate which could be directly decomposed into its components on heating.   A
procedure using di-n-butylamine in triethylamine, which could be carried out
at ambient temperature, was adopted in the U.S.A.   The ^{13}C separation factor
was 1.01.   Russian workers[37] have claimed a higher separation factor for an
n-butylamine in octane - carbon dioxide system.

The so-called Cyanex system, based on the reversible equilibrium in equa-

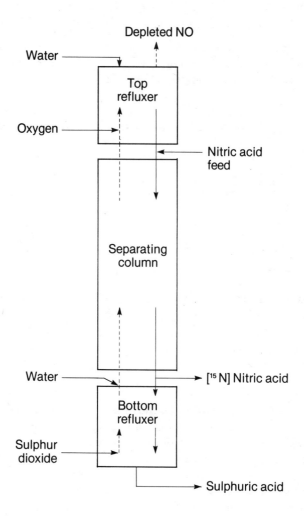

Figure 3. Separation of ^{15}N by the Nitrox process – schematic. (Rising oxides of nitrogen are shown by vertical broken arrows)

tion 4, is a liquid-liquid system in which exchange occurs on partition between
an organic solvent, such as xylene, and water.[38]   Separation

$$Et_2C(OH)^{12}CN_{(org)} \ + \ K^{13}CN_{(aq)} \ \rightleftharpoons$$
$$Et_2C(OH)^{13}CN_{(org)} \ + \ K^{12}CN_{(aq)} \qquad (4)$$

factors of 1.04 were achieved; the ^{13}C concentrated in the cyanhydrin.   Although it is claimed to be an economic process, the technical problems and
hazards involved make it unlikely to be commercially viable.

Other systems have utilised exchange between hydrogen cyanide and aqueous
sodium cyanide,[39,40] and between carbon dioxide and aqueous sodium bicarbonate.[40,41]

The exchange between sulphur dioxide and bisulphite represented by equation (5) has been adopted for the separation of ^{34}S in both the U.S.A. and
France.[42,43]

$$^{34}SO_2 \ + \ H^{32}SO_3^- \ \rightleftharpoons \ ^{32}SO_2 \ + \ H^{34}SO_3^- \qquad (5)$$
$$\text{(gas)} \qquad \text{(liq)} \qquad \text{(gas)} \qquad \text{(liq)}$$

In principle, the operation is similar to that described for the Nitrox process
but exchange is carried out at elevated temperatures (70 - $100^{\circ}C$). Acid is
added to the bisulphite in the bottom refluxer to liberate sulphur dioxide which
rises into the separating column and exchanges with the bisulphite.   Addition
of alkali to the sulphur dioxide in the top refluxer regenerates the bisulphite.
The American system uses a five-stage unit, each column being 9.1m long.   The
single stage separation factor is 1.011.

Separation by Thermal Diffusion

The prediction that a temperature gradient in a gas mixture can give rise
to diffusion was confirmed experimentally by demonstrating that in an initially
uniform mixture, the setting up of such a gradient effected a partial separation of the components.[44]   Clusius and Dickel[45] appreciated that the temperature gradient would generate convective circulation within the gas and that
this could be used to make the separation process cumulative.   Most procedures
for isotope separation use columns that consist of a water-cooled tube with an
axial heating element, frequently consisting of a hot wire, at temperatures
up to $1000^{\circ}C$; the lighter isotope normally diffuses up the temperature gradient.
Thermal diffusion has been widely employed to separate the isotopes of the
noble gases.

[^{13}C] Methane was separated at 99.8 atom% ^{13}C as early as 1954 using a
24.3m column with a hot wire as the heating element.[46]   Thermal diffusion of

nitrogen in an 82m column afforded 98.95% of the $^{14}N^{15}N$ species.[47]   In subsequent work with a 27m column, the mixed molecules were randomised by the
action of an electric discharge which led to the production of $^{15}N_2$ at 99.8
atom% ^{15}N.[48]

$[^{17}O]$ Oxygen has been prepared at 96 atom% by thermal diffusion of $[^{17}O]$
oxygen at 20 atom%.[23]   The branched separation cascade comprised 96 separation
tubes 1.5m long and having an internal diameter of 12mm.   The ^{17}O was separated
both from the lighter ^{16}O and the heavier ^{18}O.

Work in the U.S.A. on the separation of ^{34}S provides an example of thermal
diffusion in the liquid phase.[42]   In this technique, separation occurs in the
annulus between a steam-heated inner tube and a water-cooled outer one.   Whereas a gas column would typically have a spacing of 3-15mm and a temperature
differential of up to $700^{\circ}C$, in a liquid column the working space between the
vertical surfaces is in the range $200-500\mu m$ with a temperature difference of
up to $200^{\circ}C$.   As before, the lighter isotope migrates up the column while the
heavier isotope is separated at the base.   By feeding carbon disulphide to a
cascade of eight columns in series with a total length of 10.8m, $[^{34}S]$ carbon
disulphide was produced at up to 50 atom%.[42]

Other Methods of Isotope Separation

The separation of $[^{15}N]$ nitrogen from $[^{14}N]$ nitrogen has been achieved
by gas chromatography on a graphon column at 77 K.[49]   However, such techniques
are unlikely to be of any real consequence in a preparative role although they
may have some interest in an analytical context.

In recent years, considerable attention has been paid to laser isotope
separation.[50]   The concept depends upon the fact that it is possible to
excite selectively an atom or molecule of a particular isotopic composition.
Excitation of a species (X) in a mixture (X + Y) changes its physical and
chemical properties and this may be used for separation.   Early experiments
employing two-step photodissociation of ammonia achieved a partial separation
of ^{14}N and ^{15}N.   Promising results on the selective dissociation of trifluoro-
methyl iodide have demonstrated its feasibility for ^{13}C enrichment.[51]   Although work on the photopredissociation of formaldehyde has indicated enrichment
of ^{2}H, ^{13}C, ^{17}O, and ^{18}O, it has been suggested that its most promising area
of application is in the removal of ^{17}O from mixtures that have been previously
enriched by distillation.[50]

^{13}C-enrichment has also been demonstrated in the photolysis of dibenzyl

ketone in a soap solution.[52] Photolysis normally generates phenylacetyl and
benzyl radicals; the former lose CO to give benzyl radicals that recombine
to 1,2-diphenylethane. However, if the carbonyl contains ^{13}C, the magnetic
moment of the radical may reverse its electron spin and facilitate recombination to dibenzyl ketone which will become enriched in the heavier isotope.
It has been suggested that this technique, depending upon a magnetic isotope
effect, might be extended to any isotope that has such a magnetic moment.[52]

Whilst it is unlikely that these techniques will have a significant impact
on the production of isotopes of carbon, nitrogen, oxygen, and sulphur in the
near future, they illustrate some aspects of current thinking and possible
future trends.

Compounds Labelled with Stable Isotopes

In any project involving the use of stable isotopes, the initial problem
is the choice and source of an appropriately labelled compound. It will be
apparent that the form of the separated isotope does not always conform to ex-
pectations of the ideal synthetic starting point and it is therefore proposed
to provide some guidelines as an aid to the chemist in the most efficient use
of his resources.

Small-scale chemical synthesis is normally labour intensive and as such,
isotope price frequently only represents a very small proportion of the total
costs. However, it is prudent that incorporation of the isotope should be
at as late a stage in the sequence as possible. High yields are important
and the possibility of recovering unused isotope should be seriously considered.

The position or positions to be labelled in the molecule and the isotope
to be employed will be dictated by the information that is required from the
investigation and the analytical procedures that are to be used. In biological
studies, metabolic or enzymatic changes will often determine the feasibility
of alternative labelling possibilities, but in practice the final choice may
well be determined by the availability of synthetic routes.

The level of labelling required at a given centre in a molecule will be
related to the sensitivity of the detection methods employed and to the isotopic dilution that may arise in the system under study. It should be emphasised
that since the stable isotopes are naturally occurring, analytical measurements
are frequently concerned with the difference between the natural abundance of
the species and the level in the compound to be determined. Where large dilu-
tions are met, as will often occur in human clinical studies, this difference

is maximised by using as high an enrichment as possible.   On the other hand,
in some mechanistic investigations in organic chemistry, especially for example
using ^{13}C-nmr, the increase in sensitivity attainable by incorporation of even
a low level of ^{13}C will effect savings in machine and operator time that out-
weigh the increased costs of isotopic labelling.

A number of advantages may be obtained by incorporating more than one
labelled centre in the molecule.   Multiple labelling may be homogeneous or
heterogeneous and may also involve both stable and radioisotopes. . Sodium
$\left[1,2-^{13}C\right]$acetate, for example, has been employed for the detection of intact
acetate residues in the study of Pencillium multicolor.   Their existence in
the metabolites formed on feeding the doubly labelled acetate was revealed by
the presence of ^{13}C-^{13}C coupled satellites superimposed on natural abundance
singlets in the proton decoupled ^{13}C-nmr spectrum.[53]

Multiple labelling results in shifts of more than one mass number in
studies involving the use of mass spectrometry.   In turn, this will distinguish
the species under investigation from its naturally occurring isotopic analogues and considerably enhance the sensitivity.   Using heavy methane (^{13}CD$_4$)
as an atmospheric tracer, it has been detected at concentrations of 2 to 10 x
10^{-17} parts by volume;[54,55] mass 21 methane is virtually non-existent in the
atmosphere whereas ^{13}C methane comprises 1.1% of the naturally occurring compound.   Multiple labelling with deuterium has been frequently used in studies
of drug metabolism and its extension to other isotopes is assuming increasing
importance in this role.

Special attention must be paid to the enrichment level when using multiply
labelled compounds; the proportion of the fully labelled species falls off
rapidly as the level of enrichment is reduced.   At 99 atom% ^{13}C, sodium
$\left[1,2-^{13}C\right]$acetate contains 98% of the doubly labelled molecules compared with
81% when the isotopic content is 90 atom%.

Simultaneous labelling of a molecule with both ^{13}C and ^{14}C offers the
combination of ease of detection with rapid structural identification.   Circumstances will dictate if a mixture of the two singly labelled compounds is
used or whether the multiply labelled product should be synthesised from a
mixed precursor.   The use of L- and D-$\left[3-^{14}C,^{15}N,^{35}S\right]$cystine to demonstrate
that cystine was a direct precursor of penicillin provided an early example
of such mixed labelling.[56]

Compounds labelled with deuterium or the heavy isotopes of oxygen are fre-

quently prepared by isotopic exchange reactions and the possibility of further
exchange occurring either in subsequent synthetic steps or in the process
under investigation must always be considered. Although inherently less
likely with the isotopes of carbon and nitrogen, it must not be overlooked.
For example, it has been shown that exchange occurs between the two nitrogen
atoms in some 2-aminopyridines under mild hydrolytic conditions.[57]

The availability of gas mixtures with an isotopically labelled component
offers further opportunities for the exploitation of stable isotopes. The
versatility of analytical techniques may be enhanced, since, for example,
natural nitrogen and $[^{13}C]$carbon monoxide are readily distinguishable in the
mass spectrometer. Synthetic air may be obtained with the oxygen, nitrogen,
and/or the carbon dioxide labelled and finds utility in the study of a number
of biological systems while the use of a mixture containing $[^{13}C]$carbon monoxide
in laser work [58] has demonstrated how a labelled component may confer desirable
properties on a mixture in a specific context.

<u>Chemical Synthesis</u>. The preparation of a number of key intermediates labelled
with ^{13}C from $[^{13}C]$carbon monoxide is illustrated in Scheme 1. These are
compounds that are prepared in high yield on a large-scale by manufacturers
and frequently employ scaled-down versions of industrial processes or utilise
specialist equipment.

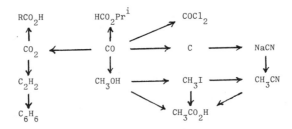

Scheme 1.   The preparation of some key intermediates labelled with ^{13}C

By way of illustration, $[^{13}C]$methanol is prepared by catalytic reduction
of $[^{13}C]$carbon monoxide or carbon dioxide using hydrogen under pressure.[59]
Although frequently obtained by carbonation of a Grignard reagent, a catalytic
reaction of carbon monoxide and methanol in the presence of rhodium chloride
at $175^{\circ}C$ may be adopted for the preparation of acetic acid on the large scale.[60]
Appropriate choice of starting materials affords the various possible labellings.
An efficient conversion of carbon dioxide to acetylene is based on techniques
developed for radiocarbon dating. Reaction of carbon dioxide with lithium

at 600°C in a stainless steel vessel affords lithium carbide; acetylene is liberated on addition of water to the cooled carbide,[61] and may be trimerised to benzene in the presence of a vanadium pentoxide catalyst.

Scheme 2 illustrates the preparation of a number of useful intermediates and gaseous products labelled with ^{15}N. Bulk conversion methods have been developed for the oxidation of $[^{15}N]$nitric oxide to nitric acid and for its reduction to ammonia.[21] Ammonia is prepared from nitric acid by catalytic hydrogenation or by reduction with Devarda's alloy in alkali.

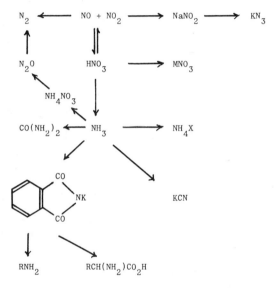

Scheme 2.   The preparation of some key intermediates labelled with ^{15}N.

With a range of procedures established for the preparation of key intermediates in high yield, the subsequent synthesis of more complex materials follows the standard procedures of organic chemistry, subject to the considerations enumerated in the previous section.

The use of isotopic exchange plays an important role in the preparation of compounds labelled with the oxygen isotopes. When many compounds, such as inorganic nitrates and carbonates and organic substances such as carboxylic acids, are dissolved in $[^{18}O]$water, equilibrium is established between the oxygen in the water and in the solute. Adjustment of pH and an elevated temperature are often necessary for the exchange to proceed to completion. When

applicable, the technique offers a simple method of oxygen labelling but unless
the labelled solvent can be reutilised the method is wasteful of isotope
especially at high levels of isotopic enrichment.

Multiply labelled compounds containing either a homogeneous or heterogeneous combination of heavy isotopes are prepared by analogous methods.   Since
all isotopically labelled compounds originate from very simple molecules, the
incorporation of more than one label in a simple intermediate such as $[^{13}C, {}^{15}N]$
potassium cyanide, $[^{13}C, {}^{2}H]$methyl iodide, $[^{13}C, {}^{18}O]$carbon dioxide, or
$[1,2-^{13}C]$sodium acetate presents few problems.

The synthesis of compounds labelled with both a radio and a stable isotope
at the same centre in a molecule is unlikely to assume any great significance
as admixture of the two singly labelled species will normally be as effective.
However, the technique has been adopted to study some aspects of the metabolism
of 4-morpholino-2-piperazinothieno$[3,2-\underline{d}]$pyrimidine(1). Potassium $[^{13}C, {}^{14}C]$
thiocyanate was prepared from a mixture of potassium $[^{13}C]$ and $[^{14}C]$cyanides
and was converted by a series of reactions into the thienopyrimidine (1)
labelled at the position shown.[62]

(1)

The use of depleted materials, especially ^{12}C and ^{14}N, must not be overlooked.   Solvents labelled with ^{12}C and deuterium have been used in ^{13}C-nmr
studies but hopes that depleted materials would provide a more economical
approach to stable isotope labelling are unlikely to be realised, although
^{12}C-labelling has been used in the study of organic reaction mechanisms.[63]
On the small scale, there are minimal effective cost savings on the preparation
of a ^{12}C-labelled compound as compared with its ^{13}C-labelled analogue.
$[^{14}N]$Ammonium sulphate has been prepared on the tonnage scale and used in
fertiliser studies.[64]

Preparation of Labelled Compounds by Biosynthesis.   Biosynthetic methods have
been applied to the preparation of a number of natural products and related
compounds.   Where such techniques are adopted, it is essential that the

labelled product should be isolable in a chemically pure state.

The large-scale photosynthetic production of ^{13}C-labelled sugars is an important example of the use of biosynthesis for the preparation of uniformly labelled natural products. $[UL-^{13}C]$starch, glucose, fructose, and sucrose were isolated from tobacco leaves after incubation with $[^{13}C]$carbon dioxide in specially designed chambers for 40 h.[65] Uniformly ^{13}C-labelled amino acids have been prepared from algae, Spirulina maxima, grown in the presence of sodium $[^{13}C]$bicarbonate.[66]

Enzymatic methods have been used to effect more specific labelling. Reduction of 2-oxoglutaric acid with reduced nicotinamide adenine dinucleotide phosphate in the presence of $[^{15}N]$ammonium chloride afforded L-$[\alpha-^{15}N]$glutamic acid.[67]

Analytical Techniques

Analysis is an essential component of any project involving the use of labelled compounds. It is required to ensure chemical identity, purity, and isotopic content of the materials used, and to determine both the qualitative and quantitative fate of the label in the experiments undertaken. This section will be devoted to a brief outline of isotopic analysis; the principal techniques will be discussed by other contributors. Checks on chemical purity use conventional chemical and physico-chemical methods. However, the differences in the physical properties and molecular weight brought about by the incorporation of a stable isotope produce effects of which the chemist must be constantly aware. It will, for example, cause shifts in the ir spectrum which will appear more complex than that of the unlabelled molecule.

Mass Spectrometry. The most widely applied method of isotopic analysis is mass spectrometry and two types of instrument are employed. Isotope mass spectrometers are designed to measure the abundance of an isotope in a gas sample compared with a known reference gas; they typically cover the range m:e 2-100. Absolute isotopic content cannot be determined and in determining natural variations in isotopic abundance it is necessary to employ standards to ensure comparability of results between laboratories. Where the enrichments measured are relatively large, the isotopic content is normally compared with the same species before introduction of the label. The high resolution analytical instrument, which covers a wider range, produces a mass:energy spectrum. The application of mass spectrometry to isotope studies has recently been reviewed.[12,68]

Quantitative conversion of the sample to a gaseous form is an essential

step in the use of the isotope mass spectrometer. ^{2}H is usually assayed as
hydrogen, ^{13}C as carbon dioxide, and ^{34}S as sulphur dioxide. The preparation
of samples as nitrogen for ^{15}N analysis is normally achieved by a Dumas direct
combustion technique or by a Kjeldahl conversion to ammonia followed by
oxidation with hypobromite. A frequently employed method for the determination of ^{18}O in water involves equilibration with carbon dioxide followed by
analysis of the carbon dioxide isotopic ratio.[69] Isotopic ratios of carbon
and nitrogen have been determined with very good precision by placing a combustion tube between a gas chromatograph and an isotope mass spectrometer and
monitoring the ions at masses 28 and 29 for nitrogen and at 44 and 45 for
carbon.[70,71] With a computer controlled mass spectrometer, enrichments as
low as 0.004 atom% excess were detected.[71]

The organic mass spectrometer, with its variety of possible sample inlet
systems, is less restricted in the form of isotopic species that it can
accept. With a gas inlet system, it can be used for the analysis of both
labelled gases and gas mixtures and for the analysis of compounds that can be
readily converted to an appropriate gaseous form. Other systems allow organic
compounds such as amino acids to be assayed after conversion to a volatile
derivative. When linked to a gas chromatograph, complex mixtures, such as
are found in biological fluids from metabolic studies, can be fractionated and
structural information derived as well as the location of the label in the
components.

Selected ion monitoring uses the mass spectrometer as a highly sensitive
specific detector for a limited number of ions. With the mass spectrometer
linked to a gas chromatograph and with computer systems to process the data
obtained, quantitative detection of specifically labelled ions can be made at
the sub-nanogram level.

Nuclear Magnetic Resonance Spectroscopy. ^{1}H (1/2), ^{2}H (1), ^{13}C (1/2), ^{14}N (1),
^{15}N (1/2), and ^{17}O (5/2) all have non-zero nuclear spins and can, therefore,
be detected by nmr spectroscopy. So far as tracer work with stable isotopes
is concerned, the overwhelming majority of the studies employing nmr have involved proton and ^{13}C. A number of reviews and monographs are available which
deal with the nmr techniques used in conjunction with stable isotopes;[72-79]
particular aspects will be dealt with in more detail by other contributors.
To the chemist involved in the synthesis of ^{13}C-labelled compounds, nmr spectroscopy is an important tool both for isotopic assay and for structural confirmation of the final product. To the research worker using ^{13}C as a tracer,
it will provide information on the location of the label in a molecule, as well

as acting as a non-specific detection method.　Sample requirements can be as low as the microgram level.

In early work, useful information was obtained from proton nmr using the so-called satellite method.　Hydrogen atoms attached to carbon-13 show spin coupling and therefore each proton resonance in the nmr spectrum is accompanied by two satellite bands.　In the spectrum of a compound with ^{13}C only at the natural abundance level, the intensity of these satellite signals is so low that they are not normally observed.　However, with ^{13}C enrichment at specific sites, signals corresponding to the labelled positions will be enhanced and in many cases will allow a quantitative estimate of the ^{13}C level to be made. The method clearly has limitations, especially if the main proton bands overlap the satellite signals.　However, it is frequently employed for enrichment checks on compounds such as sodium $[2-^{13}C]$ acetate.　In more complex molecules, the difficulties can sometimes be overcome and the method has been successfully employed in studies of biosynthetic pathways.[75,80]

Direct measurement of ^{13}C nuclei is clearly preferable.　However, the sensitivity of ^{13}C-nmr is lower than that of proton nmr by a factor of about 6000.　The inefficiency of conventional nmr derives from the fact that only one frequency is observed at any given instant.　This problem was overcome by the advent of Fourier transform nmr.[81]　Using a short pulse radiofrequency, all ^{13}C nuclei in a sample can be excited simultaneously.　The absoprtions of individual frequency components by each nucleus are detected by the receiver and abstracted by Fourier transformation using data acquisition and processing equipment.

Substitution of a hydrogen atom with deuterium produces characteristic perturbations in the ^{13}C-nmr spectrum and the carbon carrying the deuterium can be readily identified.　Work of this type has been reviewed.[76]

In nmr studies, the sample is frequently dissolved in an appropriate solvent which, in most cases will need to be deuterated.　Many solvents are now available depleted in ^{13}C (^{12}C-deuterated solvents);　the use of these solvents avoids the problems that sometimes arise when solvent signals obscure vital information in the ^{13}C-nmr spectrum under study.　It is especially important when only small samples of material are available.　With ^{13}C depleted to 0.05- 0.1 atom%, the solvents show no observable signal when examined by Fourier transform nmr.[82,83]

The use of both ^{14}N-nmr and ^{15}N-nmr has been reviewed [84,85] but these techniques have not achieved the widespread application enjoyed by ^{13}C-nmr. Although ^{15}N-nmr spectra display good resolution, their use has been restricted by poor signal to noise ratios. The nuclear quadrupole moment of ^{14}N nuclei significantly broadens their nmr resonances. ^{17}O and deuterium nmr are other developing areas that relate to the isotopes under discussion.

Optical Emission Spectroscopy. Nitrogen isotope measurements using optical emission spectroscopy depend upon the wavelength separation of the three species, ^{14}N$_2$, ^{14}N^{15}N, and ^{15}N$_2$, due to the isotopic shift. It is necessary to convert the compound to be assayed to nitrogen gas which is excited by a radiofrequency in a flow discharge tube. Although relatively simple, the method does not provide direct information on the molecular environment of the isotope. However, the technique has been routinely employed in a number of areas including agricultural studies and the estimation of protein turnover in man. Sample conversion, optical detection, and electronic calculation have now been combined in one instrument. Its use has been demonstrated in clinical metabolic studies using $[^{15}$N$]$glycine as a tracer. With a sample requirement of 120 µg, the relative accuracy was ± 1% over a range of 0.36–25 atom% ^{15}N.[86]

Infrared Spectroscopy. Substitution of stable isotopes into a molecule causes changes in the infrared spectrum. Infrared data on labelled compounds have been reviewed.[87] In certain cases it is possible to use such data for the assay of labelled compounds. The absorption peaks at 2193 cm^{-1} for ^{12}CO and at 2144 cm^{-1} for ^{13}CO are sufficiently well separated to determine the isotopic ratio of carbon monoxide. With ^{13}C enrichments of 1–20 atom%, the mean quoted was about 2% with a relative standard deviation of about 6%.[88] The carbonyl peak of acetophenone is at 1685 cm^{-1} whilst on substitution of ^{13}C at this centre it shifts to 1645 cm^{-1}.[82]

Miscellaneous Methods. Raman scattering has been investigated as a means of determining isotopic ratios but has not found widespread use. Isotopic analysis of nitrogen (^{14}N^{15}N) and oxygen (^{16}O^{18}O) can be made to accuracies of ±0.1% using an argon laser.[89]

   In activation analysis, a stable isotope is made to undergo a suitable nuclear reaction and the parent isotope determined by measurement of the resulting nuclide. The ^{18}O content of water samples as small as 1.5 µl has been determined by charged particle activation.[90]

Applications of Stable Isotopes
   It is not intended in this review to provide detailed coverage of the

many applications of stable isotopes. However, a number of areas will be indicated to provide a raison d'être for the work described in the previous sections.

Chemical Studies. Although most of the chemical applications of stable isotopes are associated with problems of biosynthesis, pharmacology, and other life sciences, they also provide opportunities for the study of reaction mechanisms.

Reaction of $[1-^{13}C]$-2-(methylthio)ethanol (2) in carbon tetrachloride with trialkyl phosphines in benzene afforded an equimolar mixture of $[1-^{13}C]$-1-chloro-2-(methylthio)ethane and the $[2-^{13}C]$ analogue (4) which supported the hypothesis that the reaction proceeded through the 1-methylthiiranium ion (3).[91]

$$MeSCH_2*CH_2OH \xrightarrow[CCl_4]{R_3P}$$

$$\begin{array}{c} H_2C \!\!-\!\!\blacksquare\!\!-\!\! *CH_2 \\ \diagdown \quad \diagup \\ S+ \\ | \\ Me \end{array} \xrightarrow{Cl^-} \begin{array}{c} MeSCH_2*CH_2Cl \\ + \\ MeS*CH_2CH_2Cl \end{array}$$

       (2)   (3)   (4)

Preparation of urea by the action of heat on ammonium cyanate labelled with ^{15}N on the ammonium nitrogen afforded a product that was almost exclusively labelled at only one of the two possible positions.[92] Randomisation clearly does not occur in Wohler's classical rearrangement.

Although the use of ^{12}C is unlikely to assume the importance of ^{13}C, it has been employed to distinguish two possible mechanisms for the formation of benzocyclopropene (6) by dehydrochlorination of the dichloro compound (5).[63] The absence of a signal corresponding to C-7 in the ^{13}C-nmr spectrum of the product (6) indicated that skeletal rearrangement did not occur.

           (5)   (6)

Applications in Biochemistry. The use of stable isotopes in the study of the biosynthesis of antibiotics has been particularly rewarding; there is efficient

incorporation of ^{13}C into bacterial and fungal metabolites and yields are high.
A common procedure has involved feeding   specifically labelled ^{14}C possible
precursors to an appropriate culture;   screening the products obtained for
radioactivity gives an indication of which compounds have been incorporated.
Repetition with appropriately labelled ^{13}C compounds and subsequent examination
of the product by ^{13}C-nmr provides information on the exact location of the
incorporated precursor without the necessity for degradative studies.   The
data have frequently been adequate to deduce the metabolic pathway.   Preliminary use of radioisotope labelling is not essential and indeed in the case of
^{15}N and ^{18}O is not possible.   Furthermore, mass spectrometry has frequently
been used for analysis, either with or without the need for chemical degradation.   Several useful reviews describing the use of stable isotopes in biosynthetic studies are available.[75,76,78,80,93-96]

     Stable isotope studies have played a significant role in elucidating the
biosynthesis of penicillins and cephalosporins.[97]   When benzylpenicillin was
formed by P. chrysogenum in the presence of L-$\left[3-^{14}C,^{15}N,^{35}S\right]$cystine, the
three labels appeared in the same proportion in the product and confirmed that
cystine was a direct precursor of penicillin.[56] ^{13}C-Nmr studies of cephalosporin C (7) obtained after incubation of the culture with either sodium $\left[1-^{13}C\right]$
acetate or sodium $\left[2-^{13}C\right]$acetate showed that most of the side chain carbon atoms
were derived from acetate residues[98,99] whilst incorporation of an entire valine
skeleton into both penicillin N and cephalosporin C was demonstrated in experiments with (2S,3S)-$\left[^{15}N-(3-methyl-D_3)\right]$valine.[100]   Shaking mycelial suspensions
of Cephalosporium acremonium in an enriched oxygen-18 atmosphere resulted in
incorporation of one oxygen atom at the C-17 position which was adduced as
evidence of the participation of hydroxylase in the biosynthesis.[101]

(7)

     Lasalocid A (8), an antibiotic produced by Streptomyces lasaliensis, is
unique in being a polyether with three ethyl groups.   Studies on the incorpor-
ation of sodium $\left[1-^{13}C\right]$acetate, sodium $\left[1-^{13}C\right]$propionate and sodium $\left[1-^{13}C\right]$buty-
rate showed that all three ethyl groups arose from complete butyrate residues.

The skeleton was additionally derived from five acetate and four propionate units.[102]

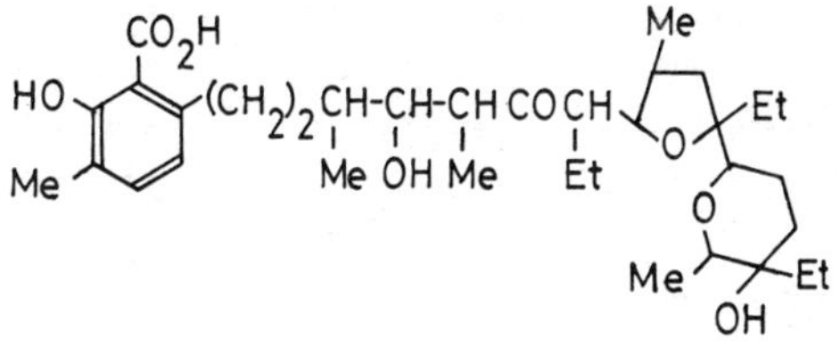

$$CO_2H \quad Me$$
$$HO-\!\!\!\bigcirc\!\!\!-(CH_2)_2CH\text{-}CH\text{-}CH\,COCH \overset{}{\underset{O}{\diagdown}} Et$$
$$Me \qquad Me\ OH\ Me \qquad Et$$
$$Me \overset{}{\diagup}\!\!\!\diagdown Et$$
$$OH$$

(8)

The wide spectrum of activity of the ansamycin antibiotics has led to a number of studies that have been reviewed.[78,95,96]

Investigations on multicolic acid (9, R = CH_2OH) and multicolosic acid (9, R = CO_2H), isolated from cultures of P. multicolor after feedings with sodium $[1,2\text{-}^{13}C]$acetate, utilised a unique method for the detection of intact acetate residues from the presence of $^{13}C\text{---}^{13}C$-coupled satellites superimposed on natural abundance singlets in the proton decoupled ^{13}C-nmr spectrum.[53] Its use established the first example of an intermediate aromatic precursor in the biosynthesis of a fungal metabolite.

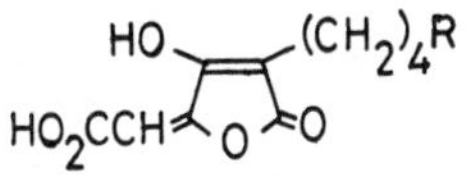

$$HO-\!\!\!\diagup\!\!\!\diagdown(CH_2)_4R$$
$$HO_2CCH \overset{}{\underset{O}{\diagdown}}\!\!\!\diagdown O$$

(9)

Biochemical applications have not been confined to biosynthesis. Indeed, extensive work on the metabolism of proteins and the intermediary metabolism of carbohydrates and lipids using stable isotopes had taken place by the mid-1940s.[1] The work, which was of immense biochemical significance, also provides some useful information on the preparation of labelled compounds.

A study of the replication of DNA in Escherichia coli by density gradient centrifugation employed ^{15}N labelling and showed that the DNA nitrogen is divided equally between two physically continuous sub-units, one of which is received by each daughter molecule. The sub-units are conserved through many

generations.[103] $[^{13}C]$Ethanol has been used to study ethanol metabolism in
rats,[104] while the potential of bile acids labelled with ^{2}H and ^{13}C in
gastroenterological research has been critically evaluated and compared with
radioactive labelling.[105]   Cerebral oxygen metabolism has been studied in
different behavioural situations using mixtures of $[^{18}O]$oxygen with natural
nitrogen.[106]   Review articles[10,12] and published proceedings of symposia[3,5-8]
indicate the continuing importance of stable isotopes to the biochemist.

<u>Applications in Pharmacology</u>.   There are three important areas where the use
of stable isotopes is of interest to the pharmacologist:

(1)   as tracers for the identification of metabolites,

(2)   as internal standards for the estimation of drugs in biological
       fluids,

(3)   in studies of the bioavailability of drugs.

Although most of the work has employed deuterium labelling, the use of ^{13}C and
^{15}N is assuming an increasingly important role.   To gain the maximum advantage,
labelling should be in a position in the molecule that is largely resistant
to biotransformation and where as many metabolites as possible will be labelled.

The application of the technique is illustrated by the identification of
urinary and biliary metabolites of nortriptyline using the dideuterium,
^{15}N-labelled drug (10).   In human studies, with a 25 mg dose, metabolites were
detected and identified from the presence of M, M+3 doublets when trifluoro-
acetylated extracts were examined by gc-ms.[107]   The use of multiple labelling
is important since the utility of M,M+1 doublets is limited by the effects of

$$CHCH_2CD_2^{15}NHMe$$

(10)

the contribution of the natural abundance species.   With multiply labelled
compounds and appropriate instrumentation, levels of 1 pg in a 1 μg sample can
be detected.[108]

The purpose of an internal analytical standard is to correct for losses
of a compound in the procedures adopted.   Chemical similarity and similarity
of behaviour must exist between the standard and the compound under examination.

When gc-ms is employed in studies of drug metabolism, an isotopically labelled compound is ideal since mass differences alone can be used to discriminate between the compound analysed and the standard.   Labelled nortriptylines have been added to crude plasma samples before gc-ms for plasma estimations of the drug.[109]   In other applications, the labelled drug has been added prior to extraction, derivatisation, and analysis.[110]   Using computerised gc-ms systems and labelled drugs as internal standards, urine, plasma, breast milk, and amniotic fluid concentrations have been analysed quantitatively in the picogram range.[111]

Labelled drugs offer important advantages in studies of bioavailability. Two experiments involving separate administration of an intravenous and an oral test dose are normally required on each subject and assumptions must be made that the drug kinetics remain unchanged between doses.   Administration of the drug orally with simultaneous intravenous injection of a labelled version overcomes these disadvantages.   The feasibility of this approach has been demonstrated on studies of N-acetylprocainamide (NAPA) and its ^{13}C-labelled version (11).[112]

$$CONH(CH_2)_2NEt_2$$

$$NH^{13}COMe$$

(11)

Clinical Applications of Stable Isotopes.   Radioisotopes have found widespread use in clinical studies, but their use is limited by the effects of the radioactivity and alternatives become increasingly important as regulations become more stringent.   Clearly there are widespread potential applications of stable isotopes in clinical investigations.

Since the heavier stable isotopes are chemically indistinguishable from their more abundant, lighter counterparts, any possible harmful effects would derive from differences in their mass.   Many instances of isotope effects have been recorded and caution must therefore be exercised in the interpretation of results.   However, these are inherently more probable with deuterium where there is a two-fold mass difference.   A number of studies have been carried out to detect possible untoward effects of ^{13}C in animals but, as yet, abnormalities have not been detected.[113-115]

Considerable attention has been paid to carbon dioxide breath tests which are based on the premise that a patient with a metabolic disorder may oxidise a specific ^{13}C-labelled substrate at a different rate from a normal subject. Administration of an appropriately labelled precursor followed by measurement of the isotopic ratio in the expired carbon dioxide may therefore provide evidence, albeit indirect, of such a disorder. Oral administration of $[1-^{13}C]$trioctanoin and monitoring the expired ^{13}C carbon dioxide has provided a sensitive test for fat maladsorption.[116] Much evidence has been devoted to similar investigations of diabetic patients after administration of $[UL-^{13}C]$ glucose; significant differences between normal and diabetic children have been observed.[117]

A knowledge of total body water is important in a number of clinical situations where fluctuations in body hydration occur. Deuterium oxide has been elegantly employed for such measurements in human studies.[118] Administration to a subject will result in exchange of H_2O molecules to give HDO, and with other labile hydrogens for which corrections can be applied. After equilibration, samples (5 μl) of urine, plasma, or saliva were reduced and the gaseous sample measured in the mass spectrometer. The accuracy of such total body water measurements is better that $\pm$ 0.5%.

^{15}N has played an important role in the measurement of total body protein turnover in man. Such work has been reviewed.[119] In most cases, excretion of ^{15}N in the urine has been measured after administration of a ^{15}N-amino acid. An alternative method involves measurement of the free amino acid in the plasma. L-$[\alpha - ^{15}N]$Lysine has been employed in this way to evaluate protein synthesis rates more directly.[120]

Urea nitrogen utilisation for albumin synthesis has been determined by a single combined injection of $[^{15}N]$urea, $[^{14}C]$urea, and ^{125}I-labelled albumin in uraemic and normal patients.[121] Clinical disorders of purine metabolism lead to elevated plasma and urinary uric acid levels and ^{15}N-labelled precursors have been used to investigate uric acid synthesis. $[^{15}N]$Glycine has also been employed in haematological investigations. These are examples of a number of clinical investigations that have been reviewed.[12]

Agricultural Studies. About half of the published work using ^{15}N-labelling in the life sciences concerns agricultural problems; the technique has become an accepted part of the investigation of the nitrogen cycle.

Less than 50% of nitrogen added to the soil is utilised in plant growth and
the economic importance of improving the efficiency of its conversion and of
its uptake from added fertiliser has prompted much of this work.[122]   Low
enrichment $[^{15}N]$ammonium salts, $[^{15}N]$nitrates, and $[^{15}N]$urea have been widely
employed in these studies.

Extensive work has been carried out on the efficiency of nitrogen incorporation by plants.[54,122]   Early studies revealed differences in the recovery
of tracer nitrogen in greenhouse and field experiments which led to the recognition of the importance of soil water in the control of nitrogen uptake.[123]
The availability of $[^{14}N]$ammonium sulphate on the tonne scale has enabled large-
scale nitrogen balance studies and field measurements to be made on the
efficiency of nitrogen utilisation and of nitrate movement.[64,124]

Biological denitrification results in losses of nitrogen to the atmosphere
and has been studied as part of the nitrogen cycle.[125]   Using calcium $[^{15}N]$
nitrate, the relative conversions of nitrate to gaseous products or to ammonium
ion were made under various conditions.[126]   Nitrogen fixation has been investigated in ^{15}N-labelled atmospheres; the limitations imposed by the confined
atmosphere have been overcome by labelling the soil nitrogen and determining
the resultant dilution of the isotopic nitrogen.[127]

The use of ^{15}N as a tracer in studies of the metabolic pathways involved
in nitrogen assimilation in plants has been reviewed.[128]   Investigations of
nitrogen utilisation by hens[129,130] and pigs[131] provide examples of the application of ^{15}N-labelling in animal nutrition.

The application of other isotopes in agriculture is more limited.
$[^{18}O]$oxygen may be employed in studies of the mechanism of photosynthesis and
^{13}C could assume an important role in plant biochemistry.   It has been suggested
that ^{34}S may prove useful in agricultural research on problems related to sulphur deficiency in the soil.[7]

Environmental Applications.   The possible consequences on the environment of
the increasing use of fertiliser nitrogen, such as the rise in inorganic nitrate
in fresh water, underline the importance of obtaining quantitative information
on the nitrogen balance of the total biosphere.[54,132]   Such studies, which
are closely related to the agricultural problems already discussed, provide
further examples of the importance of ^{15}N labelling in tracer emperiments.

The fate of pesticides is an environmental problem that requires the quantitative estimation of low-level residues in crops, soil, water, and animal tissues, together with the identification of chemical and metabolic products. The problems are similar to those of the pharmacologist studying drug metabolism and variations on the analytical techniques described earlier can be used in field studies.[54]

The potential of stable isotopes as atmospheric tracers has been referred to earlier. On releasing 84 g of heavy methane ($^{13}CD_4$) into the atmosphere, it was detected at distances of 1500 - 2500 km at concentrations as low as 2×10^{-17} parts by volume in experiments to track air motions on a continental scale.[54,55]

The importance of stable isotopes has been established in a variety of disciplines. It is hoped that this brief resumé will have provided some guidance to their potential application and to benefits that may be derived from their use.

References

1   "A Symposium of the Use of Isotopes in Biology and Medicine," 1947 University of Wisconsin Press, Madison, 1948.

2   "Proceedings of a Seminar on the Use of Stable Isotopes in Clinical Pharmacology," University of Chicago, Illinois, 1971, P.D. Klein and L.J. Roth, eds., U.S. Atomic Energy Commission, Conf.-711115, 1972.

3   "Proceedings of the First International Conference on Stable Isotopes in Chemistry, Biology, and Medicine," ·Argonne National Laboratory, Argonne, Illinois, 1973, P.D. Klein and S.V. Peterson, eds., U.S. Atomic Energy Commission, Conf.-730525.

4   "Isotope Ratios as Pollutant Source and Behaviour Indicators," Vienna, 1974, International Atomic Energy Agency, Vienna, 1975.

5   "Proceedings of the Second International Conference on Stable Isotopes," Oak Brook, Illinois, 1975, E.R. Klein and P.D. Klein, eds., U.S. Energy Research and Development Administration, Conf.-751027.

6   "Stable Isotopes - Applications in Pharmacology, Toxicology, and Clinical Research," T.A. Baillie, ed., Macmillan Press Ltd., London, 1978.

7   "Stable Isotopes in the Life Sciences," Leipzig, 1977, International Atomic Energy Agency. Vienna, 1977.

8   "Stable Isotopes, Proceedings of the Third International Conference," E.R. Klein and P.D. Klein, eds., Academic Press, New York, 1979.

9   "Separation of Isotopes," H. London, Ed., George Newnes Ltd., London, 1961.

10 N.A. Matwiyoff and D.G. Ott, Science, 1973, 181, 1125.

11 D. Staschewski, Chem. Tech., 1975, 4, 269.

12 D. Halliday and I.M.Lockhart, "Progress in Medicinal Chemistry–Vol. 15," G.P. Ellis and G.B. West, eds., North Holland, Amsterdam, 1978, p. 1.

13 D.R. Knapp and T.E. Gaffney, in Ref. 2, p. 249.

14 A.O. Edmunds and I.M.Lockhart, in Ref. 4, p. 279.

15 E.S. Robinson, in Ref. 2, p. 5.

16 W.R. Daniels, A.O. Edumnds, and I.M. Lockhart, in Ref. 7, p. 21.

17 D.E. Armstrong, T.R. Mills, B.B. McInteer, J.G. Montoya, C.A. Lehman, R.C. Vandervoort, and M Goldblatt, in Ref. 8., p. 177.

18 K. Clusius and K.Schleich, Helv. Chim. Acta, 1958, 41, 1342.

19 K. Clusius and K.Schleich, Helv. Chim. Acta, 1959, 42, 2654.

20 B.B. McInteer and R.M. Potter, Ind. Eng. Chem., 1965, 4, 35.

21 D.E. Armstrong, B.B. McInteer, T.R. Mills, and J.G. Montoya, in Ref.8, p.175.

22 I.Dostrovsky and A. Raviv, "Proceedings Amsterdam Conference on Isotope Separation," 1957, North–Holland, Amsterdam, 1958, Ch. 26.

23 D. Wolf and H. Cohen, Canad. J. Chem. Eng., 1972, 50, 621.

24 D. Staschewski, in Ref. 7, p. 85.

25 W. Spindel and T.I. Taylor, J. Chem. Phys., 1955, 23, 981.

26 W. Spindel and T.I. Taylor, Trans. N.Y. Acad. Sci. Ser. I, 1956, 19, 3.

27 W. Spindel and T.I. Taylor, J. Chem. Phys., 1956, 24, 626.

28 D. Axente, Stud. Cercet. Chim., 1971, 19, 395.

29 E. Krell, Isotopenpraxis, 1976, 12, 188.

30 E. Krell and C. Jonas, in Ref. 7, p. 59.

31 H.G. Thode and H.C. Urey, J. Chem. Phys., 1939, 7, 34.

32 G.M. Begun, A.A. Palko, and L.L. Brown, J. Phys. Chem., 1956, 60, 48.

33 G.M. Panchenkov, A.I. Kuznetsov, and A.V. Makarov, Dokl. Akad. Nauk. SSSR, 1965, 164, 1101.

34 T.I. Taylor, J. Chim. Phys. Chim. Biol., 1963, 60, 154

35 J.P. Agrawal, Sep. Sci., 1971, 6, 819.

36 J.P. Agrawal, Sep. Sci., 1971, 6, 831.

37 E.D. Oziashvili, A.S. Egiazarov, S.I. Dzhidzheishvili, and N.F. Bashkatova, in Ref. 7, p. 29.

38 L.L. Brown and J.S. Drury, J. Inorg. Nucl. Chem., 1973, 35, 2897.

39 I. Roberts, H.G. Thode, and H.C. Urey, J. Chem. Phys., 1939, 7, 137.

40 C.A. Hutchison, D.W. Stewart, and H.C. Urey, J. Chem. Phys., 1940, 8, 532.

41 A.F. Reid and H.C. Urey, J. Chem. Phys., 1943, 11, 403.

42 W.M. Rutherford and W.J. Roos, in Ref. 4, p. 295.

43   Y. Belot, J.C. Bourreau, N Dubois, and C. Pauly, in Ref. 4, p. 403.

44   S. Chapman and F.W. Dootson, Phil. Mag., 1917, 33, 248.

45   K.Clusius and G. Dickel, Naturwiss., 1938, 26, 546.

46   K. Clusius and H.H. Buhler, Z. Naturforsch., 1954, 9A, 775.

47   K. Clusius and E.W. Becker, Z. Naturforsch., 1947, 2A, 154.

48   K. Clusius, Helv.Chim. Acta, 1950, 33, 2134.

49   F. Bruner and A.D. Corcia, J. Chromatogr., 1969, 45, 304.

50   V.S. Letokhov and C.B. Moore, "Chemical and Biochemical Applications of
      Lasers," C.B. Moore, ed., Academic Press, New York, 1977, Vol. III, Ch.1.

51   V.N. Bagratashvili, V.S. Doljikov, V.S. Letokhov, and E.A. Ryabov, "Laser
      Induced Processes in Molecules," Springer-Verlag, Berlin, 1978, Springer
      Series in Chemical Physics, Vol. 7.

52   N.J. Turro and B. Kraeutler, J.Amer. Chem. Soc., 1978, 100, 7432.

53   J.A. Gudgeon, J.S.E. Holker, and T.J. Simpson, J. Chem. Soc. Chem. Commun.,
      1974, 636.

54   N.A. Matwiyoff, G.A. Cowan, D.G. Ott, and B.B. McInteer, in Ref. 4, p. 305.

55   G.A. Cowan, D.G. Ott, A. Turkevich, L.Machta, G.J. Ferber, and N.R. Daly,
      in Ref. 5, p. 605.

56   H.R.V. Arnstein and P.T. Grant, Biochem. J., 1954, 57, 360.

57   M. Wahren, Tetrahedron, 1968, 24, 451.

58   M. Poliakoff, personal communication, 1976.

59   D.G. Ott, V.N. Kerr, T.W. Whaley, T. Benziger, and R.K. Rohwer, J.Labelled
      Compd., 1974, 10, 315.

60   F.E. Paulik and F.J. Roth, Chem. Commun., 1968, 1578.

61   H.Barker, Nature, Lond., 1953, 172, 631.

62   A. Zimmer, A. Prox, H. Pelzer, and R. Hankwitz, Biochem. Pharmacol., 1973,
      22, 2213.

63   J. Prestien and H. Günther, Angew. Chem., 1974, 13, 276.

64   F.E. Broadbent, in Ref. 4, p. 373.

65   V.H. Killman, J.L. Hanners, J.Y. Hutson, T.W. Whaley, D.G. Ott, and C.T.
      Gregg, Biochem. Biophys. Res. Commun., 1973, 50, 826.

66   S. Tran-Dinh, S.S. Fermandjian, E. Sala, R. Mermet-Bouvier, M. Cohen, and
      P. Fromageot, J. Amer. Chem. Soc., 1974, 96, 1484.

67   W. Greenaway and F.R. Whatley, J. Labelled Compd., 1975, 11, 395.

68   J.T. Watson, in Ref. 6, p. 15.

69   M. Cohn and H.C. Urey, J. Amer. Chem. Soc., 1938, 60, 679.

70   M. Sano, Y. Yotsui, H. Abe, and S. Sasaki, Biomed. Mass Spectom., 1976,
      3, 1.

71   D.E. Matthews and J.M. Hayes, Anal. Chem., 1978, 50, 1465.

72   C.T. Gregg, in Ref. 2, p. 175.

73 G.C. Levy and G.I.. Nelson, "Carbon-13 Nuclear Magnetic Resonance for Organic Chemists," Wiley-Interscience, New York, 1972.

74 J.B. Grutzner, Lloydia, 1972, 35, 375,

75 U. Séquin and A.I. Scott, Science, 1974, 186, 101.

76 J.B. Stothers, "Topics in Carbon-13 NMR Spectroscopy," G.C. Levy, ed., Wiley-Interscience, New York, 1974, Vol. 1, pp. 230-286.

77 J. Feeney, N. Tech. Biophys. Cell Biol., 1975, 2, 287.

78 A.G. McInnes, J.A. Walter, J.L.C. Wright, and L.C. Vining, "Topics in Carbon-13 NMR Spectroscopy," G.C. Levy, ed., John Wiley and Sons, New York, 1976, Vol. 2, pp. 123-178.

79 N.A. Matwiyoff in ref. 2, p. 153.

80 G. Lukacs, Bull. Soc. Chim. France, 1972, 351.

81 R.R. Ernst and W.A. Anderson, Rev. Sci. Inst., 1966, 37, 93.

82 I.M. Lockhart, unpublished observations.

83 L. Pohl, W. Theyson, and R. Unger, Ger. Offen., 2,257,005, Chem. Abstr., 1974, 81, 77439, Brit. Patent, 1,405,970.

84 E.W. Randall and D.G. Gillies, Prog. Nucl. Magn. Reson. Spectrosc., 1971, 6, 119.

85 M. Witanowski and G.A. Webb, eds., "Nitrogen NMR," Plenum Press, London, 1973.

86 K. Wetzel, H. Faust, and W. Hartig, in Ref. 5, p. 400.

87 S. Pinchas and I. Laulicht, "Infrared Spectra of Labelled Compounds," Academic Press, London, 1971.

88 R.S. McDowell, Anal. Chem., 1970, 42, 1192.

89 S.D. Bloom, R.C. Harney, and F.P. Milanovich, Appl. Spectrosc., 1976, 30, 64.

90 R.A. Wood, K.A. Nagy, N.S. MacDonald, S.T. Wakakuwa, R.J. Beckman, and H. Kaaz, Anal. Chem., 1975, 47, 646.

91 D.C. Billington and B.T. Golding, J. Chem. Soc. Chem. Commun., 1978, 208.

92 P.H. Buckley and T. Frenkiel, Unpublished observations.

93 M.F. Grostic and K.L. Rinehart, "Mass Spectrometry: Techniques and Applications," G.W.A. Milne, ed., Wiley-Interscience, New York, 1971, pp. 217-287.

94 N. Neuss, Methods Enzymol., 1975, 43, 404.

95 M. Tanabe, Specialist Periodical Reports Biosynthesis, The Chemical Society, London, 1975, Vol. 3, p. 247.

96 M. Tanabe, Specialist Periodical Reports Biosynthesis, The Chemical Society, London, 1976, Vol. 4, p. 204.

97 P.A. Fawcett and E.P. Abraham, Specialist Periodical Reports Biosynthesis, The Chemical Society, London, 1976, Vol. 4, p. 248.

98 N. Neuss, C.H. Nash, P.A. Lemke, and J.B. Grutzner, J. Amer. Chem. Soc., 1971, 93, 2337 and 5314 (correction).

99   N. Neuss, C.H. Nash, P.A. Lemke, and J.B. Grutzner, Proc. Roy. Soc.
      Ser. B, 1971, 179, 335.

100   F.-C. Huang, J.A. Chan, C.J. Sih, P. Fawcett, and E.P. Abraham,
      J. Amer. Chem. Soc., 1975, 97, 3858.

101   C.M. Stevens, E.P. Abraham, F.-C. Huang, and C.J. Sih, Fed. Proc., 1975,
      34, 625.

102   J.W. Westley, R.H. Evans, G. Harvey, R.G. Pitcher, D.L. Pruess,
      A. Stempel, and J. Berger, J. Antibiot., 1974, 27, 288.

103   M. Meselson and F.W. Stahl, Proc. Natl. Acad. Sci. U.S.A., 1958, 44, 671.

104   T. Cronholm, R. Curstedt, and J. Sjövall, in Ref. 5, p. 97, and Refs.
      therein.

105   A.F. Hofmann and P.D. Klein, in Ref. 6, p. 189.

106   D. Samuel, in Ref. 3, p. 196.

107   D.R. Knapp, T.E. Gaffney, R.E. McMahon, and G. Kiplinger, J. Pharmacol.
      Exp. Ther., 1972, 180, 784.

108   M. Anbar, J.H. McReynolds, W.H. Aberth, and G.A. St. John, in Ref. 3,
      p. 274.

109   T.E. Gaffney, C.-G. Hammar, B. Holmstedt, and R.E. McMahon, Anal. Chem.,
      1971, 43, 307.

110   D.R. Knapp, T.E. Gaffney, and K.R. Compson, Adv. Biochem.
      Psychopharmacol., 1973, 7, 83.

111   M.G. Horning, J. Nowlin, K. Lertratanangkoon, R.N. Stilwell, W.G.
      Stillwell, and R.M. Hill, Clin. Chem., 1973, 19, 845.

112   J.M.Strong, J.S. Dutcher, W.-K. Lee, and A.J. Atkinson, Clin. Pharmacol.
      Ther., 1975, 18, 613.

113   C. Gregg, J. Hutson, J. Prine, D. Ott, and J. Furchner, Life. Sci.,
      1973, 13, 775.

114   H. Spielmann, H.G. Eibs, D. Nagel, and C.T. Gregg, Naunyn-Schmiedebergs
      Arch. Pharmakol. Exp. Pathol., 1975, 287, Supp. R86.

115   H. Spielmann, H.-G. Eibs, U. Jacob, and D. Nagel, in Ref. 6, p. 217.

116   J.B. Watkins, D.A. Schoeller, P.D. Klein, D.G. Ott, D.B. Neucomber, and
      A.F. Hofmann, in Ref. 5, p. 274.

117   H. Helge, B. Gregg, C. Gregg, S. Knies, I. Nötges-Borgwardt, B. Weber,
      and D. Neubert, in Ref. 6, p. 227.

118   D. Halliday and A.G. Miller, Biomed. Mass Spectrom., 1977, 4, 82.

119   P.J. Garlick and J.C. Waterlow, in Ref. 7, p. 323.

120   D. Halliday and R.O. McKeran, Clin. Sci. Mol. Med., 1975, 49, 581.

121   R. Varcoe, D. Halliday, E.R. Carson, P. Richards, and A.S. Tavill,
      Clin. Sci. Mol. Med., 1975, 48, 379.

122   "Nitrogen-15 in Soil Plant Studies," International Atomic Energy Agency,
      Vienna, 1971.

123 W.V. Bartholomew, in Ref. 122, p. 1.

124 R.D. Hauck and V.J. Kilmer, in Ref. 5, p. 655.

125 H. Broeshart, in Ref. 122, p. 47.

126 G. Stanford, J.O. Legg, and T.E. Staley, in Ref. 5, p. 667.

127 J.O. Legg and C. Sloger, in Ref. 5, p. 661.

128 S. Ivanko, in Ref. 122, p. 119.

129 I. Havassy, K. Boda, K. Kosta, J. Varady, and E. Rybosova, in Ref. 7, p. 363.

130 A. Hennig, K. Gruhn, and G. Jahreis, in Ref. 7, p. 371.

131 G. Gebhardt. T. Zebrowska, W. Souffrant, and R. Köhler, in Ref. 7, p. 383.

132 R.D. Hauck, in Ref. 122, p. 65.

ISOTOPIC LABELLING WITH CARBON-14 AND TRITIUM

Dr E Anthony Evans

The Radiochemical Centre Limited,

Amersham, Buckinghamshire, England

INTRODUCTION

The relative ease of automated measurement of radioactivity in large numbers of samples, when compared for example with the measurement of mass, and the great sensitivity with which small quantities of radioactive compounds can be accurately measured, has resulted in the widespread use of radiochemicals in research. Organic compounds labelled with carbon-14 and tritium (hydrogen-3) are of special importance in biological research. Indeed, much of the research in the life sciences has been made possible only by the commercial availability of a wide selection of organic compounds, such as amino acids, carbohydrates, drugs, nucleosides, nucleotides, vitamins and steroids, labelled with ^{14}C or ^{3}H.

In isotopic labelling a compound is labelled with an isotope of an element already present in the compound such that it is identical (apart from the isotope) with the normal unlabelled form. It should also be clearly distinguished from nonisotopic, sometimes referred to nonnuclidic, labelling in which a compound is labelled with an isotope of a "foreign" element not normally present in the unlabelled substance. Examples of this latter type are the numerous peptides and proteins labelled with radioactive iodine isotopes extensively used, for example, in radioimmunoassays.

The development of isotopic labelling methods with carbon-14 and tritium continues to challenge the ingenuity of the biochemist and organic chemist. Unlike the preparation of compounds

labelled with stable isotopes, syntheses and biosyntheses with
these two radionuclides usually involve very small scale
preparations with tens of milligrams for carbon-14 and often only
a few milligrams in the case of preparations with tritium.
However, although the chemical scale is small, such preparations
often involve large quantities of radioactivity and the safe
handling of multicuries of labelled organic compounds, for which
special training of staff and laboratory facilities are usually
necessary.

There are already a number of excellent published recent texts
concerning methods for the preparation of compounds labelled
with carbon-14 (1-3) and with tritium. (1-4) In this paper general
methods of isotopic labelling with ^{14}C and with ^{3}H are briefly
reviewed with special attention to examples of compounds likely
to be of wide interest in biological research.

PROPERTIES OF CARBON-14 AND TRITIUM

In biological research the choice of tracer radionuclide is
often between carbon-14 or tritium and sometimes the use of both
radionuclides in an experiment is desirable. However, there is
seldom a requirement to have both radionuclides in the same
molecule and a mixture of the ^{14}C- and ^{3}H-labelled forms is
usually sufficient. In such experiments using double isotopically
labelled compounds, the need to have both labelled forms of the
compound in a similar state of radiochemical purity is a
requirement to avoid spurious isotope effects.

Each of these two radionuclides have properties which may
make it the isotope of choice for a particular investigation.
Some of these properties of ^{14}C and ^{3}H are summarised in Table 1.

Both isotopes have a conveniently long half-life and there is
seldom a requirement to correct for natural decay of the radionuclide during experiments with ^{14}C- or with ^{3}H-labelled compounds.

The production of carbon-14 $^{(1,10)}$ by irradiation of aluminium
nitride in a reactor for about two years and of tritium $^{(4)}$ by
irradiation of lithium (often in the form of alloys) for several
weeks, ensures both these isotopes are available at 100 per cent
radionuclidic purity i.e. all the radioactivity is in the
specified nuclide. In addition, the production methods also ensure

TABLE 1. PROPERTIES OF CARBON-14 AND TRITIUM

PROPERTY	CARBON-14	TRITIUM
Radiation	Beta (100%)	Beta (100%)
Half-life	5730 years	12.35 years
Energy (maximum)	0.159 MeV	0.0186 MeV
Range of beta particles in photographic emulsion	100 μm	1 μm
Production method – reactor irradiation	AlN $^{14}N(n,p)^{14}C$	Li alloys $^{6}Li(n,\alpha)^{3}H$
Decay product	^{14}N (stable)	^{3}He (stable)
Available isotopic abundance	100%	100%
Nuclear spin	Zero	$\frac{1}{2}$
Maximum specific activity	62.4 mCi/mA	29.12 Ci/mA
Common specific activities for compounds	50-300 mCi/mmol	10-100 Ci/mmol
Method of measurement (efficiency)	Liquid beta scintillation (80%)	Liquid beta scintillation (40%)
Method for determination of patterns of labelling	Degradation but usually labelling pattern unambiguous from method of synthesis	Tritium nuclear magnetic resonance (tnmr) spectroscopy
Tracer uses	As tracer for carbon	As tracer for hydrogen and carbon structures
Cost (raw materials)	£2000 per Ci	£2 per Ci

that these radionuclides are available at virtually 100 per cent
isotopic abundance. However, although it is possible to prepare
^{14}C- and ^{3}H-labelled compounds of almost 100 per cent isotopic
purity i.e. near theoretical maximum specific activity, only
tritium labelled compounds can be made at very high molar specific
activities; in fact 2-3 orders of magnitude higher than for
^{14}C-labelled compounds. Both radionuclides decay to stable
isotopes as is shown in the table.

The weak beta radiation from ^{14}C and even weaker from ^{3}H,
is normally measured by liquid beta scintillation counting
(5)
with the counting efficiency for carbon-14 about twice that for
tritium using currently available equipment. The very weak energy
and hence low penetrating power of the beta particles from tritium
has made this radionuclide of unique importance for studies
(6,7)
involving the use of high resolution autoradiographic techniques.
Isotopic labelling with tritium therefore permits the precise
localisation of the tracer compound in biological tissues and
cells; such studies are not possible with stable isotopically
labelled compounds.

Measurement of radioactivity obviously follows the fate of the
radioactive atoms that are put into the experiment, hence for
many tracer applications it is necessary to know precisely the
position of the radioactive atom(s) in the labelled compound being
used as tracer. Until the development of tritium nuclear magnetic
(8,9)
resonance spectroscopy for routine use, the distribution of
isotope was not always known with certainty and imposed a serious
limitation in the uses of tritium labelled compounds as tracers
for carbon structures, especially in biological systems.
Fortunately one property of tritium is that it has a nuclear spin
which makes tritium nuclear magnetic resonance spectroscopy
possible.

Cost is always an important consideration in the design of radiochemical tracer experiments. Cost is of two main kinds; the purchase of radionuclide or labelled intermediate, and the investigator's "own time". In Universities and similar academically funded research organisations,attention tends to be focussed on actual "cash flow" and "own time" will often have its own educational value in, for example, synthesizing a labelled compound which might have been purchased. If a realistic cost is attributed to "own time", as in most commercial organisations, it usually pays to buy labelled compounds, unless there is an unusually large and recurrent need, and is often not beneficial not to purchase the final labelled compound even in this case. The cost of a compound labelled with carbon-14 or with tritium is usually a small fraction of the total investment in a research project.

Although ^{14}C-labelled compounds are normally preferred for studies of catabolism and metabolism, in pharmacology for example, the relative ease of labelling complex molecules with tritium, coupled with the properties of tritium and the great versatility of this isotope, often makes tritium labelled compounds the tracers of choice for many investigations.

LABELLING TECHNIQUES

In addition to reviewing general methods for labelling compounds, it is necessary to identify the types of labelling which can be achieved.

Types of labelling

Four main types of labelling should be distinguished as follows:

(a) Specific labelling - in which the radioactive atoms occupy known specific positions in molecules without ambiguity

(b) Uniform labelling ("U") - where the radioactive atoms are distributed throughout the labelled molecules in a statistically uniform pattern

(c) General labelling ("G") - where the radioactive atoms are distributed in a general or random pattern, not always known with any certainty, in the molecules

(d) Nominal labelling ("n or N") - is used to indicate the position of the radioactive atoms where there is uncertainty as to whether the labelling is confined to the positions specified.

Carbon-14 labelled compounds are normally either specifically labelled or uniformly labelled; tritium labelled compounds are seldom uniformly labelled but usually have specific or general labelling. Examples of the various types of labelling are given in the text which follows.

General methods

Current practical methods for isotopic labelling of compounds either with ^{14}C or with ^{3}H fall into three main categories as follows:

(a) chemical syntheses

(b) biochemical methods

(c) isotope exchange reactions

The first two general methods (a and b) are especially important for the preparation of carbon-14 labelled compounds and the first and last (a and c) are of particular importance for labelling compounds with tritium. Although all three methods can provide compounds with clearly defined patterns of labelling, this rather general statement needs cautious interpretation for tritium labelled compounds.

Characteristic of most radiochemical syntheses is the small chemical scale varying from a few milligrams to a gram or so i.e. generally about 2 to 10 millimoles. In general tritium labelled compounds are prepared on a very small chemical scale but often involving multicurie quantities of radioactivity whereas the syntheses of carbon-14 labelled compounds are on a larger chemical scale but involving millicuries of radioactivity.

Synthetic routes for the preparation of ^{14}C-labelled compounds
are frequently multi-stage whereas for ^{3}H-labelled compounds
there is seldom more than one or two stages with radioactive
products involved. In order to maximise yields the isotopic
label should be introduced at the latest practicable stage of the
synthesis.

Although in general the effects of radiation on the progress
of radiochemical syntheses has been observed to be minimal,
the possibility of such effects should always be borne in mind.
This is especially true for reactions involving materials at
very high specific activity where the effects of radiation can
result in low yields or failure of the reaction. (3,11) Such
effects are believed to be caused by free radical reactions
initiated by the radiation and some examples are found in the
preparation of ^{14}C- or ^{3}H-labelled styrene, acrylic acid,
β-propiolactone and acetylene. (11) The Claisen reaction for
coupling, for example, dimethyl oxalate with dimethylsuccinate
gives very reduced yields when either intermediate is labelled
with carbon-14 above about 10 mCi/mmol. (11)

Chemical syntheses for ^{14}C-labelled compounds. Syntheses of
^{14}C-labelled compounds tend to follow the broad lines of
classical organic chemistry albeit on a small chemical scale.
Barium [^{14}C]carbonate or [^{14}C]carbon dioxide are commonly used as
starting materials from which the labelled atom(s) is usually
derived. A number of extremely useful intermediates are
prepared by reduction reactions as follows:

Reduction to [^{14}C]methanol - [^{14}C]Carbon dioxide is reduced to
methanol with lithium aluminium hydride in tetrahydrofurfuroxy-
tetrahydropyran solvent (10) (1),

$$4CO_2 + 3LiAlH_4 \longrightarrow LiAl(OCH_3)_4 + 2LiAlO_2 \xrightarrow{ROH} 4CH_3OH + LiAl(OR)_4$$

$$\cdots\cdots(1)$$

ROH = tetrahydrofurfuryl alcohol

or by catalytic reduction with hydrogen in a Fischer-Tropsch type synthesis (12) (2):

$$CO_2 + 3H_3 \underset{\substack{200°-250°C \\ 150-600 \text{ psi}}}{\overset{\substack{Cu/Zn/Cr \\ \text{mixed catalyst}}}{\rightleftharpoons}} H_2O + CH_3OH \quad \cdots(2)$$

In the former process a small amount of $[^{14}C]$formate is produced (10) whereas the latter process gives virtually a quantitative conversion of the carbon-14 doxide to $[^{14}C]$methanol. The ^{14}C-labelled methanol is readily converted into $[^{14}C]$methyl iodide with hydriodic acid, and is another very important synthetic intermediate for further ^{14}C-labelled compounds.

Reduction to $[^{14}C]$cyanide - Barium $[^{14}C]$carbonate can be converted to barium $[^{14}C]$cyanamide with ammonia and the barium cyanamide converted into sodium $[^{14}C]$cyanide with sodium metal (13) (3):

$$BaCO_3 + 2NH_3 \xrightarrow{800°C} BaN \cdot CN + 3H_2O$$

$$BaN \cdot CN + 2Na \xrightarrow{(BaO/TiO_2)} 2NaCN + N_2$$

$$\cdots(3)$$

Alternatively $[^{14}C]$carbon dioxide can be reduced to $[^{14}C]$methane using a nickel catalyst and the methane converted catalytically into hydrogen $[^{14}C]$cyanide with ammonia (14) (4):

$$CO_2 + 4H_2 \xrightarrow[300°-350°C]{\text{Ni on pumice}} CH_4 + 2H_2O$$

$$CH_4 + NH_3 \xrightarrow[1000°-1100°C]{\text{Pt wool}} HCN + 3H_2$$

$$\cdots\cdots(4)$$

The hydrogen $[^{14}C]$cyanide is absorbed in methanolic potassium hydroxide and the potassium $[^{14}C]$cyanide extracted into liquid ammonia. The overall yield from $^{14}CO_2$ is virtually quantitative.

Reduction to $\left[^{14}C\right]$acetylene and to $\left[^{14}C\right]$benzene - A route into
the aromatic series of ^{14}C-labelled compounds is from $\left[^{14}C\right]$-
benzene via $\left[^{14}C\right]$acetylene. Heating lithium metal to 650°C in
$\left[^{14}C\right]$carbon dioxide gives lithium $\left[^{14}C\right]$carbide (15,16) which on
treatment with water yields $\left[^{14}C\right]$acetylene in good overall
yields (5):

$$Li \quad + \quad CO_2 \xrightarrow{\quad 650°C \quad} LiC \quad + \quad H_2O \longrightarrow HC\equiv CH$$

$$\dots\dots(5)$$

Trimerisation of the $\left[^{14}C\right]$acetylene can be effected by means
of suitable catalysts such as triphenylphosphine-nickel carbonyl, (17)
triethylaluminium and titanium tetrachloride (18) in hexadecane at
25°C or with niobium pentachloride. (19) The $\left[^{14}C\right]$benzene so prepared
is of course labelled uniformly in the ring.

Reduction to $\left[^{14}C\right]$urea - Heating barium $\left[^{14}C\right]$carbonate in ammonia
at 850°C yields barium $\left[^{14}C\right]$cyanamide which on treatment with
dilute sulphuric acid gives $\left[^{14}C\right]$urea (10) (6):

$$BaCO_3 + 2NH_3 \xrightarrow{\quad 850°C \quad} BaN.CN + 3H_2O \xrightarrow[0°-5°C]{H_2SO_4} H_2N.CO.NH_2 + BaSO_4$$

$$\dots\dots(6)$$

Miscellaneous reduction reactions - A number of other useful
one stage reactions with $\left[^{14}C\right]$carbon dioxide include reduction
to formate with sodium trimethoxyborohydride, (20) reduction to
elementary carbon with magnesium metal, (21) and reduction to $\left[^{14}C\right]$-
carbon monoxide with zinc metal (wool) at 410°C. (22)

Most of the basic conversions of $\left[^{14}C\right]$carbon dioxide were
worked out in the late 1940's and early 1950's (10) and, apart
from modifications to catalysts for example, there have not been
many further significant advances.

Illustrative of the multistage requirements for the synthesis of complex ^{14}C-labelled compounds is the preparative route for the isotopic labelling of cholecalciferol (vitamin D_3). This important vitamin labelled in the 4-position is prepared in nine stages from $\left[^{14}C\right]$methyl iodide. The yield of $\left[4-^{14}C\right]$cholester-ol from $\left[^{14}C\right]$methyl iodide is 30 per cent; (11,23,24) the yield of $\left[4-^{14}C\right]$vitamin D_3 is 1.5 per cent based on $\left[^{14}C\right]$methyl iodide and 5 per cent from $\left[^{14}C\right]$cholesterol. (11) The route is outlined in (7).

Chemical syntheses for ^{3}H-labelled compounds. Current emphasis is on methods which can provide tritium labelled compounds at very high molar specific activity (containing four or more tritium atoms per molecule) and specifically labelled in well defined positions. Such compounds are required for example, (6,7) in high resolution autoradiographic studies and in receptor (25) assays used to pre-screen the behaviour of drugs which can therefore often reduce the number of animals required for in vivo testing.

Radiochemical syntheses with tritium are generally one or two stage reactions and are usually much less complex than those used for isotopic labelling with carbon-14. Starting materials are tritium gas, tritiated water or tritiated metal hydrides.

Reduction reactions are of even greater significance for isotopic labelling with tritium than for carbon-14 as even complex organic compounds are labelled by simple one-step reductions. Such reactions fall into four main groups as follows:

Reduction by catalytic hydrogenation - Simple and complex organic compounds are readily labelled by catalytic addition of tritium (4) to suitable unsaturated precursors. Examples are the preparation of 9,10-dihydro$\left[9,10-^3H\right]$ergocryptine by reduction of ergocryptine using a 10 per cent palladium/charcoal catalyst in dioxan (8)

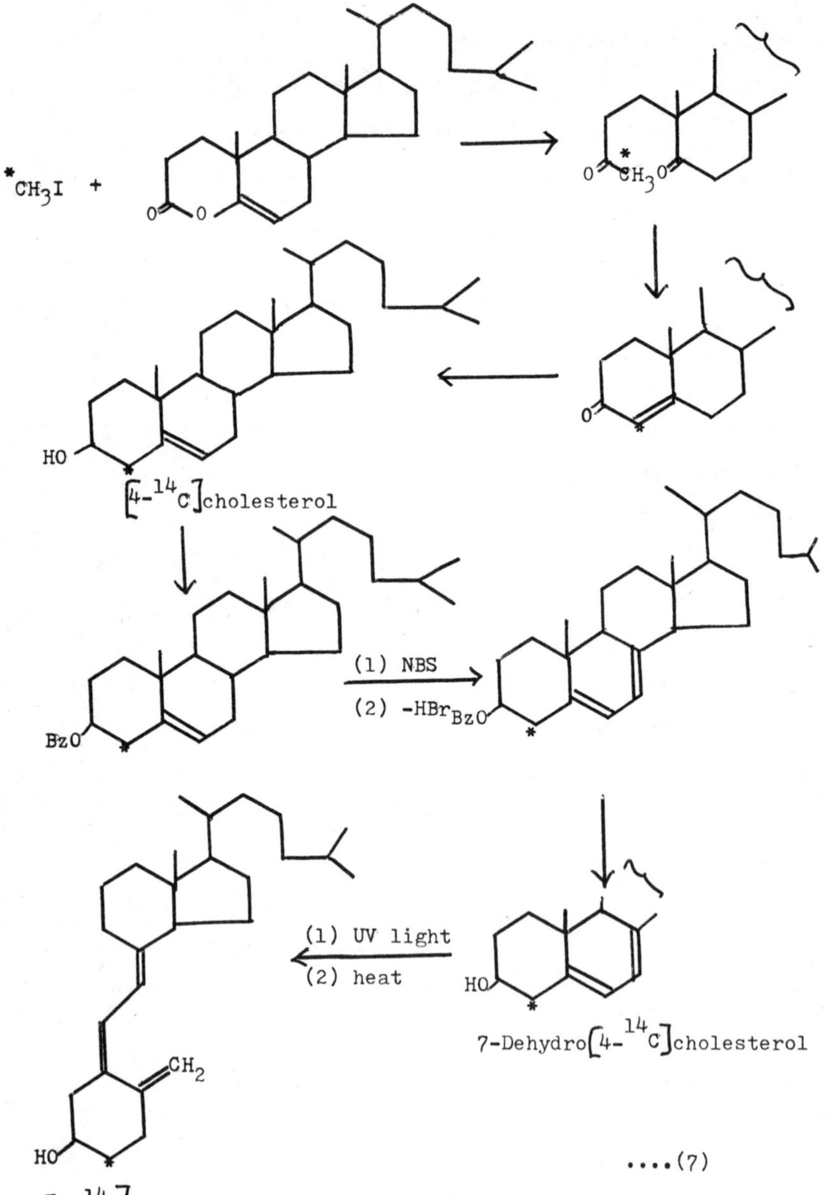
*CH3I +
0
0
0
*CH3
0
0
*
HO
*
[4-14C]cholesterol
(1) NBS
(2) -HBr
BzO
*
BzO
*
(1) UV light
(2) heat
HO
*
7-Dehydro[4-14C]cholesterol
CH2
HO
*
....(7)

[4- ^{14}C]cholecalciferol (vitamin D_3)

*Position of carbon-14 atom; NBS is N-bromosuccinimide and Bz is benzoyl

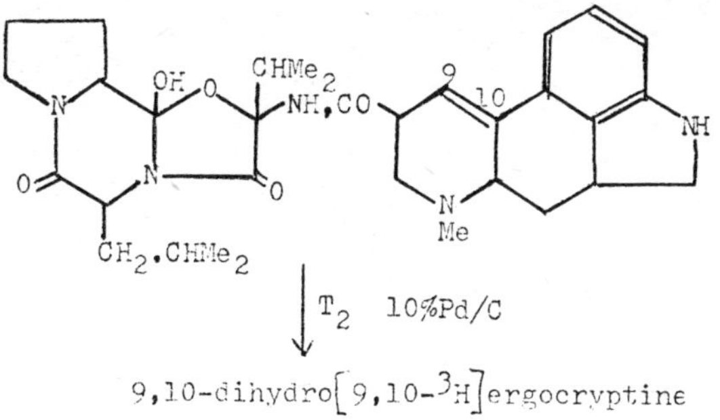

$$9,10\text{-dihydro}\left[9,10\text{-}^{3}\text{H}\right]\text{ergocryptine}$$

$$\ldots\ldots(8)$$

and the hydrogenation of 15,16-didehydronaltrexone with tritium
in ethanol solution using a 10% palladium/charcoal catalyst [26] (9)
to yield the potent narcotic antagonist $\left[15,16\text{-}^{3}\text{H}\right]$naltrexone.

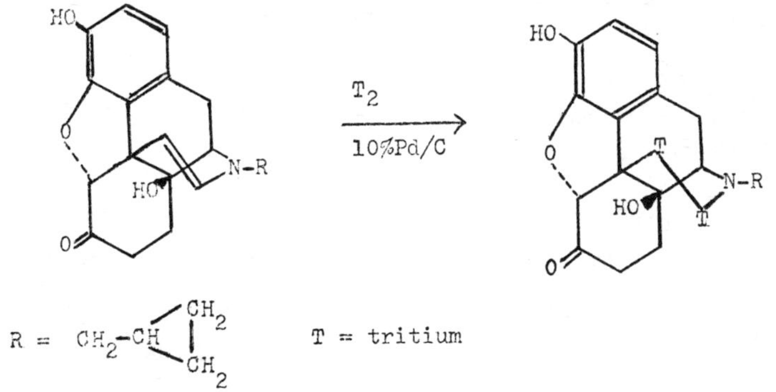

$$\ldots\ldots(9)$$

Both tritium labelled ergocryptine and naltrexone are important
tracer compounds in neurochemistry. [25]

Numerous steroid hormones are conveniently labelled with
tritium by reduction of unsaturated precursors. However, in this
case the choice of catalyst is important if a high degree of
stereospecific labelling is required. An example is the preparat-
ion of $\left[1,2\text{-}^{3}\text{H}\right]$testosterone by the catalytic reduction of

androsta-1,4-dien-3-one-17β-ol (10):

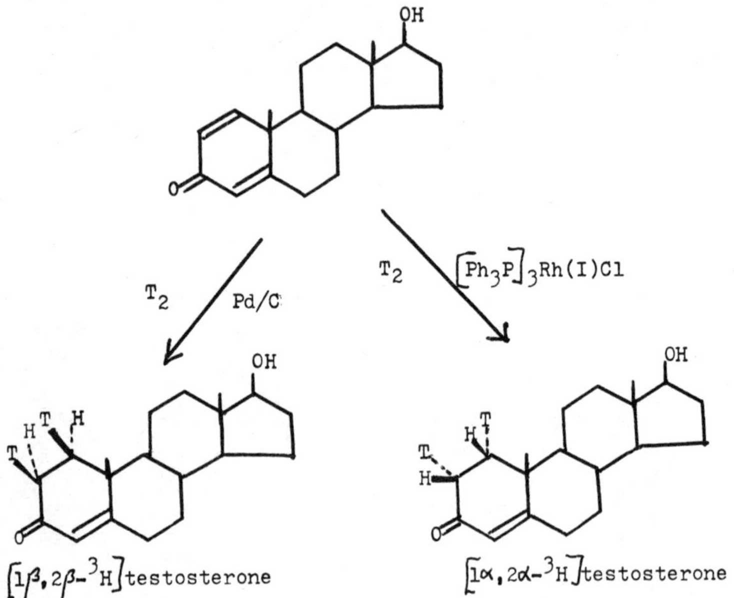

T_2 / Pd/C

T_2 \ $\left[Ph_3P\right]_3Rh(I)Cl$

$\left[1\beta,2\beta-^3H\right]$testosterone

$\left[1\alpha,2\alpha-^3H\right]$testosterone

....(10)

Using palladium/charcoal (heterogeneous) as catalyst yields
mainly $1\beta,2\beta-^3H$ labelling whereas the homogeneous (Wilkinson)
catalyst tris(triphenylphosphine) rhodium(I) chloride yields
mainly $1\alpha,2\alpha-^3H$ labelling. Neither catalyst is completely
stereospecific and the distribution pattern of labelling has
been determined by tritium nuclear magnetic resonance spectroscopy. (9)
Numerous other steroids and miscellaneous compounds are labelled
with tritium by similar reduction methods. (4)

Reduction by catalysed halogen-tritium replacement - Frequent
use is made of the catalytic replacement of a halogen atom
(usually Cl, Br or I) with tritium. An example is the labelling
of the tricyclic antidepressant drug imipramine from 3-chloro-
imipramine using a palladium supported on charcoal catalyst (11):

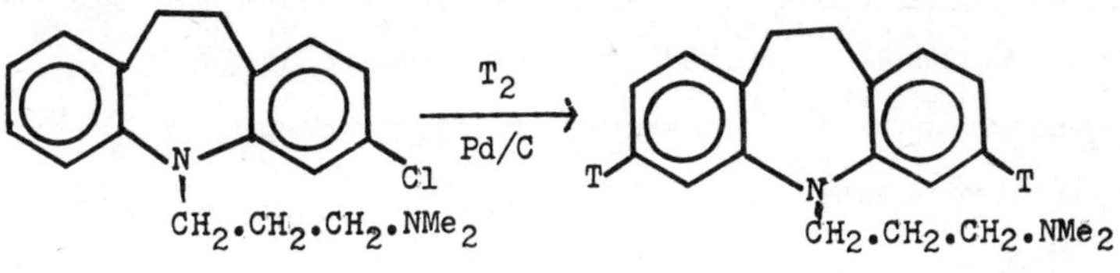

$$\ldots\ldots(11)$$

Dioxan is used as a suitable solvent and a few microlitres of an organic base (e.g. diisopropylethylamine) is added to prevent the tritium halide formed in the reaction from poisoning the catalyst. It should be noted also that in this example the 3- and 7-positions are indistiguishable after labelling and hence the labelled compound produced is $\left[3(7)-^3H\right]$imipramine.

<u>Reduction with tritiated metal hydrides</u> - Lithium, potassium and sodium borohydrides are readily prepared isotopically labelled by heating in tritium gas [4]. These tritiated metal hydrides are very important in the synthesis of numerous tritium labelled compounds by reduction of aldehydo-, keto(oxo)- and imino- groups. An example is the preparation of $\left[7-^3H\right]$noradrenaline from noradrenalone (12):

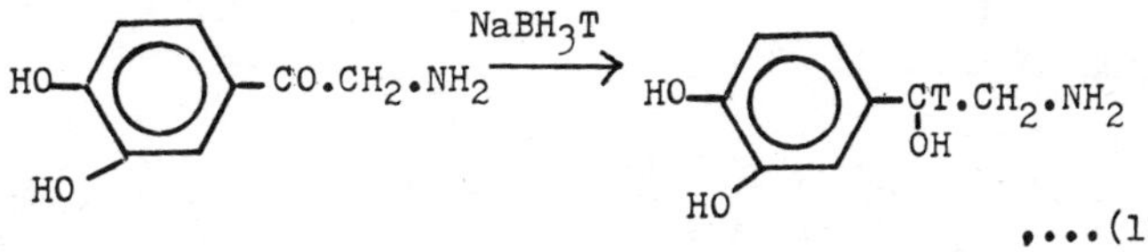

$$\ldots\ldots(12)$$

Tritiated borohydrides are especially useful in the preparation [4] of a number of tritium labelled carbohydrates.

Lithium aluminium hydride labelled with tritium (from lithium tritide and aluminium trichloride) is similarly used in many reduction reactions.

Reduction with tritiated water - Grignard reagents and organo-
lithium compounds (prepared from the corresponding halogen
compounds) can be used to prepare tritiated compounds by reaction
with tritiated water (13):

$$R.MgX \text{ (or R.Li)} \quad + \quad THO \quad \longrightarrow \quad R.T$$

....(13)

An example is the synthesis of $\left[1-^3H\right]$ naphthalene from 1-naphthyl
(4)
magnesium bromide and tritiated water.   It is interesting to
note that quite large isotope effects are observed with the
(4)
kinetics of this reaction.

Multistage chemical syntheses with tritium - Although a synthesis
involving many stages is more unusual for tritium than for
carbon-14 labelled compounds, it is of course necessary for some
compounds. A good example is the synthesis of 25-hydroxy$\left[23,24-^3H\right]$-
(27)
cholecalciferol,   The unlabelled material is formed in the liver
(14) as an important metabolite of vitamin D$_3$.

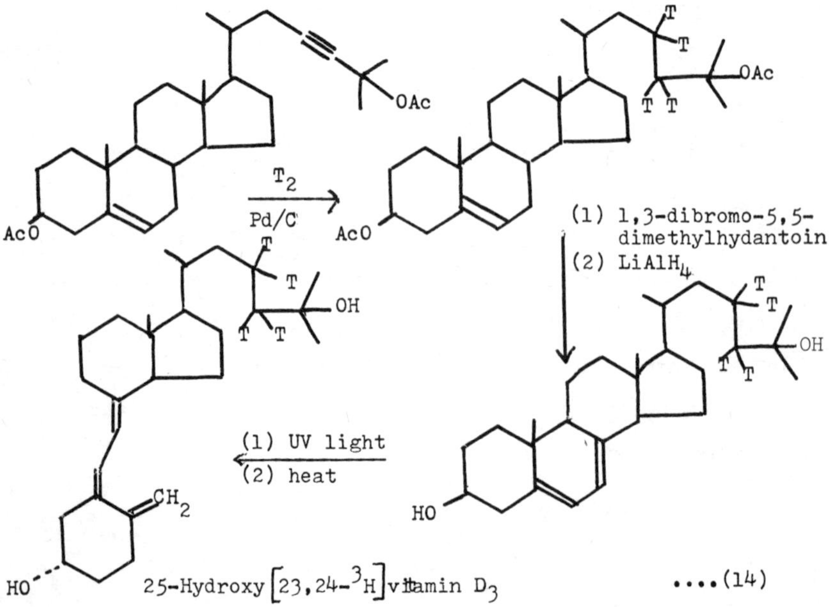

25-Hydroxy $\left[23,24-^3H\right]$ vitamin D$_3$ ....(14)

Reduction of cholest-5-en-23-yne-3β,25-diol diacetate with

tritium in tetrahydrofuran using a palladium/charcoal catalyst

introduces four tritium atoms into the molecule. Bromination

with 1,3-dibromo-5,5-dimethylhydantoin followed by dehydro-

bromination with collidine yields, after hydrolysis of the

acetoxy groups, 7-dehydro$[23,24-^3H]$cholesterol which on irradiation

with uv light and isomerisation of the pre-vitamin D_3 by heating
(27)

gives the required 25-hydroxy$[23,24-^3H]$cholecaciferol. This

five stage synthesis compares with the labelling of the same

compound in the 26(27)-methyl group which requires only a one-

step reaction of $[^3H]$methyl magnesium iodide (or $[^3H]$methyl
(28)

lithium) with 25-keto-26-norcholecalciferol-3-acetate. The

maximum theoretical specific activity in this case is 87 Ci/mmol

if all three hydrogen atoms of the methyl group are replaced

by tritium atoms, compared with over 100 Ci/mmol achieved
(11)

through the acetylenic intermediate route shown (14).

Advantages and disadvantages of chemical syntheses. The main

advantage of chemical synthesis of a labelled compound is the

ability to control the specificity of labelling. This is usually

unambiguous in the case of carbon-14 labelled compounds from the

synthetic route chosen, and also of the reduction methods using

tritiated metal hydrides or tritiated water for tritium labelled

compounds. However, in the reduction of unsaturated precursors

or in halogen-tritium replacement reactions, specific labelling

is often expected but not always achieved in practice. Non-

specific hydrogen-tritium exchange is always a possibility in the

presence of metal hydrogen transfer catalysts (such as Pt or Pd).

An example to illustrate this point is the preparation of

tritiated folic acid by catalysed halogen-tritium replacement

from 3',5'-dibromofolic acid (15). By chemical degradation of

the tritiated folic acid it was shown that only about 40 per cent

(29)

of the tritium activity resides in the 3',5'-positions.

$$\left[3',5',7,9-^3H\right] \text{Folic acid} \qquad \qquad \cdots\cdots(15)$$

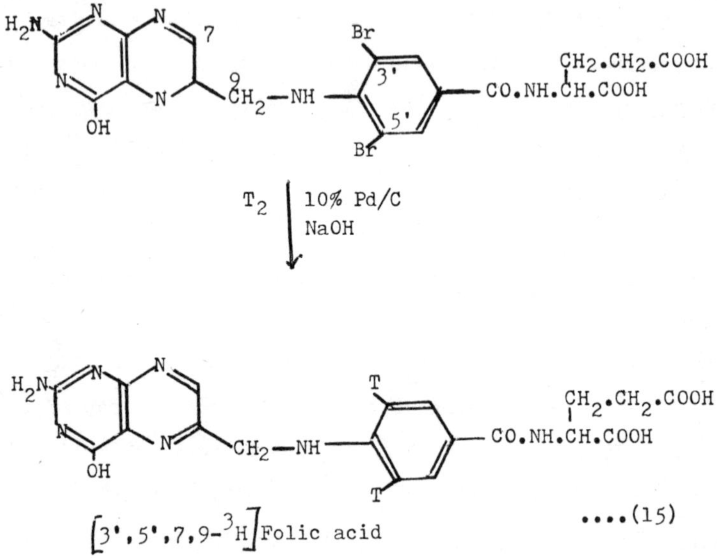

Originally it was believed that the remaining 60 per cent of the
(29)
tritium activity was located at the 9-methylene position.
However, labelling patterns in tritiated folic acid and an
analogous compound (methotrexate) by tritium nuclear magnetic
resonance spectroscopy indicated that oxidative degradation with
permanganate gave a false tritium distribution and that 25 per
cent of the tritium activity in folic acid is located at the
(30)
7-position of the pteridine ring.

This non-specific isotopic substitution could present a serious
problem in some applications of tritium labelled compounds as
tracers but the confirmation of patterns of labelling by tritium
nmr spectroscopy has now virtually eliminated this potential
disadvantage.

Another important advantage of chemical syntheses for tritium labelled compounds is the ability to achieve very high molar specific activities, often more than 100 Ci/mmol. However, for carbon-14 labelled compounds multiple or uniform labelling is limited to a few reactions in practice. In turn this is a limitation on specific activity. Fortunately tritium is relatively a low cost isotope by comparison with carbon-14 and radiochemical yields are therefore less important for tritium labelled compounds than for carbon-14 labelled compounds.

Disadvantages of chemical syntheses include difficulties with stereoisomers and the limitations of specific activity for carbon-14 labelled compounds already mentioned. Resolution of optically active compounds is difficult on a small scale and becomes very inefficient when two or more active centres are involved unless stereospecific enzymes are used.

Biochemical methods. There are many publications of biochemical methods for isotopic labelling with carbon-14 and with tritium. These methods employ either a purified or partially purified enzyme, or intact organism or cells. A few of these reactions lend themselves to commercial scale manufacture of labelled compounds.

Biosynthesis is effective and important for labelling a range of ^{14}C-labelled compounds widely used in tracer applications. The common L-amino acids, carbohydrates, nucleosides and nucleotides are readily available in their natural configurations, uniformly labelled, by growing algae on $[^{14}C]$carbon dioxide or by photosynthesis in detached plant leaves. This approach, properly used, can give high specific activities (up to 95 per cent isotopic abundance in all carbon atoms) and good yields i.e. economic in the commercial sense. The blue-green alga

<u>Anacystis</u> <u>nidulans</u>, which is even more radiation resistant than
the previously used <u>Chlorella</u>, has the specific advantage of
giving practicable yields of the deoxyribonucleosides and
(31)
deoxyribonucleotides.   Attempts to use <u>E.coli</u> for this purpose
with a substrate of $\left[\text{U-}^{14}\text{C}\right]$glucose at high specific activity
were unsuccessful, apparently because of the organism's sensitivity
(11)
to radiation for reproductive growth.

   High yields of $\left[\text{U-}^{14}\text{C}\right]$sucrose, up to 70 per cent of $^{14}\text{CO}_2$,
are obtainable in detached leaves even with carbon-14 at greater
than 95 per cent isotopic abundance because the time exposure is
short (a few hours photosynthesis) and no reproductive growth
(32)
is required.   Tobacco and <u>Canna indica</u> leaves are often used.

   DNA labelled with ^{14}C or ^{3}H is prepared by feeding
respectively $\left[^{14}\text{C}\right]$- or $\left[^{3}\text{H}\right]$-thymine to <u>E.coli</u>.   (33)   The use of such
labelled organic substrates gives a much wider choice of biosynth-
etic organisms; however, the risk of nonspecific labelling
probably discourages some uses of biosynthesis. A classical
example is $\left[3,4\text{-}^{14}\text{C}\right]$glucose, in which from 4-25 per cent of the
radioactivity may "wander" into the C_1, C_2, C_5 and C_6 positions
(34,35)
from a specifically labelled substrate.   Determination of the
patterns of labelling for carbon-14 labelled compounds is not
easy as there is no simple "non-destructive" spectroscopic method
and careful work is needed to ensure that degradation reactions
are correctly interpreted.

   Biosynthetic labelling with tritium, except by efficient
specific enzyme syntheses, has proved of limited practical use,
(4)
due mainly to the limitations imposed by radiation effects.   In
addition isotope exchange and synthetic approaches are relatively
attractive and easy with tritium so that biochemical methods
have been less studied than those for carbon-14 labelled compounds.

Frequent use is indeed made of enzymes in a purified or partially purified form for the conversion of ^{14}C - and ^{3}H -labelled substrates. One of the most important examples is the synthesis of $[^{14}C]$- or $[^{3}H]$-labelled thymidine from the corresponding ly labelled thymine using a crude enzyme transdeoxyribosylase isolated from cattle liver$^{(4)}$ (16). Deoxyuridine is normally employed as the donor of the deoxyribose moiety.

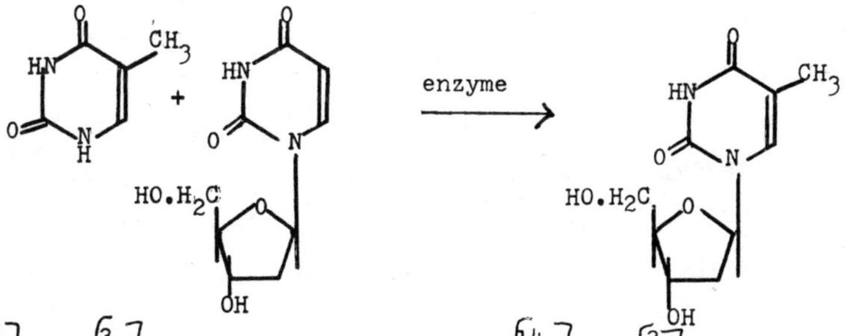

$[^{14}C]$- or $[^{3}H]$-
thymine   deoxyuridine

enzyme →

$[^{14}C]$- or $[^{3}H]$-
thymidine

....(16)

The influence of various factors, for example drugs, on cell replication and DNA synthesis is of great interest generally and especially important in the testing of potential cytotoxic drugs for possible use in the chemotherapeutic treatment of cancer. As a specific tracer for DNA, it is perhaps not surprising that isotopically labelled thymidine is currently the most widely used tracer compound.

Biological hydroxylations, of steroids for example, have also been utilized either with an enzyme in the case of the preparation of 2-hydroxy$[4-^{14}C]$oestradiol,$^{(36)}$ or with an intact organism as in the case for 15α-hydroxy$[4-^{14}C]$progesterone.$^{(37)}$

The scope of isotopic syntheses using enzymes is of course
very great; the common difficulty from a commercial point of
view is to make the reactions work consistently, in reasonable
economic yield and on the required scale.

Isotope exchange reactions. In isotope exchange reactions an atom
in a molecule is substituted by its radioactive equivalent
(say ^mZ) as illustrated in (17)

$$A^mZ \quad + \quad B^nZ \quad \rightleftharpoons \quad A^nZ \quad + \quad B^mZ$$

$$\dots\dots(17)$$

Such reactions are of course reversible. Although isotope
reactions have been known for several decades, they have very
limited practical   application to isotopic labelling with
carbon-14. However, such exchange reactions are extremely
important and are widely used for isotopic labelling with tritium.

The preparation of tritium labelled compounds by hydrogen
isotope exchange can be classified into the following types
of reactions:

Isotope exchange reactions with tritium gas - Exchange reactions
with tritium catalysed by radiation were developed by Wilzbach [38]
in 1956 and consisted simply of exposing the compound to tritium
gas at virtually 100 per cent isotopic abundance. This development
did much to promote the uses of tritium labelled compounds in the
late 1950's. There are numerous publications relating to this
technique and several modifications to the original method
including perhaps the most promising, that of tritium labelling
by microwave discharge activation of tritium gas. [39,40] However,
because of the serious difficulties in purifying compounds
labelled this way and the relatively low molar specific activities
attainable, alternative methods of labelling with tritium are now
preferred. In practice the Wilzbach method (and modifications of it)

is seldom used for commercial manufacture of tritium labelled
    (4)
compounds.

Catalysed hydrogen isotope exchange in solution with tritium
                               (41)
gas is a method developed by Evans et al   for isotopic labelling

of a wide variety of compounds with tritium. The reaction depends

upon the ability of hydrogen atoms occupying certain positions in

molecules to exchange with tritium on the surface of a metal

hydrogen transfer catalyst such as Pt or Pd. A mechanism (18)
                                                             (4)
has been proposed which assumes the formation of a carbanion   or
                       (41)
free radical   intermediate species, but the exact mechanism is

still awaiting further studies.

$$RH + M \rightleftharpoons (RM)^- + H^+$$

$$T_2 + M \rightleftharpoons (TM)^+ + T^-$$

$$(RM)^- + (TM)^+ \rightleftharpoons (RTM) + M$$

$$(RTM) \rightleftharpoons RT + M$$

$$T^- + H^+ \rightleftharpoons HT$$

M= metal catalyst; T= tritium atom   ....(18)

The anion $(RM)^-$ is protonated with a tritium atom formed by

dissociation of a molecule of tritium gas on the catalyst surface,

prior to desorption of the compound. In order to explain the

high molar specific activities obtained in practice and the fast

rate of the exchange reaction, the tritium atom must arise directly

from a molecule of tritium gas and not by dissociation of a trit-

iated water molecule on the surface of the catalyst.

This method of isotopic labelling with tritium is suitable for

labelling benzylic compounds, steroids, drugs, purine nucleosides

and purine nucleotides, and numerous other compounds. Some

examples are given in Table 2.

TABLE 2. ISOTOPIC LABELLING BY CATALYSED EXCHANGE IN SOLUTION WITH TRITIUM GAS

COMPOUND*	SPECIFIC ACTIVITY Ci/mmol	POSITION OF TRITIUM ATOMS**
L-Arabinose	10.0	1
2-Deoxy-D-glucose	10.8	1
L-Fucose	1.0	1
D-Glucose	3.9	1
Lactose	3.0	1-position of glucose moiety
Adenine arabinoside	5.6	8
Adenosine 5'-triphosphate	20.0	8
Guanosine 5'-triphosphate	1.5	8
Inosine 5'-triphosphate	2.7	8
Puromycin	3.7	8
d-Amphetamine sulphate	10.0	benzylic methylene
3,4-Dihydroxyphenylethylamine	4.5	benzylic methylene
Folic acid	30.0	7 and 9
Methotrexate	11.0	7
Oestriol	40.0	6 and 9
Oestetrol	46.0	6 and 9
L-Phenylalanine	6.7	benzylic methylene
L-Phenylalanylglycine	7.6	benzylic methylene

* For experimental details see references 4 and 41

** Determined by tritium nmr spectroscopy

Use of a very active catalyst results in the labelling of the 2-position of the adenine moiety as well as the 8-position.[11]
The method leads to products with a surprisingly high degree of specific labelling. Palladium(II) oxide is reported to be the most efficient catalyst in such exchange reactions[42] and work with

dibenzyl suggests that metal oxide reduction is the rate
determining step.
$$(43)$$

Isotope exchange reactions with tritiated solvents - Homogeneous

catalysed exchange using tritiated solvents such as water, acetic

acid, trifluoroacetic acid etc, may be base or acid catalysed.

Treatment of an oxo-compound, for example, with base/THO places

tritium in the positions adjacent to the oxo-group by
enolisation (19):
$$(4)$$

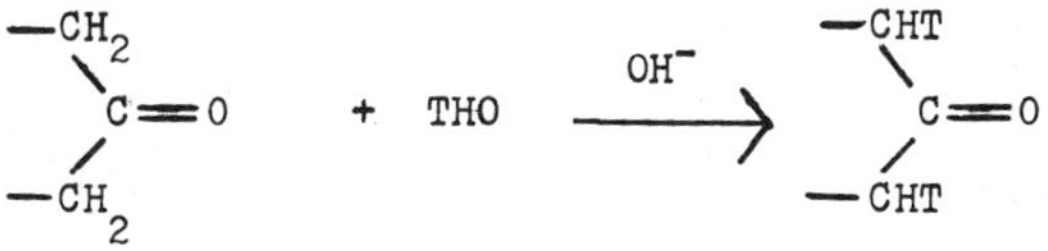

$$\ldots(19)$$

Heating kainic acid in tritiated acetic acid for 16 hours

at 85°C provides a useful method for labelling this important
compound in neurochemical research (20).
$$(25)$$

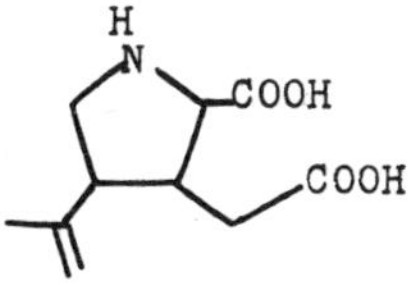

$$\ldots(20)$$

Acid catalysed reactions are especially useful with aromatic

compounds. Examples are tritiated colchicine, obtained by heating

for 7 hours at 70°C in tritiated trifluoroacetic acid, and

vinblastine (21), which is readily labelled in just 2 hours at
$$(44)$$

room temperature with tritiated trifluoroacetic acid.

Alkanes containing a tertiary hydrogen atom are usefully

labelled with ethylaluminium dichloride and tritiated water.
$$(45)$$

Heterogeneous catalysed exchange in a tritiated solvent makes

use of a metal hydrogen transfer catalyst such as Pt and Pd, and

there are many examples of the use of this method.
$$(4)$$

In all these solution exchange methods the molar specific
activity of the final product is critically dependent upon the
specific activity of the tritiated solvent used, the temperature
of the exchange reaction and the efficiency of the catalyst.
The results are often difficult to predict with any certainty
because of the variations in catalyst activity which are
unpredictable.   The methods give rise to products which are
generally (G) labelled but the use of tritium nmr spectroscopy
permits the determination of the patterns of labelling in such
products.   An example is that of tritiated vincristine[44] where
tritium nmr spectroscopy shows that the distribution of tritium
(as a percentage) is 76, 11, 6 and 7 in positions 12'+13', 11',
14' and 17 respectively.   The spectrum is shown in Figure 1.

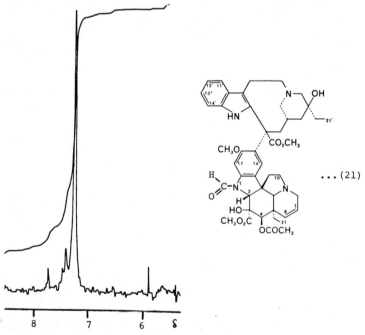

... (21)

Figure 1   Fourier-transform ^{3}H nmr spectrum with ^{1}H decoupling
of [G-^{3}H] vincristine

PURIFICATION AND ANALYSIS OF LABELLED COMPOUNDS

Methods. The complete success of any isotopic labelling method depends upon the ability to isolate the tracer compounds in a state of high radiochemical and chemical purity. Preparative chromatographic techniques are ideally suited for the purification of labelled compounds because of the small chemical weights involved. Column, thin-layer plate, paper and high performance liquid chromatography (46,47) are all especially useful methods. The difficulties of separating molecules of similar chemical structure should always be borne in mind and stringent tests applied to ensure confidence in the identity and purity of the labelled compounds.

Quality control. Radiochemical purity is normally established by chromatographic techniques (thin-layer plate, paper and high performance liquid chromatography) using several different solvent systems. Chemical identity of the labelled compound is also established by co-chromatographic behaviour with authenticated carrier (unlabelled) compound. Quality control and analysis of radiotracer compounds are thoroughly discussed by Sheppard and Thomson. (48) Especial attention is required to possible artefacts in chromatography (49) which may lead to a false radiochemical purity for a ^{14}C- or ^{3}H-labelled compound. This aspect is discussed further in the following chapter.

Having prepared the tracer compound, purified it to a high radiochemical purity and analysed the compound, it should always be remembered that self-decomposition on storage may reduce this purity to an unacceptably low level for valid experimental use. The radiochemical purity should always be determined immediately prior to use of the labelled compound in a tracer investigation to eliminate at least one possible source of error in the validity of the data obtained. (50) The topic of self-decomposition

of radiochemicals is extensively covered in the literature$^{(51)}$ and
is discussed in the following chapter.

HANDLING AND HEALTH PHYSICS

In the enthusiastic use of ^{14}C- and ^{3}H-labelled compounds for
tracer applications, one should of course not forget that such
substances are radioactive. Although both ^{14}C and ^{3}H are weak
beta emitting radionuclides normally classified as "low toxicity
isotopes", good housekeeping is essential in order to avoid
laboratory contamination. Although not necessarily presenting
a hazard to health, laboratory contamination can be a source of
error in measurements in tracer experiments. Regular monitoring
of the working areas is thoroughly recommended.

The safety aspect of radiotracer experiments is fully
discussed by Muramatsu et al,[52] Catch[53] and by Evans.[54]

SUMMARY AND CONCLUDING REMARKS

Isotopic labelling with carbon-14 and with tritium continues
to provide interesting, challenging and exciting opportunities
for the biochemist and organic chemist because of the continuing
demand for more sophisticated tracer compounds in biological
research. More than 1800 different labelled compounds are now
routinely produced by commercial manufacturers and many more
compounds labelled with ^{14}C and ^{3}H are prepared to special
demand. However, this still represents a very small fraction
(probably less than 1 per cent) of organic compounds used in
biological research of all kinds.

The routine use of tritium nuclear magnetic resonance spectroscopy for establishing patterns of labelling in complex molecules
has opened new frontiers and now enables tritium compounds to be
used with confidence for research which formerly would have
demanded difficult carbon-14 syntheses.

REFERENCES

1
J.R.Catch,"Radiotracer Techniques and Applications" Eds
E.A.Evans and M.Muramatsu, Marcel Dekker Inc., New York and
Basel, 1977, Chapter 5, p.129

2
W.J.Le Quesne, Ann.Rep.Prog.Appl.Chem.,1977, 584

3
R.J.Bayly, E.A.Evans, J.S.Glover and J.L.Rabinowitz,
"Radiopharmacy" Eds M.Tubis and W.Wolf, John Wiley & Sons Inc.,
New York, 1976, Chapter 13,p.303

4
E.A.Evans, "Tritium and Its Compounds", 2nd edn, Butterworths,
London, 1974, Chapter 4, p.238

5
C.T.Peng, "Sample Preparation in Liquid Scintillation Counting",
Review Booklet No.17, The Radiochemical Centre,Amersham, 1977

6
A.W.Rogers, "Techniques of Autoradiography",3rd Edn, Elsevier,
Amsterdam, 1979

7
A.W.Rogers,"Practical Autoradiography", Review Booklet No.20,
The Radiochemical Centre, Amersham, 1979

8
J.Bloxsidge, J.A.Elvidge, J.R.Jones and E.A.Evans, Org.Mag.Res.,
1971, 3, 127

9
V.M.A.Chambers, E.A.Evans, J.A.Elvidge and J.R.Jones,
"Tritium Nuclear Magnetic Resonance (tnmr) Spectroscopy",
Review Booklet No.19, The Radiochemical Centre, Amersham, 1978

10
J.R.Catch, "Carbon-14 Compounds", Butterworths, London,1961,p.15

11
The Radiochemical Centre, Amersham,England - unpublished results

12
cf. D.G.Ott, V.N.Kerr,T.W.Whaley, T.Benziger and R.K.Rohmer,
J.Label.Compounds,1974,10, 315

13
  P.Vercier, J.Label.Compounds, 1968,4, 91

14
  D.Banfi, S.Mlinkó and T.Palágy, J.Label.Compounds,1971,7, 221

15
  J.E.Noakes, S.M.Kim and J.J.Stipp, Proc.Int.Conf. on Radiocarbon
  and Tritium Dating, Pullman,Washington, June 1965, Springfield,
  Virginia, 1966, p.68

16
  T.W.Whaley and D.G.Ott, J.Label.Compounds, 1974,10, 461

17
  L.Pichat and C.Baret, Tetrahedron, 1957,1, 269

18
  S.Ikeda and A.Tamaki, J.Polym.Sci.Pt.B.,1966,4, 605

19
  K.Schmid, H.Fürer, and G.Dändliker, Adv.Tracer Methodol.,
  1966,3, 37

20
  R.Nystrom and D.Gurne,"Methods of Preparing and Storing Labelled
  Compounds", Ed.J.Sirchis, European Atomic Energy Community -
  Euratom, EUR 3746 d-f-e, Brussels,1968, p.103

21
  R.Abrams, J.Amer.Chem.Soc., 1949,71, 3835

22
  H.L.Huston and T.H.Norris, J.Amer.Chem.Soc.,1948,70, 1968

23
  G.I.Fujimoto and J.Prager, J.Amer.Chem.Soc.,1953,75, 3259

24
  W.G.Dauben and H.L.Bradlow, J.Amer.Chem.Soc.,1950,72, 4248

25
  J.F.Collins, Ann.Rep.Chem.Soc.,1976,73(B), 416

26
  G.A.Brine and J.A.Kepler, J.Label.Compounds Radiopharmaceuticals,
  1976, 12, 401

27
  R.R.Muccino, G.G.Vernice, J.Cupano, E.P.Oliveto and A.A.Liebman,
  Steroids, 1978,31, 645

28
  P.A.Bell and W.P.Scott, J.Label.Compounds, 1973,9, 339

29
S.F.Zakrzewski, E.A.Evans and R.F.Phillips, <u>Analyt.Biochem.</u>,
1970,<u>36</u>, 197

30
E.A.Evans, J.P.Kitcher, D.C.Warrell, J.A.Elvidge, J.R.Jones
and R.Lenk, <u>J.Label.Compounds Radiopharmaceuticals</u>, 1979,<u>16</u>
(In the press)

31
K.C.Tovey, G.H.Spiller, K.G.Oldham, N.Lucas and N.G.Carr,
<u>Biochem.J.</u>, 1974,<u>142</u>, 47

32
E.W.Putman and W.Z.Hassid, <u>J.Biol.Chem.</u>,1952,<u>196</u>, 749

33
B.Ginsberg and H.Keiser, <u>Arth.Rheum.</u>, 1973,<u>16</u>, 199

34
J.C.Bevington, E.J.Bourne and G.N.Turton, <u>Chem.Ind.</u>,1953, 1390

35
E.J.Bourne and H.Weigel, <u>Chem.Ind.</u>,1954, 132

36
P.H.Jellinck and B.J.Brown, <u>Steroids</u>, 1971,<u>17</u>, 133

37
G.Giannopoulos and S.Solomon, <u>Biochemistry</u>,1967,<u>6</u>, 1226

38
K.E.Wilzbach, <u>Chem.Engng News</u>,1956; <u>J.Amer.Chem.Soc</u>.,1957,<u>79</u>,
1013

39
W.C.Hembree, R.E.Ehrenkaufer, S.Lieberman and A.P.Wolf,
<u>J.Biol.Chem.</u>,1973,<u>248</u>, 5532

40
R.L.E.Ehrenkaufer, A.P.Wolf, W.C.Hembree and S.Lieberman,
<u>J.Label.Compounds Radiopharmaceuticals</u>, 1977,<u>13</u>, 359

41
E.A.Evans, H.C.Sheppard, J.C.Turner and D.C.Warrell,
<u>J.Label.Compounds</u>, 1974, <u>10</u>,   569

42
O.Buchman, I.Pri-Bar and M.Shimoni, <u>J.Label.Compounds
Radiopharmaceuticals</u>, 1978,<u>14</u>, 155

43
  I.Pri-Bar and O.Buchman, <u>Int.J.Appl.Radiat.Isotopes</u>, 1976,<u>27</u>,53

44
  Unpublished results, Radiochemical Centre, Amersham and
  University of Surrey

45
  J.A.Elvidge, J.R.Jones, M.A.Long and R.B.Mane, <u>Tetrahedron Letts</u>,
  1977,<u>49</u>, 4349

46
  G.L.Hawk,Ed.,"Biological/Biomedical Applications of Liquid
  Chromatography", Marcel Dekker Inc., New York and Basel, 1979

47
  A.Pryde and M.T.Gilbert, "Applications of High Performance
  Liquid Chromatography", Chapman and Hall, London, 1979

48
  G.Sheppard and R.Thomson, "Radiotracer Techniques and
  Applications", Eds E.A.Evans and M.Muramatsu, Marcel Dekker Inc.
  New York and Basel, 1977, Chapter 6, p.171

49
  G.Sheppard, "The Radiochromatography of Labelled Compounds",
  Review Booklet No.14, The Radiochemical Centre, Amersham,1972

50
  E.A.Evans, <u>Chem.Ind.</u>,1976, 394

51
  E.A.Evans, "Radiotracer Techniques and Applications", Eds
  E.A.Evans and M.Muramatsu, Marcel Dekker Inc., New York and
  Basel, 1977, Chapter 7, p.237

52
  M.Muramatsu, Y.Suzuki, I.Miyanaga and Y.Wadachi, "Radiotracer
  Techniques and Applications", Eds E.A.Evans and M.Muramatsu,
  Marcel Dekker Inc., New York and Basel, 1977, Chapter 3,p.33

53
  J.R.Catch, "Carbon-14 Compounds", Butterworths, London, 1961,
  Chapter 8, p.107

54
  E.A.Evans, "Tritium and Its Compounds", 2nd edn, Butterworths,
  London, 1974, Chapter 3, p.190

PURITY AND STABILITY OF RADIOCHEMICALS

E.Anthony Evans

The Radiochemical Centre Limited, Amersham,Buckinghamshire,
England

INTRODUCTION

The validity of the results obtained from the uses of radiochemicals such as ^{14}C- and ^{3}H-labelled compounds, which are widely
used in biological, chemical and medical research, is often
critically dependent upon the purity of the tracer compound. $(1-3)$
It should also be remembered that measurement of radioactivity in
a tracer experiment is merely a measure of the radioactive atoms
that were put into the experiment; it does not give directly
information on the type or form of the product associated with
the radioactivity.

In general, the tracer principle demands that a radioactive
compound behaves in exactly the same way, except for predictable
isotope effects, in the system it is tracing as the corresponding
nonradioactive compound under study. This basic assumption is
invalidated if the nature of the radiochemical tracer is in doubt,
that is, if it consists of more than one species or has the wrong
identity. In many experiments it is also necessary to know
accurately the quantity of radioactivity being used.

Because of self-decomposition (also referred to as self-
radiolysis) the purity of the radiotracer compound is a constantly
changing parameter. This paper therefore discusses the concepts
of purity, methods of analysis, self-decomposition and effective

methods for minimising decomposition on storage, for labelled
compounds. A clear understanding of all these topics is essential
for maintaining a tracer compound which is suitable for the
experiment being undertaken and in order to interpret the
subsequent experimental data with confidence.

PURITY

Absolute purity in any chemical substance is an ideal which
is unattainable in practice. Most nonradioactive chemicals
contain from 1 to 10 per cent of impurities, and some even more.
Exceptionally pure chemicals such as those used as analytical
standards may contain only 1 part in 10^6 of impurities. The
extreme sensitivity of radiotracer methods make it quite easy to
detect impurities of the order of 1 part in 10^{10}.

Types of Purity

With radiochemicals it is necessary to distinguish three types of
purity as follows:

Radionuclidic purity - the proportion of radioactivity in the
form of the specified nuclide (and for short half-life nuclides
at a specified date)

Radiochemical purity - the proportion of radioactivity in the
specified chemical form, which may include isomeric or stereochemical form, and may also specify biological or immunological
properties

Chemical purity - the proportion by mass in the specified chemical
form (this term normally refers to nonradioactive impurities;
radiochemical impurities, of course, are also chemical impurities
but normally their mass is very small).

For the radionuclides carbon-14 and tritium it can be assumed
that the radionuclidic purity is 100 per cent from their methods
of production. [4,5]   However, contamination of a radionuclide with

another of longer half-life $(t_\frac{1}{2})$ will of course result in a
radionuclidic purity which decreases with time, and which increases
with time for contamination with a shorter lived radionuclide.
Examples are the radionuclidic purity of ^{32}P ($t_\frac{1}{2}$ 14 days) which
steadily decreases with time when contaminated with ^{33}P ($t_\frac{1}{2}$ 25
days), whereas ^{125}I ($t_\frac{1}{2}$ 60 days) increases in radionuclidic purity
with time when contaminated with ^{126}I ($t_\frac{1}{2}$ 12.8 days).

For most practical applications the really important parameter
of purity for labelled compounds is the radiochemical purity.
Of course, it is equally important also that the chemical identity
of the labelled compound is clearly established, although the
need for extremely high chemical purity is, in general, less
critical than the requirement for high radiochemical purity. It is
worth remembering that chemical impurities can interfere, for
example, with enzyme reactions and cause inhibition, or with
chemical reactions they may also obscure the measurement of mass
in order to determine the true molar specific activity of the
labelled compound and they may cause acceleration of decomposition
on storage of the compound.

Position and patterns of labelling in a compound may also be
critically important. (6-8) This is a particular problem for tritium
labelled compounds where the methods of preparation do not always
give labelling in the expected positions (7) and must be established
by the use of tritium nuclear magnetic resonance spectroscopy. (8)
In certain cases compounds with the label in different positions
from those specified may be considered as radiochemical impurities.

The overall purity requirements for a radiochemical should be
expressed quantitatively and related to the uses to which the
compound is to be put.

Measurement of Purity

It is always advisable to check the radiochemical purity of a
tracer compound immediately before its use in an experiment.

Ordinary classical methods for the determination of <u>chemical</u>
purity, such as measurement of melting point, boiling point,
refractive index (liquids), optical rotation, ultra-violet, infrared and nuclear magnetic resonance spectroscopy, and by gas-liquid
chromatography, are usually inadequate for the determination of
<u>radiochemical</u> purity and more searching and special methods
must be employed. It is possible for example, to have a compound
which by chemical methods of analysis is 99.9 per cent pure but
having most of the radioactivity associated with the 0.1 per
cent impurity. The dilution of a radiochemically impure compound
with the nonradioactive chemically pure form of the compound
often leads to this situation. In general it is recommended that
radiochemicals are purified to the desired level prior to the
addition of any unlabelled (nonradioactive) compound.

Two basic and complementary techniques are widely used to
establish radiochemical purity; these are radiochromatography and
reverse isotope dilution analysis. However, it must be stressed
that even with these more sensitive methods one must be cautious
in the interpretation of the analytical results because of possible
artefacts.

<u>Radiochromatography</u>. The radiochemical purity of non-volatile
compounds is normally determined by chromatographic methods
(paper, thin-layer plate or high performance liquid chromatography,
hplc) and by reverse isotope dilution analysis; volatile compounds
may be assessed by radio-gas-liquid chromatography and sometimes
by hplc. All methods have their pitfalls and some examples will
illustrate the problems.

Numerous examples of the failure of radiochromatography can be
quoted. One example is the chromatographic analysis of $[^3H]$cholest-
erol where more than 9 solvent systems using both paper and thin-
layer plate chromatography indicated the material was radiochemically pure. A further check by tlc over silica-gel impregnated with

silver nitrate using chloroform-acetone (98:2) indicated about
60 per cent of the radioactivity being present as $[^3H]$cholestanol. (9)
Another example is illustrated in Figure 1. The chromatographic
analysis of tritiated cholic acid (prepared by platinum catalysed
exchange in tritiated water)(5) by tlc over silica-gel in one
solvent system (toluene:acetic acid:water 10:10:1) is shown in
Figure 1a. The radiochromatogram scan suggests that the radiochemical purity is over 95 per cent. However, Figure 1b shows the
same compound analysed in another solvent system (chloroform:
acetone 9:1) which indicates a radiochemical purity of less than
25 per cent.

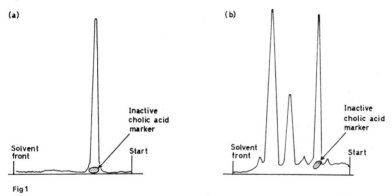

(a)

(b)

Inactive
cholic acid
marker

Inactive
cholic acid
marker

Solvent
front

Start

Solvent
front

Start

Fig 1

There are numerous examples of this kind (1,5,10) which serve
to illustrate the importance of using a number of different
solvents for the chromatographic analysis of radiochemicals and
the need to identify the solvent systems which clearly separate
impurities from the desired compound.

Artefacts in chromatography can arise in numerous ways, for
example from decomposition or irreversible adsorption on the
support medium, double peaking, spot splitting, equilibrating
mixtures (due to isomers) and overloading (2,3,10). A two-way
elution (Figure 2b) or double elution (Figure 2a) in the solvent
can often determine the reliability of the chromatographic

analysis. This is illustrated in Figure 2.

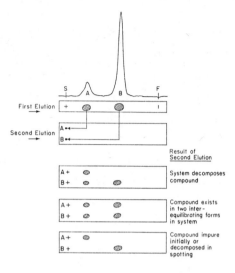

FIG. 2a.   Double elution technique.

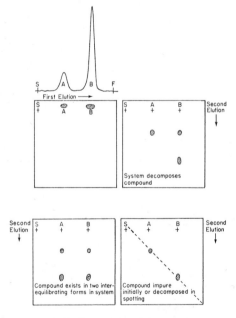

FIG. 2b.   Two-way elution technique.

A recent example of the decomposition of a radiochemical during
chromatography and electrophoresis is given by Hrnčíř *et al*. (11)
Chromatographic carriers in the paper and silica gel were
reported to cause extensive decomposition during the analysis of
L- $\left[^{14}C\right]$histidine.

Compounds which are normally regarded as being non-volatile
can produce artefacts upon analysis because of slight volatility.
The loss of material from a chromatogram during drying (or
elution) leads to false measurements. If the impurity is volatile
this will result in an overestimated value of the radiochemical
purity (RCP); if the main component is volatile the RCP values
will be underestimated. An example is provided by the homologous
series of labelled monocarboxylic acids. All members of the series
up to and including valeric acid are volatile, even as their
sodium salts because of the equilibrium between free acid and salt
in solution. Table 1 shows the RCP of some fatty acids labelled
with carbon-14 determined by reverse isotope dilution analysis
as the S-benzylthiouronium salt and by paper chromatography in an
alkaline solvent.

Table 1 **The radiochemical purity of lower fatty acids determined by
paper chromatography and by reverse isotope dilution analysis**

| | Percentage radiochemical purity by | |
Compound	Paper Chromatography*	Dilution analysis
Sodium[^{14}C]formate	80	87
Sodium[2-^{14}C]acetate	79	91
Sodium[U-^{14}C]acetate	90	97†
Sodium[1-^{14}C]butyrate	92	99
Sodium[1-^{14}C]valerate	91	96

*Lowest purity value, eluent *n*-butanol saturated with 10 per cent
ammonia.
†RCP determined by radiogas chromatography = 96 per cent.

The low RCP values found by paper chromatography indicate that
fatty acid is being lost on drying before scanning whereas the
non-volatile radioactive impurities are retained.

Many of the artefacts and problems experienced with paper and thin-layer plate chromatography are removed by the use of the newly developed methods of analysis involving high performance liquid chromatography.[12]

Reverse isotope dilution analysis depends on the ability to remove the trace chemical amounts of impurity, which may contain a high proportion of the total radioactivity present, from the compound after dilution with about 1000 times its weight of pure "carrier" (unlabelled) compound. However, the principal separation technique used with this method, recrystallisation of the material several times to constant specific activity, does not always result in a separation of impurities. An example is the analysis of mixtures of cholesterol and cholestanol.[1,9]   Suitable derivatives need to be used in such cases. The correct choice of solvent for the recrystallisation is another important considera-tion, such that decomposition, or isotope exchange in the case of tritium compounds, does not occur during the crystallisation process.[5]

The result of a reverse isotope dilution analysis is greatly influenced by the purity and the chemical form of the carrier used or derivative formed. Disagreements between RCP values obtained by chromatographic methods and by reverse isotope dilution analysis may be explained sometimes in this way.

With molecules of increasing complexity and molecular weight, recrystallisation of the diluted compound to constant specific activity is often insufficient to resolve the impurities and erroneous conclusions may be deduced if this is the only test applied for radiochemical purity of the tracer compound.

In general, radiochromatography is the preferred most powerful and versatile technique, but as already noted it does not provide a completely "failsafe" analysis and it does have the fundamental limitation that the purity value found refers only to the particular chromatography system used and its ability to effect separation of the components. Chromatographic methods are certainly not infallible and no single method of analysis is sufficient to guarantee either the RCP or the identity of the compound. In practice therefore, it is usual to apply as many different methods of analysis for the tracer compound as is practicable to achieve confidence in the value obtained for the radiochemical purity and in the suitability of the compound for the proposed investigation.

An excellent review of the quality control and analysis of radiotracer compounds has been published. (13)

SELF-DECOMPOSITION OF RADIOCHEMICALS

The purity of a radiochemical is normally at its highest value immediately following its preparation and purification, but because of self-decomposition this purity is not maintained on storage of the compound. In fact most organic chemicals can be regarded as being unstable and decompose with varying degrees of sensitivity to external factors. The presence of radioactive atom(s) in a molecule makes the compound even more unstable.

Compounds labelled with the pure beta emitting radionuclides carbon-14, tritium (hydrogen-3), sulphur-35 and phosphorus-32 are the most commonly used in tracer applications. Compounds labelled with the gamma- or x-ray emitting radionuclides such as iodine-125, iodine-131, cobalt-57, cobalt-58, mercury-197, mercury-203 and selenium-75 usually have special *in vitro* and *in vivo* applications in medicine. Compounds labelled with tritium, iodine-125 and selenium-75 are especially used in radioimmunoassay and competitive protein binding assays which are frequently associated with routine testing (screening) in medical diagnosis.

Some properties of these commonly used radionuclides are shown
in Tables 2 and 3.

Table 2
Physical properties of some beta emitting radionuclides

radionuclide	half-life	beta energy MeV		mean path length in water	specific activity max.	daughter nuclide (stable)
		max.	*mean*	*µm*	*mCi/mA**	*mCi/mmol*	
tritium	12·35 years	0·0186	0·0056	0·47	2·92 x 10⁴ 10²–10⁵	helium-3
carbon-14	5730 years	0·156	0·049	42	62·4 1–10²	nitrogen-14
sulphur-35	87·4 days	0·167	0·049	40	1·50 x 10⁶ 1–10²	chlorine-35
phosphorus-33	25·5 days	0·248	0·077	–	5·32 x 10⁶ 10–10⁴	sulphur-33
chlorine-36	3·01 x 10⁵ years	0·709	0·25	–	1·2 10⁻³–10⁻¹	argon-36
phosphorus-32	14·3 days	1·709	0·70	2710	9·2 x 10⁶ 10–10⁵	sulphur-32

*A milliatom is the atomic weight of the element in milligrams

Table 3
Physical properties of some gamma- and X-ray emitting radionuclides

radionuclide	half-life	electromagnetic transitions	specific activity max. common values for compounds	daughter nuclide (stable)
		MeV	*mCi/mA* *mCi/mmol*	
iodine-131	8·06 days	0·91(a) 0·36(b)	1·62 x 10⁷ 10²–10⁴	xenon-131
iodine-125	60 days	0·035(c)	2·18 x 10⁶ 10²–10⁴	tellurium-125
cobalt-57	270·5 days	0·13(c)	4·83 x 10⁵ 10³–10⁶	iron-57
cobalt-58	70·8 days	0·475(a) 0·81(c)	1·84 x 10⁶ 10³–10⁵	iron-58
selenium-75	120 days	0·40(c)	1·08 x 10⁶ 10–10³	arsenic-75
mercury-197	64·4 hours	0·077(c)	4·8 x 10⁷ 10²–10⁴	gold-197
mercury-203	47 days	0·21(a) 0·279(b)	2·8 x 10⁶ 10–10³	thallium-203

(a) maximum beta emission
(b) gamma photon emission
(c) electron capture photons

Factors affecting decomposition

Decomposition depends in part on the amount of energy absorbed
by the compound during storage so that it might be thought that
for a given amount of radioactivity the radiation energy emitted
could be taken as a guide to the seriousness of the problems.
If this were so the extent of self-irradiation damage should
increase as the list of pure beta emitters in Table 2 is
decended. However, this is not so in practice for several other
factors are involved. These include:

(a) <u>the fraction of energy absorbed</u>, which is much less than
unity for the more energetic beta emitters such as phosphorus-32;
on the other hand almost complete total absorption of the beta
energy occurs with tritium labelled compounds. It is not
surprising therefore that severe problems exist in the storage

of tritium labelled compounds. Gamma radiation energy is, in

general, little absorbed by the compound itself or its immediate

environs. This is seen from the effects of gamma radiation on

freeze-dried thymidine $^{(14)}$ shown in Table 4

Table 4
Effect of gamma radiation on labelled thymidine

compound	weight μg	activity mCi	Irradiation dose rads	Irradiation time mins.	radiochemical purity %
[6-³H]thymidine (freeze-dried)	92	1	nil	–	100
[6-³H]thymidine (freeze-dried)	53	2	2·5 x 10⁶	30	98
[6-³H]thymidine (freeze-dried)	53	2	2·5 x 10⁶	30	92
[6-³H]thymidine (aqueous solution)	192	1	2·5 x 10⁶	30	<5
[2-¹⁴C]thymidine (aqueous solution)	132	0·01	2·5 x 10⁶	30	<5

In solution thymidine is susceptible to attack by free radicals

produced as a result of the interaction of the radiation with

water.

(b) the specific activity of the compounds. As can be seen from

Table 2 the molar specific activity of tritium labelled compounds

is usually much higher than for compounds labelled with other

beta emitters. Notable exceptions however are L-[^{35}S]methionine

and ^{32}P-labelled nucleotides which are discussed later.

(c) the exponential decrease in absorbed energy with time, as

the radionuclide decays. This is an important factor for compounds

labelled with radionuclides of short half-life such as iodine-131,

phosphorus-32 and phosphorus-33.

Mechanisms of decomposition

    The observations of Tolbert et al published $^{(15)}$ over 25 years

ago indicated that extensive self-decomposition can occur of

compounds labelled even with the long half-life radionuclide

carbon-14, from which beta particles of quite modest mean energy

are emitted. These early observations, on compounds of low molar

specific activity (a few mCi/mmol) by current standards, certainly

shattered any illusions that radioisotopically labelled compounds

were as stable as their non-radioactive counterparts. The
significance of these observations was not fully grasped until
commercial manufacturers began extensive investigations of the
problems relating to the "off-the-shelf" supply of high quality
radiochemicals at high radiochemical purity. Over the years there
has been increasing demand for compounds at very high specific
activity, for example up to several thousand curies per millimole
for some ^{32}P-labelled nucleotides. Thus the concept of shelf-life
has emerged which must not be confused with the physical half-life
of the radionuclide. Shelf-life may be defined as "the time span
during which a labelled compound may be used validly and safely".

The modes by which the decomposition of labelled compounds
can arise were well classified by Bayly and Weigel[16] in 1960.
This classification is summarised in Table 5.

Table 5
Modes of decomposition of radiochemicals

mode of decomposition	cause	method for control
Primary (internal)	Natural isotopic decay	None for a given specific activity
Primary (external)	Direct interaction of the radioactive emission (alpha, beta or gamma) with molecules of the compound	Dispersal of labelled molecules
Secondary	Interaction of excited species with molecules of the compound	Dispersal of labelled molecules; cooling to low temperatures; free radical scavenging
Chemical and Microbiological	Thermodynamic instability of compound and poor choice of environment	Cooling to low temperatures; removal of harmful agents

Primary (internal) decomposition arises as a result of the
disintegration of the unstable nucleus of the radioactive atom.
For compounds that are labelled with radionuclides which decay
to stable isotopes, the decomposition fragments so produced will
be radioactive only if the molecules decomposing contain more
than one radioactive atom. The natural decay of tritium to stable
helium-3 for example, results in the decrease in the specific
activity of a compound by about 5 per cent per annum; it normally
contributes much less than 5 per cent to any radioactive decomposition products arising during this time even in multilabelled

molecules. Primary (internal) decomposition may become important

on storage of multilabelled polymers, labelled macromolecules or

other labelled molecules of high molecular weight, [17] but

for most practical purposes decomposition by this mode can be

ignored.

Primary (external) decomposition can be a major cause of self-

decomposition and involves the direct interaction of the particles

emitted (usually beta particles) with molecules of the tracer

compound. If these particles strike another labelled molecule and

change it in some way, then a radioactive impurity is formed

whereas if the damaged molecule is unlabelled then only a trace

of a chemical impurity is formed. The effect of primary (external)

decomposition becomes increasingly marked as the molar specific

activity of the compound is increased i.e. as the probability of

particles striking a labelled molecule is increased.

Secondary decomposition is commonly the most damaging mode and

results from the interaction of (labelled) molecules with, for

example, free radicals or other excited species produced by the

radiation.

Chemical and other causes of decomposition are often overlooked.

It should be remembered that radiochemicals are often used and

stored at very low chemical concentrations in various solvents,

and that many biochemicals are sensitive to chemical decomposition

especially under such conditions. A high chemical purity can also

be a major factor in prolonging the shelf-life of a radiochemical

and therefore factors which affect chemical decomposition, such

as heat, light (photochemical effects), microbiological

contamination, pH, and so on, must be remembered.

Methods for minimising self-decomposition

Because of the likely problems from the uses of impure

radiochemicals, it is of course important for any supplier and

user of labelled compounds to have them available in a high
state of purity for immediate use. Thus a considerable amount of
study has been done investigating methods for prolonging the useful
shelf-life of radiochemicals on storage. Unfortunately, self-
decomposition limits the useful shelf-life of labelled compounds,
the subject does not lend itself easily to rational study, and
so far no complete solution to the problems has been found.

Most methods derived for the control of decomposition are based
on empirical studies, many of which have been carried out by
the manufacturers of these compounds. The following general
methods and principles are currently employed in minimising
self-decomposition:

(a) reduction of the molar specific activity of the compound
where experimental design allows

(b) primary (external) radiation is minimised by dispersal of
the radioactive molecules which reduces the probability of direct
interaction with a (beta) particle. This is best done by using
a solvent. Suitable solvents include benzene, toluene, ethanol
and water. Aromatic solvents such as benzene or toluene are
usually excellent as these are good absorbers of energy and
produce few reactive species, but the actual choice of solvent
will depend upon the type and properties of the labelled compound
to be stored. It is essential that the solvents used are chemically
pure and free from peroxides for example.

The choice of solvent is important. For example, the observed
apparent rate of decomposition of $[^{35}S]$thiosemicarbazide in
methanolic solution is about four times faster than that observed
in aqueous solution (25) as seen from Table 6.

Table 6
Self-decomposition of [³⁵S]thiosemicarbazide[33]

Initial specific activity mCi/mmol	storage conditions	Initial radioactive concentration mCi/ml	temperature °C	storage time weeks	decomposition %	G(−M)
196	Aq. soln.	2	−30	16	8	2·2
196	Methanol	2·2	−30	16	30	9·4

Thiosemicarbazide is a reagent used in the analysis and identification (characterisation) of aldehydes and ketones, and forms thiosemicarbazones with the aldehyde(s) produced during the radiolysis of the methanol by the beta radiation, resulting in the effect observed.

(c) use of a radical scavenger, such as ethanol, if a compound is stored in aqueous solution where the (beta) radiation produces hydroxyl radicals

(d) reduce the temperature of storage, thereby raising the activation energy for radical-solute interaction

(e) store at liquid nitrogen temperature ($-196^{\circ}C$) if this is convenient, or in the vapour of liquid nitrogen (-140 to $-176^{\circ}C$)

(f) reduce the radioactive concentration of the compound in solution. In general, for compounds at high specific activity (curies per millimole) the rate of decomposition is proportional to the radioactive concentration while for compounds at low specific activity decomposition is less dependent upon radioactive concentration.

A very important fundamental point to note is that the rate of decomposition of a radiochemical may accelerate after an initial period of apparent stability.[5,17-20] This is illustrated in Figure 3 which shows the radiochemical purities of batches of $\left[2,4,6,7-^{3}H\right]$oestradiol at approximately 100 Ci/mmol. There is a marked deterioration in purity which begins after about 8 weeks storage.

Time course studies[15] on the decomposition of other compounds labelled with carbon-14, tritium and iodine-125 support this general hypothesis that some radiochemicals deteriorate markedly after a period of apparently good stability.

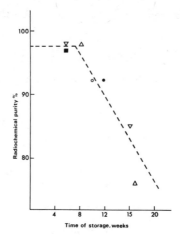

Figure 3
The variations of radiochemical purity with time of storage for batches of [2, 4, 6, 7–³H]oestradiol:
initial radiochemical purity for all batches 99%; batch 2△, batch 3○, batch 4▽, batch 5●, batch 6■.

Thus it is usually necessary to carry out purity checks at frequent intervals during the shelf-life of the radiochemical.

Two important aspects of self-decomposition of radiochemicals not always clearly understood and which merit particular attention are the effect of temperature and the effectiveness of radical scavengers.

The effect of temperature. The beneficial effects of reducing the temperature of solutions of radiochemicals for storage are clearly recognised. However, there is a need for caution; storage temperatures for radiochemicals in aromatic solvents such as benzene (freezing point +5.5°C) or toluene (freezing point -95.7°C) should not cause freezing of the solution as this can result in crystallisation of the compound and consequent removal of the protective effect of the solvent. An example of the effect of storing a benzene solution of a tritiated compound frozen is shown in Table 7.
(20)

Table 7
Effect of freezing on the decomposition of $[6,7-^3H]$ oestrone
at 42 Ci/mmol after 18 months in benzene solution

Temperature	$20^\circ C$ (unfrozen)	$+5^\circ C$	0° to $-5^\circ C$ (frozen)
Per cent decomposition	50	8	90

If aqueous solutions freeze, molecular clustering of the
solute can occur, causing an increase in the effective radiation
dose to the compound, and accelerating its rate of decomposition.
Molecular clustering is most pronounced when solutions are slowly
frozen. Fast freezing is recommended for any radiochemical which
needs to be stored frozen in solution. In general, unless a
compound can be stored at $-140^\circ C$ (or at lower temperatures) it
is best to store at the lowest temperature that is practicable
without causing the solution to freeze. The weak beta radiation
from tritium, and the consequent high local concentration of
hydroxyl radicals, means that tritiated compounds are especially
susceptible. This effect, first reported in 1963, [21] is clearly
seen from some selected examples shown in Table 8.

Table 8
Decomposition of tritiated compounds in frozen solutions

compound	specific activity mCi/mmol	storage conditions	storage temperature °C	storage time months	decomposition %
5-bromo-2'-deoxy [6-³H]uridine	588	a	+2*	7	1
	588	a	−40	4·5	5
[3',5',9-³H]folic acid	16,000	b	+2*	3	16
	16,000	b	−20	3	30
L-[methyl-³H]methionine	8,600	c	+2*	1·5	27
	8,600	c	−40	1·5	47
2-methyl-[6-³H]naphtho-1,4-quinol diphosphate	29,000	d	+2*	3	25
	29,000	d	−80	3	40
2-methyl-[5,6,7-³H]naphtho-1,4-quinol diphosphate	76,000	e	+2*	3	20
	76,000	e	−80	3	30
[6-³H]thymidine	2,500	f	+2*	8	10
	2,500	f	−40	12	28
[6-³H]thymidine	3,600	g	+2*	16	18
	3,600	g	−40	10	20
[methyl-³H]thymidine	44,000	g	+2*	1·2	4
	44,000	g	−20	1·2	17
[G-³H]uridine	3,000	h	+2*	12	17
	3,000	h	−40	10	25
[5,6-³H]uridine	55,000	g	+2*	2·5	6
	55,000	g	−20	2·5	25

*+2°C unfrozen
a aqueous solution at 0·4mCi/ml
b phosphate buffer pH 6·9 at 1·3mCi/ml
c aqueous solution at 5·5mCi/ml
d tetrasodium salt in isotonic saline under N2 at 390mCi/ml
e tetrasodium salt in isotonic saline under N2 at 360mCi/ml
f aqueous solution 6·9mCi/ml
g aqueous solution at 1mCi/ml
h aqueous solution at 20mCi/ml

Many examples of the beneficial effects of storing radiochemicals
at liquid nitrogen temperatures ($-140^{\circ}C$ to $-196^{\circ}C$) have been
published. (17,25)

The effect of radical scavengers. Many radiochemicals are soluble
only in hydroxylic solvents and water is necessarily used as the
solvent. This fact applies to numerous tracer compounds of interest
to biochemists and biologists including amino acids, carbohydrates
and nucleic acids.
                                                                                      (22)
The action of ionizing radiation on water is well known;
the primary entities believed to result from the radiolysis of
water are hydroxonium ions, hydrated electrons, hydrogen atoms,
hydroxyl radicals, hydroperoxy radicals, molecular hydrogen and
hydrogen peroxide. It is these reactive species which cause
self-decomposition of the radiochemical through secondary effects.
In solution ionisation occurs along the tracks of the (beta)
particles in discrete pockets known as spurs. The weaker the
radiation the closer together are these spurs, as illustrated
diagrammatically in Figure 4 .

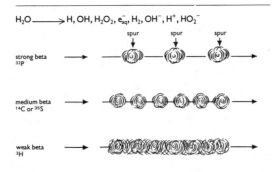

$H_2O \longrightarrow H, OH, H_2O_2, e_{aq}^-, H_2, OH^-, H^+, HO_2^-$

Figure 4
Proximity of 'spurs' or 'pockets' of reactive species in solutions of radiochemicals labelled
with beta emitting radionuclides.

The weak beta radiation from tritium results in the spurs being
very close together during the self-radiolysis of tritium
labelled compounds in solution. This effect has important

consequences when considering frozen aqueous solutions of
tritiated compounds and other compounds labelled with soft beta
emitters, as already discussed.

In aqueous solutions of radiochemicals, it is believed from
current evidence [5] that the hydroxyl radical is the most damaging
species. Hydroxylated products are found on storage of radiochemicals in aqueous solution. An example is the formation of
tyrosine on storage of aqueous solutions of [14]C- or ^{3}H-labelled
phenylalanine. The addition of hydroxyl radical scavengers to
solutions of radiochemicals should therefore result in a protection of the labelled compound from self-decomposition by secondary
effects. This is indeed found to be so in practice and ethanol,
benzyl alcohol, sodium formate, glycerol, cysteamine and
mercaptoethanol have all been used with varying degrees of
success. [5,17,18] In order to be really effective in affording
protection for a radiochemical, the hydroxyl radicals should react
faster with the scavenger than with the labelled compound. The
radical scavenger must also react preferentially and rapidly
with the reactive species and yet the products of the scavenger-
radical reaction must not themselves react with the labelled
compound.

Ethanol is widely used for the protection of many different
classes of labelled compound; [5,17,18] it has the advantage that
it can be removed easily by lyophilising the solution and at the
low concentrations employed (2-3 per cent) has resulted in very
few difficulties in the uses of the compounds. Higher concentrations of ethanol (10 per cent) are usually necessary to give
protection to compounds which react faster than ethanol with
hydroxyl radicals.

It must not be assumed that ethanol, or indeed any other free
radical scavenger, will provide protection for all labelled

compounds stored in solution. For example, ethanol affords very little protection for solutions of ^{14}C- or ^{3}H-labelled methionine because of the very rapid reaction of methionine with hydroxyl radicals; mercaptoethanol is a better protective scavenger in this case.

Some examples of the use of ethanol and other scavengers in minimising the self-decomposition of radiochemicals are given in Table 9.

Table 9 Effect of radical scavengers on the storage of radiochemicals in aqueous solution at +2°C

Compound	Specific activity mCi/mmol	Storage conditions	Radioactive Concentration mCi/mmol	Storage time Months	Decomposition %
DL-[7-³H]Noradrenaline	3680	No scavenger	1	3	6
		+1% ascorbic acid	1	3	15
L-[7-³H]Noradrenaline	2340	No scavenger	1	3	18
		+1% ascorbic acid in 0·1N acetic acid	1	3	2
L-[3,3′-³H]Cystine	1680	0·1N HCl no scavenger	1	20	18
		+2% ethanol	1	20	12
L-[2,5,6-³H]Dihydroxy-	13,000	No scavenger	1	4·5	50
phenylalanine	28,700	+1% ethanol	1	3·5	5
L-[U-¹⁴C]Isoleucine	240	0·1N HCl no scavenger	0·5	6	4
	310	2% aqueous ethanol	0·5	18	2
L-[*methyl*-³H]Methionine	8600	No scavenger	5·5	1·5	22
		+1% sodium formate	5·5	1·5	17
L-[U-¹⁴C]Phenylalanine	282	No scavenger	0·18	2	8
		+3% ethanol	0·18	2	<0·5
[6-³H]Thymidine	3000	No scavenger	1	6	22
		+1% benzyl alcohol	1	6	13
[6-³H]Thymidine	20,000	No scavenger	1	2	6
		+10% ethanol	1	3·75	2
[*methyl*-³H]Thymidine	44,000	No scavenger	1	1·25	4
		+2% ethanol	1	5·75	6
		+10% ethanol	1	5·75	3
L-[3,5-³H]Tyrosine	48,000	No scavenger	1	3	35
		+1% ethanol	1	3	<1
		+0·1% sodium formate	1	4·5	4*
[5,6-³H]Uridine	55,000	No scavenger	1	4	9
		+2% ethanol	1	4	3
		+10% ethanol	1	5·5	N.D.
[G-³H]Uridine	2440	No scavenger	2	13	30
		+1% benzyl alcohol	2	13	8

*Accompanied by complete racemisation.

N.D. = None detectable.

Decomposition of labelled macromolecules and polymers

Macromolecules include peptides, proteins, polysaccharides, polynucleotides and other polymers. It is easy to appreciate that macromolecules will decompose at a much faster rate than simple molecules and that the primary (internal) effect becomes important in such materials. Take D-$[$U-^{14}C$]$glucose, for example, and consider 500 labelled molecules. If two events occur to change irreversibly two of these labelled molecules then 0.4 per cent radiochemical impurity is the result. If these 500 molecules were used to make 5 polysaccharide molecules, each therefore of 100 D- $\left[U-^{14}C\right]$glucose units, two events occurring in different polymer units now cause a change to two out of 5 units or 40 per cent radiochemical impurity. Of course it will depend upon the use to which the tracer molecule is to be put whether such events will affect the data from a use of the labelled material.

An interesting observation illustrating the effect of specific activity on the decomposition rate of (tritiated) polymers was made during the storage of labelled polymethylmethacrylate ("Perspex"), a polymer used in the preparation of tritiated sources for autoradiography. (5) A sheet of the tritiated polymer at 316 mCi/g stored for 15 months and absorbing about 45 M rads of radiation, turned brown and readily crumbled to a fine powder when touched with a pair of forceps. Another example is that of algal protein labelled with carbon-14 at over 50mCi/mA, that is, at over 80 per cent isotopic abundance. The insoluble protein and cell walls become soluble in water after a few months of storage. (12)

Isotopic exchange during storage of tritium labelled compounds in solution

The formation of tritiated water on storage of labelled compounds in aqueous solutions is a problem peculiar to tritiated compounds (5,28) and often depends upon the pH of the solution. Chromatographic analysis by paper or thin-layer plate chromatograpgraphy seldom indentifies this effect and a direct measurement of tritium activity in a lyophilised sample is normally required. For example, $\left[5-^{3}H\right]$orotic acid at 4.6 Ci/mmol stored in aqueous solution (1 mCi/ml) was found to contain 66 per cent labile tritium (as tritiated water) after 32 weeks at +2°C.

Paper chromatographic analysis of the tritiated orotic acid
indicated a RCP of more than 90 per cent but the ultraviolet
light absorption data showed that the chemical concentration
of orotic acid had markedly decreased by self-decomposition.
In this case the impurities which formed exchanged their
tritium with the hydrogen of the aqueous solution. (5,25)

Decomposition of L-[^{35}S]methionine

The most widely used compound labelled with ^{35}S is L-methionine,
a tracer compound for studies of protein synthesis.
L-[^{35}S]Methionine is prepared and used at specific activities
approaching the theoretical maximum of 1500Ci/mmol and it has
therefore provided an interesting challenge in developing
suitable conditions for minimising its self-decomposition on
storage.

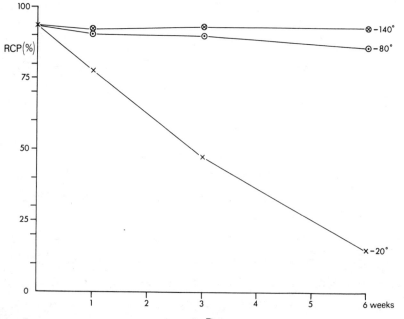

Figure 5   Decomposition of L-[^{35}S]methionine at 1200Ci/mmol

The results obtained from a stability study $^{(12,29)}$ on
L- $\left[^{35}S\right]$methionine at 1200Ci/mmol are summarised in Figure 5.
The compound was stored under nitrogen in aqueous solution
containing 0.1 per cent 2-mercaptoethanol and 20 millimolar
potassium acetate at a radioactive concentration of 4.6mCi/ml.
The rate of decomposition is greatly influenced by the temperature of storage. At -140°C the rate is less than 1 per cent per
week and about 1 per cent per week at -80°C. At -20°C the rate
of decomposition accelerates to about 14 per cent per week.
The temperature of storage recommended is therefore -80°C or
lower. No significant change in the optical purity was observed
during storage over a 6 weeks period and the main decomposition
product was identified as L- $\left[^{35}S\right]$methionine sulphoxide.

Studies on the decomposition of other compounds labelled with
^{35}S have been published. $^{(25)}$

Decomposition of ^{32}P-labelled nucleotides

Nucleotides labelled with ^{14}C, ^{3}H and ^{32}P are widely used in
studies of nucleic acid biosynthesis. The ^{32}P-labelled nucleotides
are especially useful for such studies because of the very high
molar specific activities that can be achieved (more than
1000Ci/mmol). All nucleotides, and especially the nucleoside
5'-triphosphates, are chemically unstable; considerable care is
needed in choosing the conditions used for their storage and for
their analysis. Care is needed for example, in keeping solutions
of ^{32}P-labelled nucleotides at near neutral pH (7.0 to 7.5) and
free from bivalent metal ions which can catalyse the rate of
hydrolysis. $^{(17,25)}$

The results of extensive stability studies $^{(12)}$ have shown that:
(a) the best storage conditions for ^{32}P-nucleotides are in
50 per cent aqueous ethanol at -20°C or lower temperatures.
Under these conditions rates are normally less than 3 per cent

per week, and frequently less than 1 per cent per week. At room temperature (20°C) in 50 per cent aqueous ethanol, decomposition is variable for different preparations and may be as high as 20 per cent per week.

(b) storage in aqueous solutions at -20°C or lower temperatures also results in decomposition rates which are normally less than 3 per cent per week. At room temperature (20°C) in aqueous solution decomposition is consistently 20–30 per cent per week.

(c) the rate of decomposition appears to be virtually independent of specific activity except at room temperature. For example, $[\alpha-^{32}P]$adenosine 5'-triphosphate at both 10Ci/mmol and 2000–3000Ci/mmol shows similar decomposition rates at -20°C and -80°C.

Table 10 Decomposition rates of ^{32}P-nucleoside 5'-triphosphates

compound	specific activity (Ci/mmol)	solution		decomposition rate (%/week) at −20°C	at −80°C	at −140°C	at 20°C
[α-³²P]ATP	~250	aqueous ethanol	(1:1v/v)	1	1	3	21
[α-³²P]CTP	~250	aqueous ethanol	(1:1v/v)	2	2	2	19
[α-³²P]GTP	~250	aqueous ethanol	(1:1v/v)	2	2	1	6
[α-³²P]UTP	~250	aqueous ethanol	(1:1v/v)	1	1	1	6
[α-³²P]dATP	~250	aqueous ethanol	(1:1v/v)	3	3	1	19
[α-³²P]dCTP	~250	aqueous ethanol	(1:1v/v)	2	3	1	12
[α-³²P]dGTP	~250	aqueous ethanol	(1:1v/v)	2	3	1	4
[α-³²P]dTTP	~250	aqueous ethanol	(1:1v/v)	1	1	1	15
[α-³²P]ATP	~10	aqueous ethanol	(1:1v/v)	1	1	−	1
	~10	water		1	1	−	20
[α-³²P]GTP	~10	aqueous ethanol	(1:1v/v)	1	1	−	1
	~10	water		1	1	−	20
[γ-³²P]ATP	~3000	aqueous ethanol	(1:1v/v)	1	1	−	2
	>10	aqueous ethanol	(1:1v/v)	1	1	−	1
	~3	aqueous ethanol	(1:1v/v)	1	1	−	−
	~3	water		1	1	−	20
[γ-³²P]GTP	>10	aqueous ethanol	(1:1v/v)	2	1	−	3

Table 10 shows some typical data for the decomposition rates of ^{32}P-nucleoside 5'-triphosphates stored in solution at a radioactive concentration of 1mCi/ml.

The decomposition products of nucleoside 5'-triphosphates labelled with ^{32}P are the result of both chemical hydrolysis and self-radiolysis. In general at -20°C they comprise the corresponding 5'-mono- and diphosphates, and orthophosphate, together with smaller amounts of unknown products eluting near the parent triphosphate on high performance liquid chromatography. The position of the label in the parent nucleoside 5'-triphosphate will determine which of the products is radioactive. The unknown products appearing near the parent nucleoside 5'-$[^{32}$P$]$triphosphate on hplc are common to both $[\alpha-^{32}$P$]$ATP and $[\gamma-^{32}$P$]$ATP, and hence demonstrate the integrity of the triphosphate group. (25)

The pattern of decomposition products of $[\gamma-^{32}$P$]$ATP at 3Ci/mmol in aqueous solution at 20°C was found to be unusual in that large amounts of pyrophosphate were found. Thus after 1 week at 20°C, approximately 50 per cent of the decomposition product was $[^{32}$P$]$pyrophosphate. The remainder comprised the normal mono- and diphosphate, orthophosphate and unknown modified triphosphates as mentioned previously. The mechanism of decomposition is not understood.

An interesting observation is that if vials of ^{32}P-nucleotides are stored in carbon dioxide at $-80^{\bullet}$C, enough carbon dioxide can leak into the vial and dissolve in the solution to lower its pH. This can cause significantly increased decomposition, particularly for the ^{32}P-labelled deoxyribonucleotides which are especially sensitive to low pH. (25)

Summary of decomposition rates

As a guide the typical rates of decomposition of some radiochemicals under optimum conditions of storage are for:

^{14}C- labelled compounds   1 to 3 per cent per year

^{3}H- labelled compounds   1 to 3 per cent per month

^{35}S- labelled compounds   2 to 3 per cent per month

^{32}P- labelled compounds   1 to 2 per cent per week

^{125}I- labelled compounds   5 per cent per month

Of course, the actual rate of decomposition will depend on the
labelled compound and investigators should always refer to the
latest information on best methods of storage normally provided
by manufacturers on the batch analysis data sheet.

There are a number of comprehensive reviews which discuss
further the mechanisms and control of the self-decomposition of
radiochemicals labelled with both beta and gamma emitting
radionuclides, for those investigators who wish to delve more
into this subject. (5,17-27)

EFFECTS OF IMPURITIES IN TRACER EXPERIMENTS

Whether radioactive impurities in a tracer compound interfere
and result in artefacts or misleading data depends largely upon
the nature and behaviour of these impurities in the experimental
system. However, it is important to remember that, in many
biological experiments for example, particularly at the cellular
level, enough of the tracer compound is used to saturate the
system and only a small fraction of the labelled compound may be
taken up into the system or cells. <u>Selective</u> uptake of radioactive
impurities by various tissues could thus give rise to very
misleading results and an incorrect distribution pattern in
various tissues or excretion pattern from various organs.

Spurious incorporation or uptake of labelled precursors into
an experimental system can occur in three ways as follows:

(a) incorporation of impurities initially present in the precursor

(b) degradation of the precursor to yield radioactive products
    which are then incorporated

(c) incorporation of radioactivity from the precursor into
    impurities in, or associated with, the experimental material.

The controls normally used in such tracer experiments do not
always correct adequately for any spurious incorporation of
radioactivity, and failure to identify the product, or the true
precursor, can lead to false inference of, for example,
macromolecular synthesis (proteins, DNA, RNA or polysaccharides).
Some problems in the radiometric assessment of protein and
nucleic acid synthesis are discussed in detail by Oldham [30]
and by Monks et al [31].

There are few publications which have described or have
specifically recognised the significance of radiochemical
impurities in a biological experiment using a tracer compound.
Unfortunately, there is usually very little known about the nature
or identity of the radiochemical impurities which are present
in solutions of radiochemicals. Consequently their effects in
a particular study can only be determined by direct experimentation
with isolated impurities. One such investigation by Wand et al [32]
involved the study of the uptake by Tetrahymena pyriformis of
impurities produced by the self-radiolysis of tritiated thymidine
in solution. Measurement of the uptake of radioactivity into the
cells by liquid beta scintillation counting and by autoradiography
indicated that the cells were heavily labelled by both pure
tritiated thymidine and by the sample of the isolated radiochemical
impurities, but only in the former could the label be displaced
from the cells on treatment with DNAse. A detailed examination
of the autoradiographs showed that the impurities failed to label
the nucleus of the cells while the pure tritiated thymidine
showed the characteristic heavy labelling of DNA in the nucleus.
However, the tritiated impurities showed heavy labelling of the
cytoplasm increasing with time. Neither DNAse nor RNAse removed
the label in this case. Similar results have been found by Diab
and Roth [33] on the uptake of impure thymidine labelled with tritium

in mouse intestinal crypt cells. Numerous other examples of the effects of impurities in tracer experiments, and especially biological tracer experiments, have been published.[5,13,25,30]

This example illustrates the need to check for the presence of radiochemical impurities and their effects, in order to interpret the results with confidence. It is indeed exceedingly fortunate that many investigations using impure radiochemicals have yielded a qualitatively correct result in that the impurities have either failed to incorporate into the biological system or have been removed during the treatment of the samples. The increasing awareness of such problems is very encouraging. However, one is still left wondering just how many results in the published literature are erroneous from this cause. Until work is repeated this will not be known.

REFERENCES

1
   E.A.Evans, J.Microscopy,1972,96(Pt 2),165

2
   E.A.Evans, Meth.Cell Biol.,1975,10,291

3
   E.A.Evans, Chem. and Ind.,1976,394

4
   J.R.Catch, "Carbon-14 Compounds", Butterworths, London, 1961,

   Chapter 2, p.6

5
   E.A.Evans, "Tritium and Its Compounds", 2nd Edn, Butterworths,

   London, 1974

6
   J.R.Catch, "Patterns of Labelling", Review Booklet No.11,

   The Radiochemical Centre, Amersham, 1971

7
   E.A.Evans, "Tritium and Its Compounds", 2nd Edn, Butterworths,

   London, 1974, Chapter 4, p.238

8
   V.M.A.Chambers, E.A.Evans, J.A.Elvidge, and J.R.Jones,

   "Tritium Nuclear Magnetic Resonance (TNMR) Spectroscopy",

   Review Booklet No.19, The Radiochemical Centre, Amersham,1978

9
   J.R.Catch,"Purity and Analysis of Labelled Compounds", Review

   Booklet No.8, The Radiochemical Centre, Amersham, 1968

10
   G.Sheppard, "The Radiochromatography of Labelled Compounds",

   Review Booklet No.14, The Radiochemical Centre, Amersham,1972

11
   Š.Hrnčíř, J.Kopoldove, K.Vereš, V.Dědkové, V.Hanuš and

   P.Sedmera, J.Label.Compds Radiopharmaceuticals,1978,15,47

12
   The Radiochemical Centre, Amersham, England - Unpublished results

13
   G.Sheppard and R.Thomson, "Radiotracer Techniques and Applications"

   Eds E.A.Evans and M.Muramatsu, Marcel Dekker Inc.,New York

   and Basel, 1977, Chapter 6, p.171

14
   E.A.Evans and F.G.Stanford, Nature,Lond.,1963,199,762

15
   B.M.Tolbert, P.T.Adams, E.L.Bennett, A.M.Hughes, M.R.Kirk,

   R.M.Lemmon, R.M.Noller, R.Ostwald and M.Calvin, J.Amer.Chem.Soc.,

   1953,75,1867

16
R.J.Bayly and H.Weigel, Nature,Lond.,1960,188,384

17
E.A.Evans, "Radiotracer Techniques and Applications", Eds
E.A.Evans and M.Muramatsu, Marcel Dekker Inc., New York and
Basel,1977 Chapter 7, p.324

18
G.Sheppard, Atomic Energy Review, 1972,10,3 (Vienna: IAEA)

19
L.E.Geller and N.Silberman, Steroids,1967,9,157

20
L.E.Geller and N.Silberman, J.Label.Compounds, 1969,5,66

21
E.A.Evans and F.G.Stanford, Nature Lond.,1963,197,551

22
J.K.Thomas, Adv.Radiation Chemistry, 1969,1,103

23
R.J.Bayly and E.A.Evans, J.Label.Compounds,1966,2,1

24
R.J.Bayly and E.A.Evans, J.Label.Compounds,1967,3(Suppl.1),349

25
E.A.Evans, "Self-decomposition of Radiochemicals", Review
Booklet No.16, The Radiochemical Centre,Amersham, 1976

26
E.A.Evans, Nature,Lond.,1966,209,169

27
L.Pritasil, J.Filip, J.Ekl and Z.Nejedly, Radioisotopy, 1969,
10,525

28
W.R.Waterfield, J.A.Spanner and F.G.Stanford, Nature,Lond.,
1968,218,472

29
Technical Bulletin 77/4, The Radiochemical Centre, Amersham,
November 1977

30
K.G.Oldham, Analyt.Biochem., 1971,44,145

31
R.Monks, K.G.Oldham and K.C.Tovey, "Labelled Nucleotides in
Biochemistry", Review Booklet No.12, The Radiochemical Centre,
Amersham, 1971

32
M.Wand, E.Zeuthen and E.A.Evans, Science, 1967,157,436

33
I.M.Diab and L.J.Roth, Stain Technol.,1970,45,285

^{13}C NMR Spectroscopy in Medicinal Chemistry

David M. Rackham

Lilly Research Centre Ltd., Erl Wood Manor,
Windlesham, Surrey, GU20 6PH.

1. Introduction

This review is intended as a brief, introductory survey of
some techniques in carbon-13 (^{13}C) NMR spectroscopy that may prove
of interest to workers in the pharmaceutical industry and in
medicinal research who are not yet aware of the potential of this
remarkable method. It is in no way written as a full account of
applications of ^{13}C NMR spectroscopy[1,2] or of Fourier Transform
NMR[3], both of which have been comprehensively, and excellently,
described elsewhere.

2. Historical Background

Organic spectroscopy deals with the behaviour of organic
materials in the presence of electromagnetic radiation. We are
probably aware that ultraviolet, visible and infrared radiation
can yield information about the bonds which join atoms into
chemical groups and the linking and configuration of these groups
to form chemical molecules. Less familiar are the high energy
(X-ray and γ-ray) and low energy (radar and radio wave) ends of
the electromagnetic spectrum (Figure 1). It is the last of these -
close to VHF wavebands - that contains the radio wave radiation
used for NMR experiments.

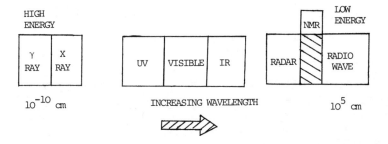

Figure 1. The Electromagnetic Spectrum

Bloch[4] and Purcell[5] and their co-workers were the first to demonstrate the NMR phenomenon in 1945 when studying the properties of protons in water and in paraffin wax. It soon became apparent that the nuclei (in a chemical sample) which possessed non-zero magnetic moments could be regarded as small spinning bar magnets when the sample was placed in a powerful magnetic field. The most important of these nuclei include the proton (^{1}H), deuteron (^{2}H), triton (^{3}H) and carbon (^{13}C) isotopes. The application of the radiofrequency raises some of the ground state nuclei (present in a small excess) into an excited state resulting in a net absorption of energy (Figure 2).

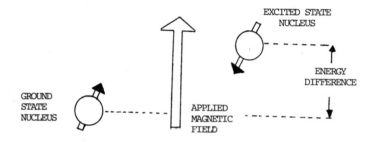

Figure 2.   Application of the Magnetic Field

For a so-called spin ½ nucleus (like ^{1}H or ^{13}C) the relationship between the absorption of energy (ΔE) and the radiofrequency is given by the equation

$$\Delta E = \frac{\gamma h}{2\pi} \times B$$ γ is called the magnetogyric
                                  ratio for the nucleus studied.

= constant x applied magnetic
       term   field

Since the sensitivity of the NMR experiment is related, *inter alia*, to ΔE, there are obvious advantages to building a system with a very high magnetic field (e.g. a superconducting spectrometer). The need for such equipment has pushed the magnetic field strength of commercial spectrometers (originally about 10,000 Gauss or 1 Tesla) through the barrier for electromagnets (2.5 Tesla for ^{1}H frequencies of 100 MHz) to superconducting magnets of more than 10 Tesla (over 400 MHz operation).

3. Continuous Wave (CW) and Fourier Transform (FT) NMR

   The plot of absorption of radiofrequency energy against the
change in applied magnetic field yields the conventional NMR
spectrum, this mode of operation being known as continuous wave
(CW) NMR spectroscopy (Figure 3). Much of the instrument time is
unavoidably wasted when running through the major areas of the
spectrum which do not contain absorption peaks.

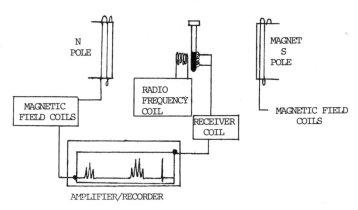

Figure 3. Simplified diagram of CW NMR equipment

   The first commercial equipment for the study of proton (^{1}H)
NMR spectroscopy became available 25 years ago.

   For small amounts of material or for nuclei like carbon-13
(^{13}C), the Fourier Transform (FT) NMR method is most often
employed. It is, in fact, used almost exclusively when studying
the ^{13}C nucleus which is both less abundant (1.1%) and less
sensitive than the near ideal proton case and less receptive than
the proton by an overall factor of about 5,700.

   In FT NMR spectroscopy, a powerful pulse of the radiofrequency energy is applied over a time period of a few micro-
seconds. The receiver coil samples the magnetisation decay in a
direction at right angles to the field derived from the electro-
magnet or superconducting magnet. This decay of magnetisation
(the Free Induction Decay, or FID) can be converted from the
'time domain' to a conventional NMR spectrum (said to be in the
frequency domain) by the process of Fourier Transformation. In
contrast to the Continuous Wave procedure, all of the spectral
information is acquired in a few seconds. The application of the

pulse, the power levels of the observing and any decoupling
radiofrequencies and the mathematical data manipulation
(including Fourier Transformation) are carried out using a
dedicated mini-computer (8-40K). Addition of data in the
computer provides an improvement of the signal/noise of the
spectrum which is directly proportional to the square root of
the number of spectral scans totalled (Figure 4).

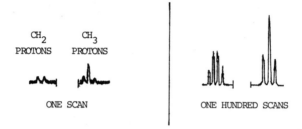

Figure 4. The NMR spectrum obtained for a) one and b) one
hundred accumulated scans for ethyl benzene

   Commercial equipment for FT NMR has now been generally
available for ten years.

   Although the study of ^{1}H NMR predominated for the first two
decades, the development of FT equipment has enormously
enhanced the awareness of the practising spectroscopist to the
attractions of the ^{13}C nucleus. Study of three recent copies of
Organic Magnetic Resonance shows this very clearly; 66% of the
published papers were concerned with ^{13}C NMR (sometimes in
combination with ^{1}H NMR) whereas 17% described work using ^{1}H NMR
alone and 17% other nuclei. The recent commercial availability
of broad band frequency synthesisers now allows ready
observation of many nuclei other than ^{1}H and ^{13}C (in particular
^{15}N and ^{31}P) but these are outside the scope of this chapter
(see references 6 and 7).

4.   Information derived from the ^{13}C NMR experiment

   ^{13}NMR very greatly extends the structural and interactional
information about an unknown material beyond that available from
^{1}H NMR alone. Some of the valuable parameters that can be
obtained from both techniques are discussed and compared below.

A. The Chemical Shift. This parameter measures the position of the NMR absorption peak on the spectral chart. It is partly related to the chemical nature of the proton or the carbon nucleus under study (e.g. alkyl, aryl, alkenyl, alkynyl, heterocyclic). Most signals for protons are found within a range of 10 parts per million (ppm) of the applied magnetic field. Figure 5 shows some typical chemical shift ranges for commonly encountered protons.

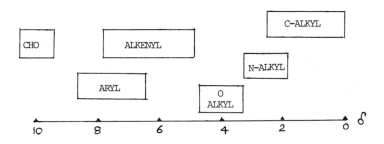

Figure 5. ^{1}H Chemical Shifts

In contrast, carbon signals occur over a chemical shift range of about 200 ppm (Figure 6).

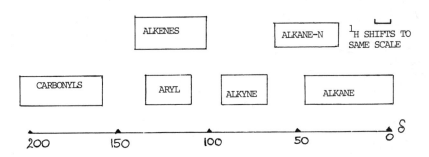

Figure 6. ^{13}C Chemical Shifts. The range of Proton Shifts is given on the same scale

There is, thus, a much wider spread of carbon signals, and molecules with overlapping, unresolved proton signals can often give discrete lines in the <u>carbon</u> spectrum. Figure 7 shows a comparison of spectra for 3-heptanone, $CH_3CH_2COCH_2CH_2CH_2CH_3$. Single lines are seen in the ^{13}C spectrum corresponding to the seven carbon atoms in the molecule.

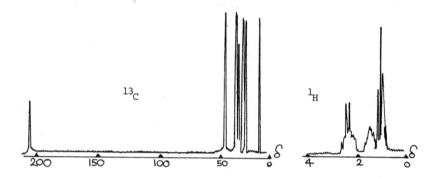

Figure 7. Comparative ^1H and ^{13}C spectra for 3-heptanone

· It is often possible to calculate ^{13}C chemical shifts from additive parameters in Tables (e.g. for alkanes[8]). Hence structural information can be derived from very much more complex synthetic and bio-molecules by analysis of the ^{13}C NMR. Thus warfarin (a) and its sodium salt (b) can be positively identified as the hemi-ketal and the open chain ketone by the ^{13}C NMR peaks at 100δ in (a) and 216.5δ in (b) (Figure 8).

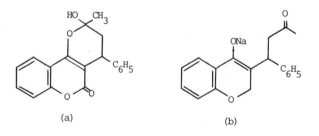

(a)   (b)

Figure 8. Proposed structures for Warfarin and its hemiketal

B. **Spin-Spin Splitting.** This phenomenon, occurring between
magnetic nuclei, gives rise to the multiplet nature of many
peaks in the NMR spectrum. The number and separation of peaks
in the multiplet can lead to the number and orientation of nearby
coupling nuclei and hence to detailed structure and stereochemistry. This important feature of ^{1}H NMR is sacrificed in the
most common type of ^{13}C NMR experiment (the broad band decoupled
spectrum) in which all ^{13}C - ^{1}H interactions (and hence
splittings) are deliberately removed. In return, the ^{13}C NMR
spectrum (signal intensified through the Nuclear Overhauser
effect) will consist of single stick lines - one for each
chemically and/or magnetically different carbon atom in the
molecule (including carbonyl, nitrile and tetra-substituted
olefins for which there are no ^{1}H NMR signals). Figure 9 shows
the broad band decoupled spectrum of vinyl acetate.

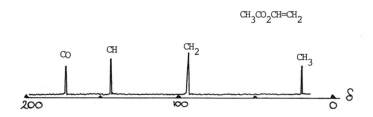

$$CH_3CO_2CH{=}CH_2$$

Figure 9. Broad Band Decoupled ^{13}C NMR spectrum of Vinyl Acetate

Under slightly different conditions of carbon spectrum
accumulation (the off resonance decoupled experiment) one can
recover the peak multiplicities (e.g. doublets for CH, triplets
for CH_2 and and quartets for CH_3 carbons). Structural and
stereochemical information is thus regained but at the expense
of signal enhancement (Figure 10 shows the ^{13}C NMR spectrum of
vinyl acetate recorded under these conditions).

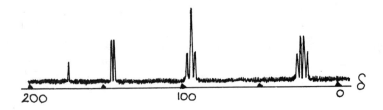

Figure 10. Off Resonance Decoupled ^{13}C NMR Spectrum of vinyl acetate

C. Integrated Peak Areas. Electronically integrated peak areas can be related to the relative numbers of nuclei in different chemical environments. The ease with which integrated areas can be obtained in ^{1}H NMR has led to its widespread use for quantitative analysis in many disciplines (e.g. in the assay of pharmaceutical formulations,[9,10] and on a more general analytical front[11]). The problems associated with peak quantitation in ^{13}C NMR are more fully discussed in Sections 4D and 5B.

D. Spin–Lattice (T_1) Relaxation Measurements. The occurrence of non-equal peak heights and areas for equal numbers of carbon nuclei is very often related to differences in their spin-lattice relaxation times (T_1). These are a measure of the time taken for the magnetisation of a nuclear spin to return to its ground state condition via loss of energy to the molecular lattice. T_1 is often measured by application of a 180° pulse to invert the nuclear spins followed after time T by a 90° pulse to place them in the observing plane of the spectrometer. Variation of T gives a collection of spectra with peaks inverted, zero intensity or the right way up which can yield T_1. As an approximate method, $T_1 \simeq$ Tzero $\div$ 0.7 where Tzero is the time delay T when the peak achieves zero intensity. A montage of such spectra is given for ethyl benzene in Figure 11.

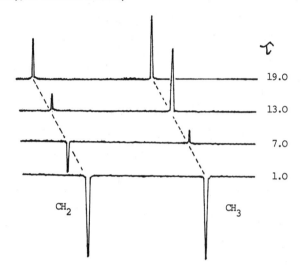

Figure 11. T_1 montage for Ethylbenzene with change in T

Protonated (-$\overset{|}{\text{C}}$H and -CH$_2$) carbons frequently have shorter
relaxation times (and hence show larger spectra peaks) than non-
protonated (-$\overset{|}{\underset{|}{\text{C}}}$-) or methyl (-CH$_3$) carbons. The protonated
carbons are also signal intensified by the Nuclear Overhauser
effect in decoupled spectra. The effect is clearly shown for
phenyl acetylene in Figure 12.

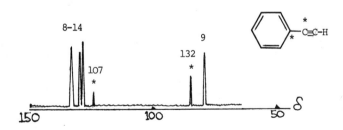

Figure 12. ^{13}C Spectrum and T_1 values for Phenylacetylene
(*). Low intensity peaks are due to the non-protonated carbons

Variations of T_1 are also related to the number of protons on carbons adjacent to a quaternary atom; e.g. in the case of mescaline the T_1 values are progressively shorter for quaternary carbons with none, one or two adjacent protons.

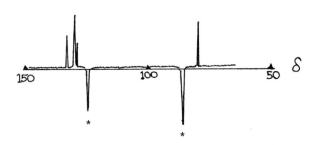

Mescaline

(T_1 in seconds)

Although a nuisance from the viewpoint of quantitative correlations, the peak intensity variation can thus be most useful evidence of structure and of intermolecular interaction. It is also possible to set up spectral acquisition conditions (partial relaxation) under which the protonated carbon atoms appear as peaks of positive magnitude whereas the non-protonated and methyl carbons (with longer values of T_1) are seen as peaks below the baseline. This is illustrated in Figure 13 for phenyl acetylene.

Figure 13. Partially relaxed spectrum of phenyl acetylene (Time delay T = 10 seconds)

E. Effects of Chemical Exchange and Rate Processes. The loss of peaks from a proton NMR spectrum when a drop of D_2O is added to the test solution is due to the process X-H → X-D and is particularly useful for identifying labile protons (X = O, N or S). Similar exchange processes (e.g. in mixtures of rotameric amides) can be readily studied by ^{1}H and ^{13}C NMR. The greater chemical shift differences in ^{13}C NMR mean that coalescence temperatures are higher and that many rate processes can be more conveniently studied nearer to room temperature for this nucleus.

Figure 14 shows the ^{13}C NMR spectrum for N,N-dimethylacetamide at temperatures below 30°C, at 80°C, and above 100°C, the coalescence temperature for merging of the N-methyl signals. The rate constant and the free energy of activation, $\Delta G^\dagger$, can be readily derived for the rotamer equilibrium in this simple case.

$$CH_3CON(CH_3)_2$$

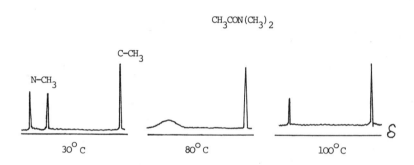

Figure 14. ^{13}C NMR spectrum of NN-dimethylacetamide at 30, 80 and 100°C

F. <u>Isotopic and Chemical Derivatisation</u>. Useful chemical shift changes can often be brought about by judicious chemical reaction, e.g. catalytic deuteration. The change CH_2 to CD_2 causes a dramatic decrease in the intensity of the attached carbon signal due to a large increase in its relaxation time and to loss of the Nuclear Overhauser enhancement when proton decoupling is used. A similar, but smaller, effect can be seen on vicinal carbon atoms. Because the deuterium nucleus possesses a higher spin number than the proton, the carbon signal is further split (into a quintet). These factors make it relatively easy to identify the position of deuteration at any carbon site in the molecule. Figure 15 shows the difference in the ^{13}C NMR spectra of N-phenylpiperazine and its d_3 (trideuterophenyl) analogue.

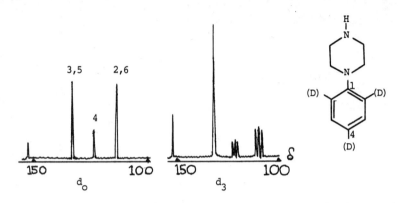

Figure 15. ^{13}C NMR spectrum of d_o and d_3 N-phenylpiperazine

The presence of a hydroxyl group can be employed in a study of acetylation shifts. The magnitude and sign of the ^{13}C chemical shift changes upon acetylation of cyclohexanes or steroids can be related to the proximity of the hydroxyl group and to its stereochemical orientation. Other chemical changes (e.g. ketones or aldehydes to ketals or acetals) are also particularly easy to characterise by ^{13}C NMR.

5. Recent Progress in ^{13}C Methodology

Several recent developments in the techniques and reagents used in ^{13}C NMR spectroscopy have significantly expanded its applicability and its appeal to analysts in the pharmaceutical industry. Five of these topics are briefly discussed below.

A. Lanthanide Shift Reagents. The commonest of these organometallic reagents are the ten or eleven carbon β-diketonates of the lanthanides europium and praseodymium. Much work since their discovery in 1970 has demonstrated their outstanding value in simplifying complex ^{1}H spectra to a degree unachievable by the most powerful, commercial, superconducting magnet systems.[12,13] Organic molecules possessing heteroatom donor groups like OH, NH, C=O or P=O can form weak donor-acceptor complexes in solution with the shift reagent. Large upfield (Pr) or downfield (Eu) chemical shifts are often induced in the donor, the magnitude of these being related to the geometry of the donor-acceptor complex and, thus, to the stereochemistry of the donor. Figure 16 shows the

remarkable simplification of the ^{1}H spectrum of pentanol upon
addition of the europium shift reagent.

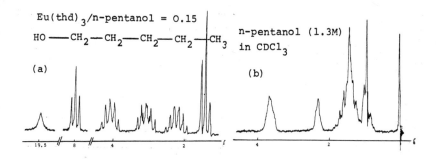

Figure 16. The ^{1}H spectrum of n-pentanol, $CH_3(CH_2)_4OH$
(a) with and (b) without addition of the europium shift reagent,
$Eu(thd)_3$

    The simplification advantage has not been as widely
exploited in ^{13}C NMR for two reasons: i) the 'stick' ^{13}C spectrum
is inherently simpler than the corresponding ^{1}H spectrum
(because of the extended range of chemical shifts) and thus
easier to interpret and ii) the shifts induced by the lanthanide
reagents often have a significant 'contact' contribution which is
not simply related to the stereochemistry of the donor. Figure 17
shows the ^{13}C spectrum of the pesticide Dieldrin.[14] Addition of
a fluorinated Eu shift reagent, $Eu(fod)_3$, causes downfield
shifts in the order 4 > 12 > 3 > 2, i.e. in order of their
proximity to the lanthanide. This behaviour clearly distinguishes
Dieldrin from its stereoisomer, Endrin.

Dieldrin

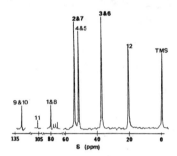

Figure 17.   ^{13}C NMR spectrum of Dieldrin (from reference 14 with permission)

B. __Quantitative Analysis by__ [13]__C NMR__. In spite of the
universal acceptance of ^{1}H NMR for quantitative analysis using
the integrated areas of spectral peaks, there are two problems
which beset the analyst attempting a similar procedure by ^{13}C NMR.
These are the variations in signal intensity caused by i) the
very different relaxation times of non-protonated and protonated
carbon atoms (Section 4D) and ii) decoupling procedures resulting
in Nuclear Overhauser enhancement for protonated carbons. It is
now possible to minimise these difficulties by one of two
procedures, although neither is ideal. i) Long delays between
pulses to allow for relaxation of all carbon atoms together with
a 'gated' decoupling procedure to remove the Nuclear Overhauser
enhancements. In the latter technique the decoupler is only
switched on during the radiofrequency acquisition pulse (too
short a time for the Overhauser enhancement to build up) and then
switched off during pulse delay. The requirement of a long pulse
delay can mean usage of an excessive amount of instrument time to
achieve a satisfactory spectrum with an adequeate signal/noise.
ii) Addition of a paramagnetic species, the most popular reagent
being the shiftless relaxation reagents Cr (acetylacetonate)$_3$ or
Cr (tetramethylheptanedionate).[15] These reagents eliminate the
Overhauser enhancements and level off relaxation times. However,
it may not always be easy to remove the chromium complex from the
solution after NMR analysis. Figure 18 shows the effects of both
these techniques on the ^{13}C NMR spectrum of acenaphthene.[16] The

area of the methylene carbon has been normalised to 2.00.

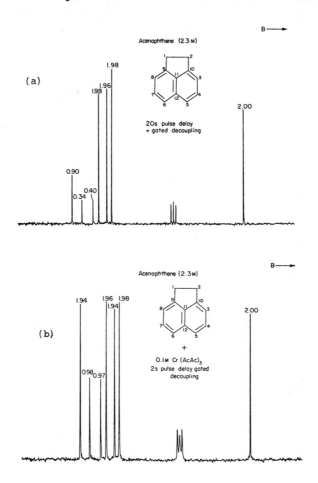

Figure 18. ^{13}C NMR spectra of acenaphthene (a) with gated decoupling, and (b) addition of 0.1 M Cr(acetylacetonate)$_3$. From reference 16, with permission

Shoolery[16] has compared the two procedures and recommends method (ii) where possible, particularly for repetitive analyses.

One must also take care that certain instrumental parameters associated with the FT technique are carefully established, in particular that the peaks are adequately described by sufficient data points and that the pulse power does not vary significantly over the range of peaks being compared.

When choosing peaks for quantitative comparison and assay it is obviously better to select those from carbons which are i) protonated and ii) close together on the chemical shift scale. It is also possible (but usually less satisfactory) to avoid either technique. Results will be reproducible (even if quaternary carbon peaks are very low in intensity) and the assay can be calibrated using solutions of known concentrations as external standards. Alternatively, conditions of high concentration, high viscosity and elevated temperature ($80°$) can sometimes give satisfactory results. O'Neill *et at*[17] have recently reported the assay of L-hydroxyproline in hydrolysed meat using ^{13}C NMR in the latter mode and with an internal standard (relative standard deviation = 5.3%).

C.Analysis of Small Quantities. Development of the Fourier Transform technique has made possible the examination of microgram quantities by ^{1}H NMR spectroscopy. The carbon nucleus has, of course, received similar benefit but the inherently low sensitivity and low abundance of the ^{13}C isotope has meant that tens of milligrams were routinely needed. The development of dedicated carbon microprobes (utilising 1-2 mm capillaries to concentrate the sample within the receiver coil) has reduced the sample requirement even further. Some representative accumulation times for a well defined ^{13}C spectrum on a budget priced FT NMR system (ca. 80 MHz for ^{1}H and 20 MHz for ^{13}C) are:-

   i)   cholesterol (molecular weight = 387)

        5 milligrams   17 minutes accumulation time
        *1 milligram   10 hours

   ii) ethyl vanillin (molecular weight = 166)

        0.5 milligrams (500 micrograms)   16 hours
        0.1 milligram   5 days

   *^{13}C spectrum of 1 milligram of cholesterol in 10 µl
   $CDCl_3$ is shown in Figure 19.

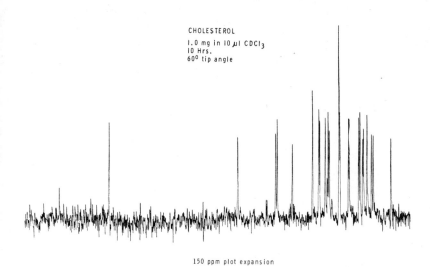

CHOLESTEROL
1.0 mg in 10 μl CDCl₃
10 Hrs.
60° tip angle

150 ppm plot expansion

Figure 19. ^{13}C spectrum of 1 milligram of Cholesterol (with
permission from Dr. J. Shoolery, Varian Associates)

D.   Two Dimension (2-D) NMR Spectroscopy. Software control of
pulse sequences by the dedicated mini-computer in FT spectrometers
has recently led to some interesting methods for spectrum
simplification. Using one of these sequences it has been
possible to resolve overlapping multiplets in a conventional
(1-D) ^{1}H spectrum into two dimensions provided that they possess
different chemical shifts. Thus Ernst et al[18] have resolved the
^{1}H signals of a mixture of five amino acids in 2-D. An added
bonus is the 'homonuclear broad band decoupled' spectrum
consisting of single peaks (with no ^{1}H-^{1}H couplings) for each
proton in the mixture of molecules.
    A different pulse sequence[19,20] can be utilised to produce
a chemical shift correlation map with ^{1}H and ^{13}C chemical shifts
on orthogonal axes, i.e. information from both nuclei in a single
multi-pulse sequence experiment.

E. ^{13}C NMR Spectra of Solids. Until recently, high resolution
NMR (with peak definition better than 1 Hz) was almost entirely
confined to the solution and liquid states. Solid organic
materials gave ^{1}H NMR spectra of effectively infinite line widths
(over 10,000 Hz) and thus of no interest for routine organic
analysis. In the case of the ^{13}C nucleus, modern developments
in instrumentation have led to solid state ^{13}C spectra with line
widths comparable to those in solution. This has been made
possible through the minimisation of three important factors
which contribute to line broadening effects, *viz*:-

   i)   chemical shift anisotropy - minimised by high speed
        rotation (2,000-3,000 Hz) about an axis which is at the
        'magic angle' (54°44') to the direction of the applied
        magnetic field.

   ii) dipolar broadening - minimised by study of the 'rare'
        nucleus (like ^{13}C) with very high power ^{1}H decoupling
        (which eliminates interactions with the 'abundant'
        protons).

   iii) long spin-lattice relaxation times (T_1) - which can be
        by-passed by transferring ^{1}H spin polarisation to the
        [13]C spins. A specific pulse sequence is used to bring
        about this 'cross polarisation' which eliminates the
        usual dependence of pulse delays on the (long) T_1 values.

The method of solid state NMR[21] seems applicable to a wide
range of polymer samples, powders, gels, biological membranes and
coal and oil shale samples. It is certain to find increasing
acceptance in the next few years in a similar manner to the rise
in popularity of ^{13}C solution NMR in the last ten. Comparisons
with X-ray crystallographic studies are already being carried out
and look particularly interesting.

    In Figure 20 the solid and solution state spectra of camphor
are compared.

6. Study of Biomolecules by ^{13}C NMR

    The extension of NMR analysis from low molecular weight,
laboratory synthesised organic molecules to biopolymers and
other biomolecules of high molecular weight has been made possible
by the developments in Fourier Transform methodology previously
described and by the availability of superconducting magnet
systems at ever increasing radiofrequencies. The FT technique
has opened up an area of study of trace quantities of material

(a)   (b)

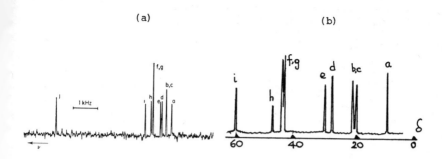

Figure 20. (a) Solid state and (b) solution spectra of camphor.
From reference 21, with permission

for ^{1}H NMR and the examination of other nuclei (e.g. ^{13}C, ^{31}P) in
low isotopic abundance and/or sensitivity. For the larger
molecules, the superconducting magnet systems, with the
appropriate range of software control of pulse sequences, have
the power to simplify the ^{1}H spectra of biopolymers (proteins,
enzymes, nucleic acids, etc.) to the point where valuable
interpretative information becomes available. The laboratory
with the more modest budget and the less powerful FT machine
(90 MHz or less) is often restricted to study of less complex
bioactive molecules. However, the increasing availability of
multinuclear probes on this equipment and the easy access to ^{13}C
T_1 relaxation times using standard software means that the
latter groups can study the interactions of small, bioactive
species, or metal ions, with the larger biopolymers from the
viewpoint of the smaller molecules.

A. The use of ^{13}C labelled materials in studies of biosynthetic pathways. As a result of the greatly increased
availability of ^{13}C labelled materials (see the chapter by
Dr. Lockhart) it has become simpler to synthesise molecules of
biochemical interest by biosynthetic pathways and to identify
the labelled carbon positions by ^{13}C NMR. There are obvious
advantages to this *in situ*, non-destructive analytical method
over multi-stage degradative chemistry which might be needed for
location of a ^{14}C label by tracer studies.

An application to the anti-fungal avenaciolide is
illustrated below and included here from the ^{13}C NMR interest.
Applications of isotopes in biosynthesis are fully described by
Dr. Young in Chapter 8.

Tanabe *et al*[22] have used 90% [13]C labelled sodium acetate
1-[13]C (CH$_3$C$^\Delta$O$_2$Na) and 60% [13]C labelled sodium acetate 2-[13]C
(ĊH$_3$CO$_2$Na) in separate growing culture experiments. These yield
the avenaciolide bis-lactone with enhanced ^{13}C signals in the
positions of the incorporation of the label (Figures 21a from the
1-^{13}C label and 21b from the 2-^{13}C).

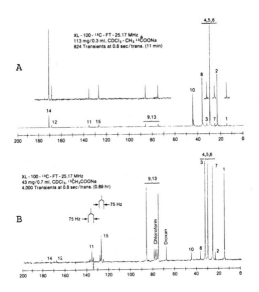

Figure 21. ^{13}C broad band decoupled NMR spectra of Avenaciolide
labelled from A, 1-^{13}C acetate (Δ) and B, 2-^{13}C acetate (•)

The observed enhancements are in agreement with the biosynthetic mechanism for the formation of avenaciolide from acetyl
Co-enzyme A (via 3-ketododecanoic acid) which predicts that the
label will occur (• or Δ) in the carbon positions shown.

Only in one instance are there two adjacent carbons bearing
the same acetate derived labels (11 and 15). This is clearly
shown by the extra doublets at the foot of these ^{13}C signals and
are due to ^{13}C-^{13}C coupling, normally not seen in unlabelled
materials.

A further attractive feature of using ^{13}C labelled materials
is also apparent in drug metabolism studies (see Chapter 7)
where ^{13}C peaks are prominently seen with little interference
from background excipients which have this isotope in natural
abundance only and, therefore, give weak peaks. Some of the
most exciting work of current interest involves feeding ^{13}C
labelled materials to enzyme systems or living cells.
R. G. Shulman (Bell Laboratories) and A. I. Scott (Texas A and M
University) have described[23] their studies on the Krebs
tricarboxylic acid cycle, glycerol to glucose conversion in
mammalian cells and the metabolic pathway leading to uro-
porphyrinogen.

B. Structural Analysis of Biomolecules.[24] The wider spread
of chemical shifts in ^{13}C NMR greatly assists the interpretation
of these complex molecules, e.g. the alpha carbon atoms of amino
acids show a ^{13}C chemical shift range of 1.0-1.5δ whereas the alpha
hydrogen atoms show a ^{1}H chemical shift spread of only 1-15δ.
Thus the ^{13}C NMR signals for all carbon atoms in relatively short
polypeptides can be assigned and many in enzymes (like lysozyme
and ribonuclease) and other proteins. From the variation in ^{13}C
shift with pH the pK_a values can be determined for ionisable
groups within these biomolecules. Allerhand and co-workers have
pioneered detailed analysis on hen egg white lysozyme. In their
recent work[25] they have used the technique of noise-modulated
off-resonance proton irradiation which yields narrow aromatic ^{13}C
spectral lines for the non-protonated carbon atoms in lysozyme.
Figure 22 shows the aromatic region of the ^{13}C spectrum of 9 mM
denatured protein in H_2O (20 mm probe, 40 hours accumulation
time). The individual ^{13}C signals can be identified for the
arginine, phenylalanine, tyrosine, histidine and tryptophan
residues.

There have also been many important applications of ^{13}C NMR
to the study of the phospholipid content of membranes. T_1
measurements on sonicated bilayered dipalmitoyl lecithin vesicles
show that the slowest motion occurs in the branched (glycerol)

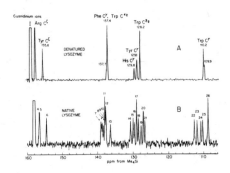

Figure 22. ^{13}C NMR spectrum of Lysozyme in Water, from
reference 25, with permission

end of the molecule (shortest T_1) with increasing mobility
towards the lipophilic alkyl chain end (from >1 to about 4
seconds).[26] The influence of cholesterol on the chain mobility
has also been studied.

The determination of pK_a by 'titration' of ^{13}C chemical
shifts has been used in several studies of nucleotides and
coenzymes like NAD to measure protonation of both the adenine
N-1 and the diphosphate group (pK_a = 4.0 and <1.0 respectively)[27].

C. Interactions of Bioactive Materials with Small Molecules.
The sensitivity of T_1 relaxation times to inter-molecular
interactions and the line broadening effects of paramagnetic
metal additives have proved of outstanding value to the study
of these phenomena. Four representative examples are listed
below:-

i) Small ^{13}C labelled molecules with proteins. As
    discussed in Section 6A, the ^{13}C signals for these
    enriched materials are seen prominently with little
    interference from the weak background peaks of the
    macromolecule. Differences in ^{13}C chemical shifts
    for ^{13}CO bound to haemoglobins reflect the biological
    source and the number of binding sites of the latter.[28]
    ^{13}C relaxation times have been used to measure the
    binding of a labelled sugar (^{13}C -α-methyl-D-gluco-
    pyranoside) to the lectin Concanavalin-A from the
    viewpoint of the conformation of the sugar ring and its
    separation from the paramagnetic manganese atom in the
    Zn/Mn lectin.[29]

ii) <u>Drug molecules with phospholipids.</u> Sonicated vesicles
are used to study the interactions of drug substances
at membrane receptor sites. The ^{13}C relaxation times
for epinephrine seem to vary with the nature of the
phospholipid, being shorter for the anionic phosphatidyl-
serine vesicles than for those derived from zwitterionic
choline or ethanolamine bases.[30] Two further papers
of interest are concerned with the binding of
penicillins[31] and of phenothiazines related to
chlorpromazine.[32]

iii) <u>Paramagnetic metal ions with drug substances.</u> [13]C and
^{1}H NMR studies by Everett and co-workers[33] have shown
that paramagnetic and diamagnetic salts are bound to the
tetracyclic antibiotic, tetracycline, through the A ring.

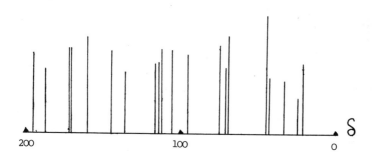

Tetracycline

(see Figure 23)

Figure 23. [13]C stick spectrum of Tetracycline in D_2O

The ^{13}C spectrum shows clearly that the carbon atoms in,
and attached to, the A ring are those which are most effectively
shifted and/or broadened by the lanthanide ions Nd^{3+} and La^{3+}.

iv) Diamagnetic metal ions with Macrocyclic Ionophores and
     Proteins. Methanol solutions of the ionophore anti-
     biotic valinomycin form stable complexes with alkali
     metal ions like K^+ and Cs^+. The ^{13}C carbonyl signals
     are moved by about 5 ppm in the metal complexes
     showing co-ordination of the ester carbonyl groups and
     intramolecular H bonding.[34] The different behaviour
     of the signals for the Na^+ complex relates to the very
     high K^+/Na^+ selectivity for valinomycin. Other
     ionophore complexes related to nigericin (Lilly A23187
     and Roche X537A) assist in the transport of metal ions
     into sonicated phospholipid vesicles.[35]

The broadening effects of a paramagnetic metal ion
relaxation probe are inversely related to the sixth power of the
separation of the probe and the relevant carbon signal (i.e. they
fall off very rapidly with distance). This relaxation
broadening has assisted in understanding the binding of the
enzyme lysozyme (e.g. gadolinium ions are strongly bound to
aspartine-52, refs 36 and 37). An interesting extension of this
paramagnetic broadening effect applies to proteins which
themselves contain a suitable metal ion. Thus the blue copper
protein, azurin, has been examined as the Cu(I) and the Cu(II)
complexes. The former is diamagnetic and shows eighteen low
field carbon signals as expected. The latter complex, however,
shows only eleven of the eighteen carbon signals indicating that
the paramagnetic Cu(II) ion closely approaches many of these
carbon atoms causing extensive line broadening.[38]

7. References

1. R. J. Abraham and P. Loftus, 'Proton and Carbon-13 NMR
    Spectroscopy', Heyden, London, 1978.

2. J. Feeney in R. H. Pain and B. J. Smith (Eds.), 'New
    Techniques in Biophysics and Cell Biology', John Wiley,
    London, Vol.2, 1975, p.287.

3. D. Shaw, 'Fourier Transform NMR Spectroscopy', Elsevier,
    Amsterdam, 1976.

4. F. Bloch, W. W. Hansen and M. Packard, Phys. Rev., 1946,
    69, 127.

5. E. M. Purcell, H. C. Torrey and R. V. Pound, Phys. Rev.,
    1946, 69, 37.

6. R. K. Harris, Chem. Soc. Rev., 1976, 1.

7. R. K. Harris and B. E. Mann (Eds.) 'NMR and the Periodic
    Table', Academic Press, London, 1978.

8. L. P. Lindeman and J. Q. Adams, Anal. Chem., 1971, 43,
    1245.

9. D. M. Rackham, Talanta, 1970, 17, 895.

10. D. M. Rackham, Talanta, 1976, 23, 269.

11. F. Kasler, 'Quantitative Analysis by NMR Spectroscopy',
    Academic Press, London, 1973.

12. A. F. Cockerill, G. L. O. Davies, R. C. Harden and
    D. M. Rackham, Chem. Rev., 1973, 73, 553.

13. J. Reuben in J. W. Elmsley, J. Feeney and L. H. Sutcliffe,
    'Progress in NMR Spectroscopy', Pergamon Press, 1973,
    Vol.9, Part 1.

14. R. L. Roberts and G. L. Blackmer, J. Agr. Food Chem.,
    1974, 22, 542.

15. G. C. Levy and J. D. Cargioli, J. Mag. Res., 1973, 10,231.

16. J. Shoolery,in J. W. Elmsley, J. Feeney and
    L. H. Sutcliffe, 'Progress in NMR Spectroscopy',
    Pergamon Press, 1977, Vol.11, p.79.

17. M. L. Jozefowicz, I. K. O'Neill and H. J. Prosser, Anal.
    Chem., 1977, 49, 1140.

18. K. Nagayama, K. Wüthrich, P. Bachmann and R. R. Ernst,
    Naturwissenschaften, 1977, 64, 581.

19. G. Bodenhausen and R. Freeman, J. Mag. Res., 1977, 28,
    471.

20. R. Freeman and G. A. Morris, J. Chem. Soc. Chem. Commun.,
    1978, 684.

21. R. K. Harris and K. J. Packer, European Spectroscopy News,
    1978, 37.

22. M. Tanabe, T. Hamasaki, Y. Suzuki and L. F. Johnson,
    J. Chem. Soc. Chem. Commun., 1973, 212.

23. J. L. Fox, Chem. and Eng. News, 1979, 27.

24. T. L. James, 'NMR in Biochemistry', Academic Press,
     New York, 1975, Ch.7.

25. A. Allerhand, Accounts of Chemical Research, 1978, 11,
     469.

26. Y. K. Levine, N. J. M. Birdsall, A. G. Lee and
     J. C. Metcalfe, Biochemistry, 1972, 11, 1416.

27. B. Birdsall and J. Feeney, J. Chem. Soc. Perkin II,
     1972, 1643.

28. R. B. Moon and J. H. Richards, J. Amer. Chem. Soc.,
     1972, 94, 5093.

29. C. F. Brewer, H. Sternlicht, D. M. Marcus and
     A. P. Grollman, Proc. Natl. Acad. Sci., U.S., 1973, 70
     1007.

30. G. G. Hammes and D. E. Tallman, Biochim. Biophys. Acta,
     1971, 233, 17.

31. J. M. Padfield and I. W. Kellaway, J. Pharm. Sci., 1973,
     62, 1621.

32. J. Frenzel, K. Arnold and P. Nuhn, Biochim. Biophys. Acta,
     1978, 185.

33. J. Gulbis and G. W. Everett, J. Amer. Chem. Soc., 1975,
     97, 6248.

34. E. Pretsch, M. Vasak and W. Simon, Helv. Chim. Acta,
     1972, 55, 1098.

35. G. R. A. Hunt, FEBS Lett., 1975, 58, 194.

36. I. D. Campbell, C. M. Dobson, R. J. P. Williams and
     A. V. Xavier, Ann. N.Y. Acad. Sci., 1973, 222, 163.

37. K. Kurachi, L. C. Sieber and L. H. Jensen, J. Biol.Chem.,
     1975, 250, 7663.

38. K. Ugurbil, R. S. Norton, A. Allerhand and R. Bersohn,
     Biochemistry, 1977, 16, 886.

Deuterium and Tritium Nuclear Magnetic Resonance Spectroscopy

J.A. Elvidge
Chemistry Department, University of Surrey, Guildford GU2 5XH

1. Introduction

1.A. Proton magnetic resonance spectroscopy

Following the first observation of nuclear magnetic resonance
in matter in 1945 [1,2] and the discovery of chemical shifts a few
years later, [3-6] the first commercial n.m.r. spectrometers became
available in the late 1950's. They were made at first for proton
resonance. Reasons obviously included the favourable nuclear
magnetic properties of the proton, and the great importance of
hydrogen compounds and the wide-spread ongoing interest and
activity in their chemistry. The enormous field of organic
chemistry was doubtless seen to be full of potential users. Being
the earliest established form of n.m.r. spectroscopy and susceptible to interpretation with simple reasoning, as well as being
so widely applicable, ^{1}H n.m.r. now has the greatest fund of
detailed information concerning the correlation of spectroscopic
features with molecular structure. ^{1}H N.m.r. is very widely
used, appreciated, and taught - indeed chemists regard the technique as indispensable - and there are many excellent texts
available, comprehensive, [7-9] theoretical, [10] and introductory. [11-20]
Fortunately, the bulk of this vast store of information is also
applicable to the interpretation of the n.m.r. spectra of the
isotopes of hydrogen. The chemist or biochemist does not start
in the fields of ^{2}H and ^{3}H n.m.r. with a correlation problem as
he may still for example with ^{13}C n.m.r.

1.B. Characteristics of ^{1}H n.m.r. spectroscopy and notes on the
interpretation of spectra

A high resolution ^{1}H n.m.r. spectrum of a compound in solution generally comprises lines and groups of multiplets. The relative intensities of the various signals are given by an integral trace or a printout. Sometimes signals may be so close as to give rise to envelope-type absorption, and individual signals may be broadened.

The origin positions of the signals, the chemical shifts, give information about the kinds of groupings containing hydrogen which are present in the molecules of the compounds. Chemical shifts are measured from the signal derived from an internal standard such as TMS (Me_4Si), or e.g. DSS ($Me_3Si[CH_2]_3SO_3Na$) in the case of aqueous solutions. It is convenient to define chemical shifts by the expression [(frequency of signal/Hz-(frequency of reference/Hz]/frequency of reference/MHz) where the denominator is often equated to (operating frequency/MHz). Chemical shifts are then numbers, δ p.p.m., independent of the spectrometer operating frequency.

The intensities of the signals, as provided by electronic integration, give direct information about the relative numbers of nuclei responsible for the signals. This obviously helps in the making of structural assignments from correlation tables based on chemical shifts.

The multiplicities of signals give information about the numbers of neighbouring ^{1}H nuclei, in simple cases, through the (n + 1) rule. In such 1st order cases, where the chemical shift between coupled nuclei (in Hz, i.e. δ x operating frequency/MHz) is large compared with the coupling constant (J), the multiplicity of each signal is n + 1, where n is the number of protons of a kind in the neighbouring group. The internal relative intensities of the lines of such a simple multiplet arising from coupling between spin 1/2 nuclei are given by the coefficients of

the binomial expansion. Thus the signal from the methylene protons in an ethoxy group, being equally coupled to the three protons of the methyl group, appears as a 1:3:3:1 quartet. The
magnitude of the splitting involved (J/Hz) gives information
(from correlation tables) about the relative proximity or stereochemistry of the groupings. Sometimes the multiplicity of ^{1}H
n.m.r. signals results from the presence in the molecule of one
or more (n) magnetic nuclei of another sort (with spin $\underline{I}$), e.g.
isotopic nuclei or those of a different element, when the (2n$\underline{I}$+1)
splitting rule applies. When two or more sets of distinct nuclei
are coupled to a group, then the resultant splitting of the signal
arising from that group is often best understood by taking each
coupling in turn and drawing in separate stages, according to the
(2n$\underline{I}$+1) splitting rule, a splitting diagram (to scale) to
ascertain whether lines will overlap or not. In this way, in 1st
order cases, the appearance of such multiplets can be predicted.
More complicated non-first order cases can be solved by quantum
mechanical analysis for which many computer programmes are generally available.

Besides using these guide lines, the chemist involved in
spectral interpretation will bear in mind the respective shielding
and deshielding effects of electron donating and electron withdrawing groups on neighbouring protons, and the long range effects
of nearby anisotropic groupings such as phenyl, carbonyl, triple
bonds, etc. With a good data source or with correlation
tables [8,9,17,19,21] and spectral compilations,[22-24] and with
practice,[14,15,17,25] the chemist or biochemist can make complete
or at least highly significant structural deductions with considerable assurance and a minimum of effort and investigative
chemistry. ^{1}H N.m.r. thus enables one to monitor structures
easily during a multistage synthesis, to identify unknown

compounds and deduce new structures, to assess purity and to
effect assays of mixtures (with moderate precision in a percentage
range), even to determine chiral purity with the aid of chiral
shift reagents, and to examine dynamic phenomena such as tauto-
merism and conformational equilibria.

1.C. Nuclear properties of the proton

Ordinary hydrogen, or protium ($_1^1H$; 99.9844% abundance) has
a nucleus, the proton, with a high sensitivity to n.m.r. detection
(1.00 with respect to other nuclei at the same field and in equal
number). This is because the proton has a high magnetogyric
ratio γ_p = 4.2577 x 10^7 Hz T^{-1} which results (γ_p = $\mu/\underline{I}.h.2\pi$) from a
high intrinsic nuclear magnetic moment μ and a spin quantum
number $\underline{I}$ = 1/2. The resonance frequency ν, given by the
expression

$$\nu = \gamma B$$

is therefore high for a given field B, being e.g. 90 MHz for a
magnetic field of 2.11 Tesla. The spectral dispersion in Hz per
unit chemical shift (δ p.p.m.) is consequently good, although
because of the small shielding of the proton (1 extranuclear
electron) the range of chemical shifts is small (10 - 20 p.p.m.)
compared with most other magnetic nuclei in the Periodic Table.
The spin quantum number 1/2 means that there are just 2 possible
orientations of the nuclear magnetic moment vector in an applied
magnetic field [i.e. (2$\underline{I}$ + 1) orientations or energy states] but
also implies spherical nuclear symmetry from which it follows
that the magnetic moment is the only nuclear property dependent
upon orientation. Consequently, the relaxation of proton spins
from their upper spin state can only be induced by magnetic
fields and these must have components perpendicular to the
applied magnetic field and which are oscillating with the Larmor

(resonance) frequency. Such oscillating magnetic fields will sooner or later arise from the magnetic moments of nuclei in the surroundings and in other molecules undergoing thermal motion in the solution, but the probability is low because of the preceding stringent requirements. Relaxation times for protons are then long (seconds) - indeed very long compared with the lifetimes of excited states in electronic or vibrational spectroscopy - and so, from the uncertainty principle, ^{1}H n.m.r. lines are very sharp (e.g. 0.2 Hz width at half height). Spin multiplets are normally highly resolved as already implied and the coupling constants are often extractable merely by inspection, so stereochemical deductions can usually very easily be made, as well as deductions concerning the proximity of other hydrogen-containing functional groups.

1.D. Main methods of producing n.m.r. spectra

There are three main techniques for generating n.m.r. spectra. In two, which are called continuous wave (CW), either the frequency is kept constant while the magnetic field is swept or the field is kept constant while the frequency is swept through the resonances. Typically a CW spectrum takes 500 - 1000 s to acquire. By accumulation of repeated spectra in a computer, signal-to-noise can be improved (as the square root of the no. of accumulations) but this is obviously slow and requires extreme instrumental stability. The third technique is that of pulse, Fourier transform spectroscopy [26,27] in which the solution sample is subjected in a constant magnetic field to repeated brief (μ second) pulses of rf radiation at the fixed characteristic n.m.r. frequency for the nuclei under study. A very short pulse (t_p) generates Fourier components covering a range of frequency ($\pm 1/t_p$) about the carrier frequency. Hence all the nuclei of a kind in the

sample are simultaneously excited, in contrast to one at a time
in CW n.m.r. The relaxation of the nuclei during a delay time
(e.g. 1 - 3 s) between pulses generates from the various decaying
induced magnetic moments an exponentially decaying interferogram
(a free induction decay - f.i.d.) which is stored in a computer
and added to subsequent f.i.d.s following each pulse. Signal-
to-noise improves (as before) as the square root of the number of
f.i.d.s stored but the cost in time is clearly much less than
with CW accumulation. Fourier transformation of the accumulated
f.i.d. signal provides the characteristic n.m.r. absorption
spectrum of lines and multiplets with their correct intensity
relationships.

2. ^{2}H N.m.r. Spectroscopy

2.A. Deuterium

Deuterium ($^2_1$H or D), is a stable isotope of hydrogen, present
at low abundance (0.0156%) in all natural hydrogen and its compounds. Commercially it is concentrated by electrolysis of water
(as in Norway) and is available as gas or as heavy water (D_2O),
or specifically incorporated into simple compounds. Methods of
hydrogen labelling are discussed in Chapters 2 and 10.

2.B.   Nuclear properties of the deuteron and characteristics of ^{2}H n.m.r.

The deuterium nucleus, the deuteron, is magnetic and so
capable of n.m.r. detection. Although studied by P. Diehl and
coworkers since 1964,[28] the potentialities of ^{2}H n.m.r. for the
analysis and investigation of deuterium labelled compounds, and
the possibilities of ^{2}H n.m.r. at natural abundance, have only
recently become more widely appreciated.[29] This is a direct
result of the development of modern instrumentation, in particular
pulse n.m.r., and of very high field n.m.r. made possible by

solutions of problems involved in the construction of practical
superconducting solenoids.

The disadvantages initially apparent in ^{2}H n.m.r. were low
sensitivity and poor spectral dispersion, further compounded by
broadened lines lacking spin coupling multiplicity. These
features are inherent, deriving from the magnetic properties of
the deuteron, but can now in part be mitigated. The low sensitivity of the deuteron to n.m.r. detection (10^{-2} that of ^{1}H or
1.5×10^{-6} at the natural abundance of 1 part in 6410 of ordinary
hydrogen) can be offset by the rapid multipulse signal-accumul-
ation technique of Fourier transform spectroscopy. Inevitably
there is cost in time, because 10^4 accumulations are needed to
give a ^{2}H n.m.r. signal of strength equivalent to that of a one
pulse spectrum from the same number of ^{1}H nuclei. The situation
may be very much better in reality because the signal-to-noise
characteristics of the spectrometer may be so high, especially
if it is a high field instrument, that even a single pule ^{2}H
n.m.r. spectrum (from a sample labelled at high abundance) may be
adequate. Use of wide-bore sample tubes to increase the number
of nuclei inside the rf sample coil of the spectrometer helps
(typically 4X, in going from a 5 mm to a 10 mm diameter tube).
This is so, provided the magnet construction can maintain the
necessary near perfect field homogeneity over the wider sample.
Again, here, wide-bore superconducting solenoids can offer sub-
stantial advantages.

The low sensitivity of the deuteron to n.m.r. detection
derives from the low magnetogyric ratio $\gamma_d = 0.6536 \times 10^7$ Hz T^{-1},
in turn deriving from a low nuclear magnetic moment and the spin
$\underline{I} = 1$. The resonance frequency is therefore low, viz. 13.8 MHz
at 2.11 Tesla (as compared with 90 MHz for ^{1}H) so that spectral
dispersion is poor. In terms of frequency, ^{2}H n.m.r. chemical

shifts are compressed or reduced, with respect to ^1H chemical
shifts at the same field, in the ratio of the magnetogyric constants, <u>i.e.</u> by a factor of 1/6.5 (= 13.8/90 = γ_d/γ_p). Although
when expressed in p.p.m. on the δ scale chemical shifts of
hydrogen isotopes are virtually identical, the frequency com-
pression merges close lines in a ^2H n.m.r. spectrum which would
be resolved in the corresponding ^1H n.m.r. spectrum measured on
the same instrument. To achieve a spectral dispersion of ^2H
chemical shifts, comparable to that long accepted by chemists in
60 MHz (1.4T) ^1H n.m.r. spectroscopy, at least a 6 fold increase
in the applied magnetic field (to 8 - 9T) is required, inevitably
meaning the use of a superconducting magnet.

     A further consequence of the low γ_d/γ_p ratio is that ^2H
coupling constants to other nuclei are smaller than the corresponding ^1H coupling constants by the factor 1/6.5. Coupling
constants are, of course, field-independent so that no
amelioration is possible. Consequently only geminal coupling of
deuterons to protons is resolved ($|^2\underline{J}_{H,D}|$ = 2.3 Hz) but not
vicinal coupling ($|^3\underline{J}_{H,D}|$ = 0 - 1.5 Hz) except perhaps that across
a double bond (<u>ca.</u> 2 - 3 Hz). When observed, the multiplicity of
a deuterium signal, resulting from coupling to ^1H or other nucleus
having spin $\underline{I}$ = 1/2, is exactly as expected from experience in
pure ^1H n.m.r. Thus the ^2H n.m.r. signal from [1-$^2H_1$]methanol
(CH_2D-OH) appears as a 1:2:1-triplet and the signal from the
deuterium in [2-$^2H_2$]acetonitrile ($CH_3 \cdot CD_2 \cdot CN$) appears as a
1:3:3:1-quartet. This last case of resolved vicinal coupling is
exceptional, two features contributing : one is the intrinsically
high internuclear coupling and the other is the absence of a
molecular electric field gradient at the methylene group so that
deuteron relaxation is relatively slow and the ^2H n.m.r. lines
are sharp.

The spin quantum number $\underline{I}$ = 1 for the deuteron indicates
($2\underline{I}$ + 1), *i.e.* 3, equally spaced energies or orientations for the
nuclear magnetic vector in an applied field. Consequently,
according to the ($2n\underline{I}$ + 1) first-order splitting rule, a single
deuteron will split the n.m.r. signal from an adjacent magnetic
nucleus or group of equivalent nuclei (*e.g.* the ^{1}H signal from a
CDH_2 group) into 3 equi-intense lines : these will be observable
in the n.m.r. spectrum of that nucleus or group. Two equivalent
deuterons, as in a CD_2H labelled methyl group, would split the
proton signal into 5 lines. These would ideally have internal
relative intensities of 1:2:3:2:1, as construction of a simple
splitting diagram will show (Figure 1). In practice, because of
the breadth of the lines (say 0.6
Hz or more) and the small magni-
tude of the $^2\underline{J}_{H,D}$ coupling constant (*ca.*2 Hz) there may well be
merging of lines and production
of an envelope signal. Geminal
coupling of a proton to carbon-13
is 120 - 200 Hz (depending on bond
hybridisation) and so the corresponding deuterium coupling con-
stants $|^1\underline{J}_{D,^{13}C}|$ (= 18 - 30 Hz)
will give rise to resolvable
splitting of both the ^{2}H and the
^{13}C n.m.r. signal and be

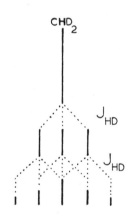

Figure 1. Ideal quintuplet
signal from the proton in a
CHD_2 labelled methyl group

observable, given sufficient isotopic abundance. The ^{13}C signal
from a ^{13}CD group will be a 1:1:1-triplet and the ^{2}H signal will
be a 1:1-doublet.

The spin $\underline{I}$ = 1 for the deuteron also implies an aspherical nuclear charge distribution. Thus the deuteron is a quadrupolar nucleus, so-called because for mathematical manipulation an arbitrary charge distribution needs to be resolved into a multipole and corresponding moments, here 4 poles (and 2 moments). As it spins, a quadrupolar-nucleus produces a highly fluctuating electric field and this interacts with the molecular electric field gradient at the nucleus to provide an additional mechanism for energy transfer and so for loss of upper spin state energy to the lattice. A consequence is shortened relaxation times for deuterons and so broadened n.m.r. lines, and often loss of multiplicity where spin coupling to protons is expected. Line broadening in multiplets occurs when $(1/2\pi.\tau_d) \approx \underline{J}_{H,D}$, where τ_d is the lifetime of the deuteron in an upper spin state and $\underline{J}_{H,D}$ is the proton-deuteron coupling constant : broadening does not occur to the same extent for each line of e.g. a triplet or quadruplet signal, so the effect can be distinguished from exchange broadening. When $(1/2\pi.\tau_d) << \underline{J}_{H,D}$ then the multiplet will be well resolved : this implies a longer relaxation time and so a high molecular symmetry about the quadrupolar nucleus so that there is zero molecular electric field gradient at the nucleus. When on the other hand $(1/2\pi.\tau_d) >> \underline{J}_{H,D}$ then a singlet is observed because now the deuteron relaxation is so rapid as to effect decoupling from the adjacent protons. This is most often so. Because of the generally rapid deuteron relaxation and its mechanism, there are no nuclear Overhauser enhancements even when ^{1}H spin decoupling is applied. Accurate integration of [2]H n.m.r. signals is thus assured.

2.C.   Applications of [2]H n.m.r.

The [2]H n.m.r. spectrum of a deuteriated compound in solution will, then, in general comprise a series of broadened lines, one

from each magnetically distinct (i.e. chemically different)
deuteriated atom incorporated into the molecule. Typical is the
^{2}H n.m.r. spectrum (at 15.28 MHz and 2.34 T) of [^{2}H]griseofulvin
(1) in chloroform, shown in Figure 2A. The compound was labelled
by biosynthesis from sodium [2-^{2}H]acetate (CD_3CO_2Na) using a culture of <u>Penicillium urticae</u>.[30] The line from natural abundance
[^{2}H]chloroform served as a reference, and field-frequency locking
was to the fluorine resonance from added hexafluorobenzene. The
spectrum (Figure 2A) shows broadened lines and no observable

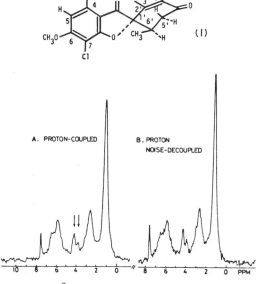

Figure 2. ^{2}H N.m.r. spectra of griseofulvin
biosynthesised from [2-^{2}H]acetate

splitting, illustrating the lack of spectral dispersion. Effectively, the various ^{2}H signals appear as though largely decoupled
from protons adjacent to the deuterons : there is only slight
sharpening of the lines in the ^{2}H n.m.r. spectrum taken with ^{1}H

decoupling (Figure 2B). Although the ^{1}H n.m.r. spectrum of
griseofulvin was obviously available, nevertheless it was felt
desirable in order to obtain assignments unambiguously, to compare
the ^{2}H n.m.r. spectrum with the ^{2}H n.m.r. spectra of several
specifically deuteriated species of the molecule. Thus griseofulvin was hydrolysed to griseofulvic acid and remethylated with
[^{2}H$_{2}$]diazomethane to obtain [2'-OCHD$_{2}$]griseofulvin, which gave
the ^{2}H n.m.r. spectrum in Figure 3D. By heating griseofulvic
acid in chloroform containing D$_{2}$O, the 3'-position was labelled
by exchange ; methylation with diazomethane in [O-^{2}H]methanol
then gave [2'-OCH$_{2}$D,3'-D]griseofulvin with the ^{2}H n.m.r. spectrum
E (Figure 3).

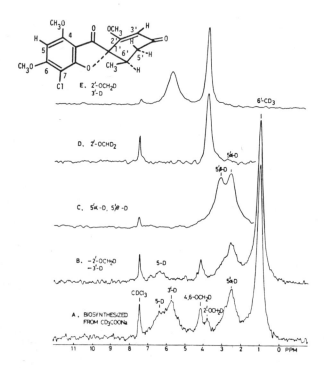

Figure 3. ^{2}H N.m.r. spectra of deuteriated griseofulvins

By exchange under basic conditions with sodium s-butoxide in [O-^{2}H]s-butanol, [5',5'-D$_2$]griseofulvin was directly obtained: this gave the ^{2}H n.m.r. spectrum C (Figure 3). Spectrum B (Figure 3) resulted from the biosynthesised [^{2}H]griseofulvin after removal of deuterium from the 2'- and 3'-positions by hydrolysis (and exchange at the 3'-position) to griseofulvic acid and then remethylation with diazomethane in ether-methanol and separation from isogriseofulvin.

From the assignments thus achieved (somewhat laboriously) in spectrum A (Figure 3), the pattern of incorporation of the deuterium from [2-^{2}H$_3$]acetate into griseofulvin was revealed (Table 1). The methyl group of acetate supplies carbon atoms 5, 3', and 5', and the 6'-methyl group as well as the methoxyls at the 4-, 6-, and 2'- positions. The results agreed with previous work employing [2-^{3}H,^{14}C]acetate and chemical degradation and counting but did not provide so much information as the tedious conventional approach because it happens that so many of the carbon atoms of the griseofulvin molecule do not carry hydrogen.

Table 1. ^{2}H N.m.r. relative incorporation data for griseofulvin biosynthesised from [2-^{2}H$_3$]acetate.

Position	Intensity (%)
6'-CD$_3$	44
5'α-D	23
2'-OCH$_2$D	3.3
4,6-OCH$_2$D	6.3
3'-D 5-D	24

However, a point of detail to emerge directly from the ^{2}H n.m.r.
study was the incorporation of the isotope into the 5'α- position
to the exclusion of the 5'β- position, but no inferences were
drawn. The three 2'-, 4-, and 6-methoxy groups retained only
a fraction of the original complete deuterium labelling (Table 1).
That the remaining deuterium in those groups (only 9.6% of total
deuterium incorporated) was now present as OCH_2D was indicated by
mass spectrometry and by the change, on ^{1}H spin decoupling, of the
height of the two relevant ^{2}H n.m.r. signals : there were 21-27%
height increases (Figure 2, see arrows) as a result of line
width decrease on removal of the coupling to adjacent protons.
Loss of deuterium had also taken place from the 6-methyl group,
the remaining isotope there now being present mainly as CHD_2.
Again, this was demonstrated by the mass spectrum and by the
relative intensity of the ^{2}H n.m.r. signal as compared with that
from the 5'α-position, and also by the 17% signal height increase
on ^{1}H decoupling.

It is possible in principle to make biosynthetic deductions
from incidental loss of hydrogen isotope during a biosynthesis, as
indicated in other studies,[31] but before this can be attempted it
is vital to ensure that the incorporation has been optimised by a
careful set of preliminary experiments which also define the
incidence of trivial exchange and conditions for its minimisation.
It is not sufficient to aim only at high incorporation merely by
use, e.g., of a replacement culture technique.

In a related study of the microbial hydrogenation of
dehydrogriseofulvin (2) and the concomitant formation of
5'-hydroxygriseofulvin (3), [5'-^{2}H]dehydrogriseofulvin (4) was
fed to a shaken culture of <u>Streptomyces</u> <u>cinereocrocatus</u>.[32] The
^{2}H n.m.r. spectrum of the resulting mixture of griseofulvin and
its 5'-hydroxy derivative showed a signal which was attributed to

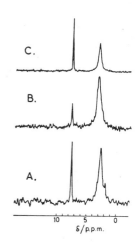

Figure 4. ^{2}H N.m.r. spectra of

A [5'α-^{2}H]griseofulvin +

5'-hydroxygriseofulvin

B [5'α, 6'β-^{2}H]griseofulvin

C [5'α-^{2}H]griseofulvin

deuterium in the 5'α-position of griseofulvin (Figure A).
Providing the assignment was correct, it had been very simply
demonstrated that the 5'-hydroxy group derives by direct replacement of the 5'α-hydrogen in griseofulvin (without inversion) and
that the latter compound is in turn formed by trans-diaxial
addition of hydrogen to the 5', 6'-double bond of dehydrogriseofulvin. For confirmation, a replacement culture was grown on a
medium containing dehydrogriseofulvin (2) and 50% of heavy water.
The isolated [^{2}H]griseofulvin gave one ^{2}H n.m.r. signal (Figure
4B). As this signal partly persisted (Figure 4C) after the
compound had been treated with neutral alumina under conditions
known to effect complete back exchange at the 5'β-position, it was

concluded that the 6'-position as well as the 5'β-position bore

label, as expected from the earlier experiment. Thus possible

ambiguity in assigning the very close ^{2}H n.m.r. signals from the

5'α-, 5'β-, and 6'-positions was circumvented. At the conven-

tional field strength of 2.34 Tesla of a 100 MHz ^{1}H n.m.r.

spectrometer, the poor spectral dispersion of ^{2}H n.m.r. chemical

shifts is something of a limitation in an otherwise elegant

approach

Another biochemical reduction recently investigated with the

aid of deuterium labelling and ^{2}H n.m.r. spectroscopy was that of

2',7-dihydroxy-4'-methoxyisoflavone (5) to the pterocarpan

phytoalexin (7) in fenugreek seedlings.[33] Special conditions

designed to effect accumulation of this product were used. In

this way the isoflavone precursor (6), labelled at the 2-position

with deuterium (96 - 97% ; assayed by mass spectrometry and ^{1}H

n.m.r.) was converted into a [^{2}H]pterocarpan, isolated as the

acetate (8), which showed a single ^{2}H n.m.r. line at δ 3.92 (in

benzene), measured with ^{1}H decoupling from the natural abundance

[^{2}H]benzene signal as reference. From the established ^{1}H chemical

shifts of the pterocarpan (7) in benzene, the deuterium signal

could at once be assigned to the 6-pro(R)position in (8). Hence

the biological reduction involved in the conversion (5) → (7)

proceeds trans-diaxially across the original 2,3-double bond in

(5). Dilution of the label had occurred, evidently because of

(5) R=H

(6) R=D

(7) R=R'=H

(8) R=D, R'=Ac

endogenous synthesis, so that the product contained 58% of
deuterium in the 6-position, as measured by comparison of ^{2}H
signal intensities with that of the natural abundance ^{2}H signal
from the benzene solvent. This finding was equally obtainable
from measurement of the ^{1}H signal intensity at δ 3.92 in the ^{1}H
n.m.r. spectrum. However, as pointed out, at higher isotopic
dilution (e.g. 10 x or more) only the ^{2}H n.m.r. signal intensity
(derived by appropriate accumulation) would give a reliable
analytical figure.

In a study of the mode of addition of one molecular proportion of hydrogen chloride to trans-1,3-pentadiene (9), deuterium
chloride was employed, along with ^{2}H n.m.r. (at 15.28 MHz).[34]
Unfortunately, the ^{2}H n.m.r. signals from the mixture of products,
(10) and (11), overlapped but the relative proportions were
estimated with the aid of a curve resolver, and compound (10) was
found to predominate. Corroboration was obtained by reductive
removal of the combined halogen in the mixture, with lithium
triethylborohydride, to give a mixture of compounds (12) and (13),

for which the ^{2}H n.m.r. signals were fully separated. That 1,2-
addition predominates over 1,4-addition suggests the inter-
mediation of ion-paired species (14) rather than a symmetrical
allylic carbonium ion (15), a conclusion of relevance to tne
mechanism of electrophilic addition to the norbornene system.

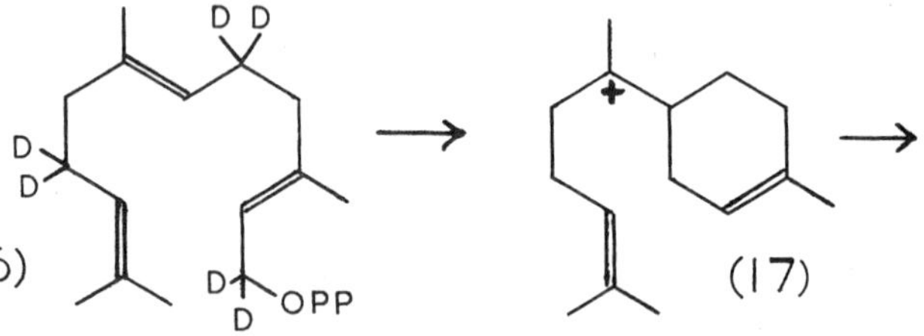

From incorporation experiments with [3,4-^{13}C]mevalonate on
the biosynthesis of the antibiotic ovalicin (19) by Pseudeurotium
ovalis, it was concluded [35] that after initial condensation of
mevalonic units to farnesyl pyrophosphate (16), there was
cyclisation to a bis-abolyl cation (17) which underwent 1,3-mi-
gration of the C_8 side chain to β-trans-bergamotene (18) followed
by oxidative cleavage. Subsequent isolation of the postulated
intermediate (18) provided strong support.[36] To distinguish
between possible mechanisms for the oxidative cleavage (18) → (19),

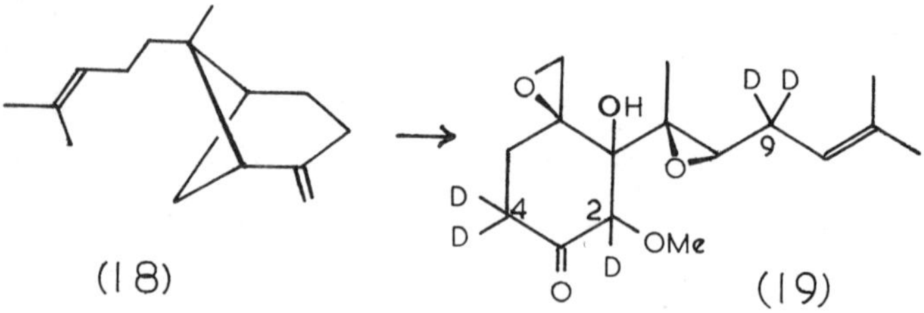

further incorporation experiments were undertaken[37] utilising

[5,5-^{2}H$_2$]mevalonate containing [2-^{14}C, 5-^{3}H]mevalonate as an

internal standard. Measurement of the radioactivity of the iso-

lated ovalicin indicated 4.5% incorporation of precursor, only

trivial loss (1/6) of tritium, and so 7.9% isotopic enrichment

with deuterium at each labelled site. By ^{2}H n.m.r. the pattern

of labelling in the ovalicin was revealed directly, as shown in

Figure 5.

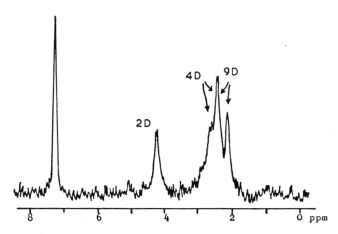

Figure 5. ^{2}H N.m.r. spectrum of ovalicin (2084 transients)

The indicated assignments were made from knowledge of the ^{1}H

chemical shifts determined from the ^{1}H n.m.r. spectrum of

ovalicin at 270 MHz, in part by specific proton spin decoupling

and in part by exchange deuteriation at the 2- and 4-positions.

The ^{2}H n.m.r. results, in terms of formula (19), support the pro-

posed biosynthetic route. In demonstrating that five out of six

of the hydrogen atoms at the 5,5-positions of three mevalonic

units are incorporated directly into ovalicin via compound (18),

the ^{2}H labelling results specifically exclude dehydrobergamotene

(20)[38] and the compound (21)[39] as possible intermediates, as well

as oxidative conversion of the methylene bridge of bergamotene (18)
into carbonyl.

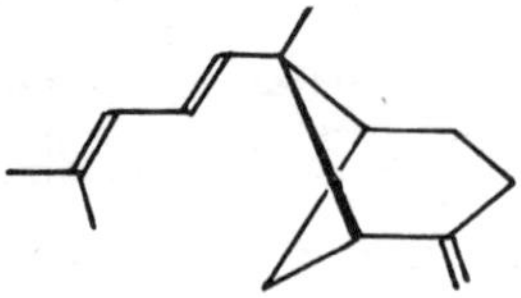

(20) 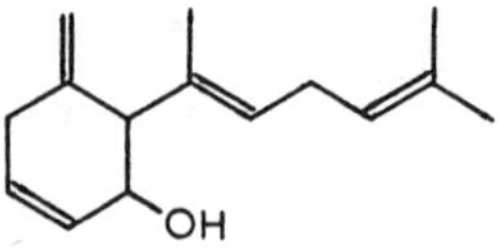   (21)

Deuterium labelling and ^{2}H n.m.r. revealed [40] the stereochemistry of the biological S_N2' ring closure reaction involved
in the biosynthesis of rosenonolactone (23). This was of
particular interest because of controversy surrounding the S_N2'
reaction generally.

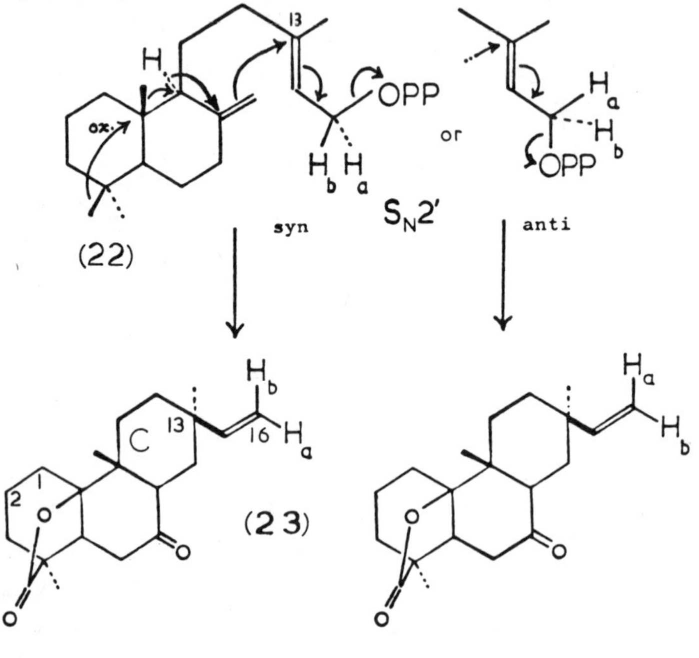

Figure 6. Cyclisation of labdadienyl pyrophosphate to
rosenonolactone

Previous biosynthetic studies [41-45] had established that

labda-8(17),13-dien-15-yl pyrophosphate (22) undergoes S_N2'

cyclisation to form ring C in rosenonolactone, involving allylic

displacement of the terminal pyrophosphate group and concomitant

hydride and methyl shifts. There is also oxidative formation of

the lactone ring.

That cyclisation of ring C in (22) occurs on to the front or

β-face of the allyl system was known from the absolute configur-

ation [46,47] of rosenonolactone. It was not known whether the

pyrophosphate group leaves <u>syn</u> or <u>anti</u> to the original 13,14-

double bond. The direction necessarily determines which of the

prochiral hydrogen atoms at the 16-position becomes <u>Z</u> (<u>cis</u>) and

which <u>E</u> (<u>trans</u>) in the resulting terminal group (Figure 6).

Since the 16-position derives from the 5-position of mevalonate,

as shown <u>e.g.</u> by biosynthesis of rosenonolactone from

[5-^{2}H]mevalonate and the ^{2}H n.m.r. spectrum of the product

(Figure 7A), it was possible to determine the stereochemical

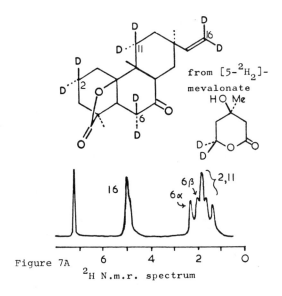

Figure 7A ^{2}H N.m.r. spectrum

course of the biological S_N2' ring-closure reaction by feeding
stereospecifically 5-labelled mevalonate.   Figures 7B and C show
the resulting ^2H n.m.r. spectra and the assignments.   Those
assignments for the vinyl group were derived from ABX analysis
of the vinyl system in the ^1H n.m.r. spectrum of rosenonolactone
at 270 MHz.   The terminal methylene signals are only 0.07 p.p.m.
apart and are not clearly resolved in the ^2H n.m.r. spectrum
(Figure 7A) even at 41.44 MHz (6.3 Tesla) with ^1H decoupling.
Nevertheless, careful comparison of the spectra (Figure 7) shows
that the 16Z(<u>cis</u>)- and 16E(<u>trans</u>)- hydrogen atoms of rosenonolactone are derived respectively from the 5-<u>pro</u>-(<u>R</u>) and 5-<u>pro</u>-(<u>S</u>)-
hydrogen in mevalonate.   The S_N2' cyclisation which generates
ring C in rosenonolactone (23) therefore occurs with overall <u>anti</u>
stereochemistry.

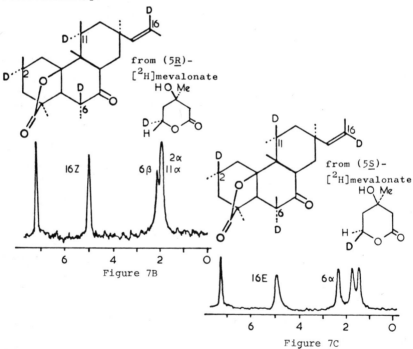

Figure 7B

Figure 7C

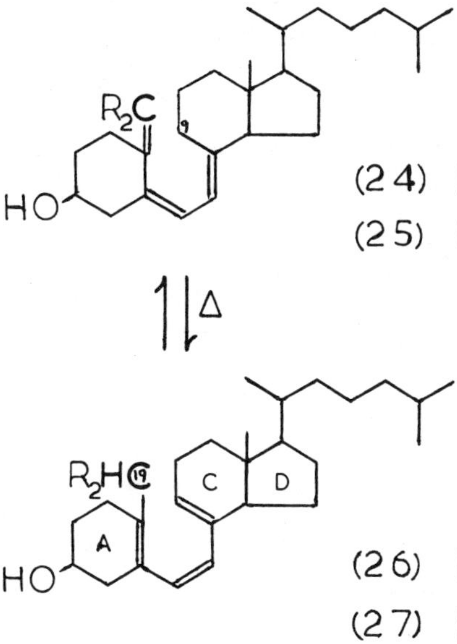

(2 4) R=H

(2 5) R=D

(2 6) R=H

(2 7) R=D

Deuterium labelling and ^{1}H and ^{2}H n.m.r. were used to follow
the equilibration of vitamin D$_3$ (24) with its structural isomer,
previtamin D$_3$ (26).[48] The vitamin was labelled with deuterium in
the terminal vinyl group by synthesis from the appropriate ketone
employing the Wittig reagent Ph$_3$P=CD$_2$. Heating the labelled
vitamin D$_3$ (25) in solution at 80°C for 2h equilibrated it with
previtamin D$_3$ (27) but unexpectedly the ^{1}H n.m.r. spectrum of the
vitamin showed virtually no change. In agreement, the ^{2}H n.m.r.
spectrum of the isolated previtamin D$_3$ showed that deuterium was
present only in the 19-methyl group. Thus under the above
conditions, the equilibration involved 1,7-protium migration to
the exclusion of deuterium migration. The latter was in fact
induced by very prolonged heating (14h), and the large isotope
effect was measured by mass spectrometric analyses as k_H/k_D = 45.
The ^{1}H n.m.r. spectrum showed a 40%increase in the 19-methyl
signal, and a decrease in that at δ 2.7 from the 9β-hydrogen of

the   vitamin,   but gave no more information because of overlap

of signals.   However, the ^{2}H n.m.r. spectrum at 41.4 MHz revealed

the full detail of the 1,7-migration in that signals appeared at

δ 1.68 and 2.70 from deuterium in the 9α- and 9β-positions,

respectively, with relative intensities of 2:1.   This showed that

the transition state for the vitamin D_3- previtamin D_3 equil-

ibration involves hydrogen transfer between the 19-position and

the 9α-position preferentially to the 9β-position.   It follows

that the preferred transition state is the one in which the cis-

triene system is twisted slightly in a right-handed sense to allow

ring A to lie below the plane of the C,D rings.   Presumably the

directing effect on this non-enzymic transformation must originate

from the chiral substitution in rings C and D.   It might further

be surmised that in Nature the transformation involves the fore-

going transition state exclusively.

2.D.   Summary

   The preceding examples illustrate well the usefulness and

also the limitations of ^{2}H n.m.r. for the analysis of deuterium

labelled products.   On the debit side there is the poor sensit-

ivity, lack of spectral dispersion, and general absence of spin

coupling information.   On the credit side, there is the absence

of intrusion from the natural abundance of deuterium (0.015%) with

compounds labelled at 1% or more, the rapid accumulation feasible

because of the relatively short relaxation times for deuterons,

the accurate quantitation possible because of the absence of

nuclear Overhauser effects, and the virtual identity of deuteron

with proton chemical shifts, making assignments straightforward

(referencing has mainly been to the natural abundance ^{2}H signal

from the solvent).   The identification and integration of

individual signals may of course be hindered by the poor

dispersion even at high field strengths, as has been indicated.

The claim that use of deuterium eliminates problems associated

with radioactivity needs putting in perspective : not infrequent-

ly, studies utilising deuterium labelling have involved parallel

experiments with [^{14}C] and [^{3}H]labelled or [^{14}C, ^{3}H] doubly

labelled precursors, in order to delineate the incorporation of

the corresponding [^{2}H]precursor.[37,40,49] When tritiated pre-

cursors are to be used anyway, it would seem logical to examine

and analyse the products by ^{3}H n.m.r., particularly in view of

the superior magnetic properties of the tritium nucleus for high

resolution n.m.r. and the very high sensitivity at 6.3 Tesla.

2.E. N.m.r. applications of ^{2}H as a quadrupolar nucleus.

Where deuterium has a unique advantage over protium or

tritium is in applications making use of the quadrupolar properties

of the nucleus. These enable molecular dynamics and orientation

to be investigated.

One approach is through measurement of spin-lattice relax-

ation times, T_1, by the inversion-recovery Fourier transform

method using the 180°-t-90° pulse sequence. Relaxation times are

simpler to interpret for ^{2}H than for ^{1}H and ^{13}C because deuteron

relaxation occurs entirely through an intramolecular quadrupolar

mechanism, as already outlined. Deuterium in a group that spins

freely will not interact efficiently with the local electric field

gradient and so will have a relatively longer relaxation time.

Slower internal motion will result in more rapid relaxation, as

time becomes available for the interaction (see Table 2).[50] Thus

in [^{2}H$_8$]toluene, deuteron relaxation is faster for the ring

positions than for deuterium in the spinning methyl group. The

rate of relaxation of the ring deuterons will be controlled in

part by the anisotropic tumbling of the molecule. Rotation about

Table 2. Dependence of deuteron spin-lattice relaxation time on internal motion of deuteriated compounds (a neat, b in CCl_4).

Compound	Group		T_1/s
$CD_3-CCl=CCl_2$ a	CD_3		2.7
CD_3-CCl_3 b	CD_3		2.0
CD_2Cl-CD_2Cl a	CD_2		1.3
CD_2Br-CD_2Br a	CD_2		0.64
$CDCl_2-CDCl_2$ a	CD		0.46
C_6D_6 (benzene) a	=CD		1.5
C_6D_5Cl b	=CD	av	0.89
$C_6D_5-CD_3$ b	CD_3		4.3
	=CD	av	0.86
Ph-C≡C-D b	≡CD		0.25

	Group		T_1/s
	=CD o		0.58
	=CD p		0.39

	Group		T_1/s
	CD_3 cis		3.0
	CD_3 trans		1.6
	CD		0.95

axes in the plane of the aryl ring is favoured, especially that axis through a substituent and the p-position. Hence in $[2,4,6-^2H_3]$nitrobenzene (Table 2) the 4-deuteron has the shortest relaxation time. The 2- and 6-deuterons will change their positions relative to the applied field much more, will experience a better average electric field gradient and so will have a longer relaxation time. Reasoning of this type enables assignments of otherwise ambiguous chemical shifts to be made from T_1 measurements on deuteriated analogues. In [$^2H_7$]dimethylformamide (see Table 2) the deuterons on the methyl groups with chemical shifts of δ 2.88 and 2.97 have T_1 = 3.0 and 1.6 s respectively. The high field methyl with the long relaxation time must experience effective averaging of quadrupolar effects : this methyl must therefore move relatively much more than the low field methyl and so lie off the major O-N rotation axis of the molecule. This identifies the high field methyl group as that <u>cis</u> to the oxygen atom,[50] in agreement with other results.[51,52]

Many revealing studies have been made of complex formation,[53] the motion in polynucleotide chains[54] and in peptides such as oxytocin,[55] and the changes in relative motions which occur during interactions between substrates and enzymes,[56,57] from ^2H relaxation measurements, both of T_1 and of T_2 the spin-spin relaxation time. Also of great interest are studies of the relaxation times of heavy water (D_2O) as a measure of the interaction of water with solutes, including electrolytes[58-61] and biopolymers.[62-65] In this way D_2O can be used as a probe for studying the aqueous phase of cells in living organisms.[66,67] The T_1 and T_2 results indicate increased organisation of some of the water in tissue material, presumably because of strong interactions with the biomolecules present : depending on the tissue, there may be external water and bound water[66] and also sometimes water in spaces within macromolecular material.[65]

A second area of investigation, making use of the quadrupolar nature of the deuteron, is in wide-line ^2H n.m.r. studies of deuterium labelled solids or liquid crystals. The n.m.r. lines from solids or orientated molecules are very broad because of rapid spin-spin relaxation (time T_2) arising from the efficient

interactions between the aligned spins. When a nucleus has a
quadrupole, there is an additional splitting of the energy levels,
independent of the applied field. From the magnitude of this
quadrupolar splitting of the n.m.r. line from a deuteron, the
average orientation of the C-D bond with respect to the molecular
field gradient axis can be determined and so the nature of the
molecular motion in orientated molecules can be assessed. The
interaction of the electric quadrupole moment of the deuteron with
a molecular electric field gradient (at the nucleus) having
cylindrical symmetry - as in a rod-like molecule or lipid chain -
gives rise to two deuteron transitions, $\underline{I}$ = 0 to $\underline{I}$ = 1 and $\underline{I}$ = -1
= 0, with an energy separation of

$$\Delta\nu_q = 3 \ e^2 qQ \ (3 \ \cos^2\theta - 1)/4h$$

where $e^2 qQ/h$ is defined as the quadrupolar coupling constant (with
eq as the local field gradient and eQ the quadrupole moment, e   e
being the nuclear charge). In liquids or solutions, the rapid
reorientation of molecules averages the term $(3 \ \cos^2\theta-1)$ to zero,
so quadrupolar splitting of the ^{2}H n.m.r. signals only appears in
solids or liquid crystals, where the splitting is dependent on
orientation between e.g. the C-D bond in question and the axis of
the molecular field gradient (or director). For deuteriated
paraffin chains in the solid, the quadrupolar coupling constant
is 170 kHz : this corresponds to maximum ordering of the hydrocarbon chains in the extended all-trans conformation. From that
value and the observed reduced quadrupolar splitting, $\Delta\nu_q$, for a
deuteriated paraffinic liquid crystal or lipid layer (e.g. in a
membrane or on the surface of a vesicle), the time averaged
orientation term $(3<\cos^2\theta>-1)$ or order parameter $(= S_{CD})$ can thus
be obtained. Thence the average orientation of chain segments or
of C-D bonds to the molecular axis can be derived. In this way,

the dynamic structure of fatty acyl chains in a phospholipid
bilayer (from aqueous deuteriated dipalmitoyl lecithin) has been
determined through ^{2}H n.m.r. measurements at several temperatures.[68] The observed quadrupolar splitting led to the conclusion that the restricted motion of the chains at 50°C resulted
in their being ca. 11.2 Å shorter than for the fully extended
state, and to an interpretation of this shortening in terms of a
set of particular preferred conformations.[68]

Many other investigations of liquid crystal [69-72] and of
model membrane systems[73-75] have been made and the studies have
been extended to biological systems. Suitable membrane material
can be derived from e.g. micro-organisms cultured on specifically
deuteriated fatty acids [76,77] and animals fed with deuteriated
choline.[78]

Like deuteriated water, deuteriated precursors may distort
normal metabolism (D_2O is toxic at 30% displacement of H_2O) so
realistic studies with living organisms may not necessarily be
possible with the deuterium isotope at really high levels.
Pulsed wide-band spectrometers operating at the highest attainable magnetic field strengths overcome the sensitivity difficulty
but there are other problems peculiar to the pulse technique, the
main one being a finite receiver dead time (following the pulse)
which leads to loss of the initial part of the free induction
decay and so distortion of the Fourier transformed spectrum.[79]

Finally in this Section, mention needs to be made of a
valuable use of deuterium in assigning ^{13}C n.m.r. signals.
Particular help was thus obtained in interpreting the ^{13}C n.m.r.
spectra of carbohydrates.[80] The signal from ^{13}C attached to
deuterium (100% labelling) either disappears from the ^{13}C n.m.r.
spectrum or is visible as a triplet (in the case of a ^{13}C-D group)
with an origin position shifted 0.1 to 0.5 p.p.m. to higher field

by the isotope effect. Progressively smaller upfield shifts are observed for β-carbons (0.02 to 0.1 p.p.m.) and for γ-carbons (0.01 p.p.m.). Splitting of these signals may also be observed. With the aid of two C-deuteriated derivatives in each case, unambiguous assignments of the ^{13}C n.m.r. signals from glucose, mannose and galactose were made, and for other common sugars and their methyl glycosides,[80] in some cases with revision of earlier conclusions.

Another example of this application of deuterium comes from an investigation of the biosynthesis of scytalone (28) from [^{13}C]acetate. In this there arose the problems of individually assigning ^{13}C signals at δ 38.8 and 47.1, both of which were from CH_2 groups - necessarily those at the 2- and 4-positions because the signals appeared as triplets in the off-resonance ^{1}H decoupled ^{13}C n.m.r. spectrum. Treatment of scytalone with alkaline D_2O exchanged the protons in the 2-, 5-, and 7-positions with deuterium. In the ^{13}C n.m.r. spectrum of the product, taken with ^{1}H decoupling, the signal previously observed at δ 47.1 had gone. This was then assigned to C-2. The signal strength from ^{13}C carrying deuterium is severely reduced because the relaxation time is now very long, ^{13}C relaxation being domin-ated by a dipole-dipole mechanism. Moreover, the absence of attached spin 1/2 protons also means there will no longer be any nuclear Overhauser enhancement of the ^{13}C signal.[81]

(28)

3. ^{3}H N.m.r. Spectroscopy

3.A.   Tritium [82]

Tritium ($^3_1$H or T) is a moderately short-lived radioactive isotope of hydrogen which is formed in the upper atmosphere as a result of nuclear reactions induced by cosmic rays. It therefore occurs in the hydrogen of rain water and so in the environment generally, although only at extremely low abundance ($< 10^{-16}$%), but not in fossil water. Tritium is manufactured by neutron bombardment of lithium compounds in an atomic pile (Equation 1) and is available commercially as pure gas, diluted tritiated

$$\underset{3}{\overset{6}{}}Li \ + \ \underset{o}{\overset{1}{}}n \longrightarrow \underset{1}{\overset{3}{}}H \ + \ \underset{2}{\overset{4}{}}He \qquad \ldots 1$$

water, and incorporated into a wide variety of compounds of biological and chemical interest, both specifically and generally at various specific radioactivities. The half-life of tritium is 12.35 years so that the rate of disintegration is high : the maximum specific radioactivity of tritiated compounds is 29.15 Ci mmol^{-1}. Hence 1 g of pure T_2O would have a radioactivity of 29.15 x 2 x 1000/22 = 2650 Ci (= 98 TBq = 9.8 x 10^{13} Bq, where 1 Ci = 3.7 x 10^{10} Bq) and there would be considerable gas evolution, partly because of disruption of chemical bonds and partly because tritium disintegrates into helium-3 (Equation 2) by β-particle emission. This last has a mean energy of 5.66 keV

$$\underset{1}{\overset{3}{}}H \longrightarrow \underset{2}{\overset{3}{}}He \ + \ \beta \qquad \ldots 2$$

and is classed as soft β-emission, the mean range in air being only 4.5 mm, so that normal glass apparatus provides complete protection from the radiation (as indeed does the skin).

Whilst tritium is one of the least toxic of radionuclides as such, tritiated compounds should nevertheless always be manipulated with the precautions normal to the handling of potentially dangerous materials.[82] If tritiated compounds are absorbed or ingested, more or less of the tritium is quickly incorporated into the body water by metabolic processes and excreted, so that the biological half-life is about 12 days. Chemical operations with tritiated compounds, especially if volatile, should be conducted

in good fume cupboards. Spill trays should be used as a matter of
routine and safety spectacles and disposable gloves must be worn.
Thus the usual procedures and provisions appertaining to a low
level tracer laboratory are very adequate because at the most only
tens of millicuries of radioactivity will be employed. Moreover,
at this level, problems arising from radiation decomposition will
not be encountered. The emphasis in tracer work should be on
cleanliness and tidiness, and procedures should be chosen such
that in the unlikely event of an accident, contamination is
properly and safely contained. Those who work regularly with
tritiated compounds should have their urine samples monitored at
weekly intervals. Instruments for the assay of tritium in
labelled compounds should be kept separate from the tracer laboratory.

In ^{3}H n.m.r. work, the n.m.r. tubes are loaded and sealed
in the radiochemical laboratory and then mounted inside a stout
metal can (suitably identified). The samples can then safely
be conveyed to the n.m.r. spectrometer. If volatile tritiated
compounds are being examined, then it is wise to exhaust the air
from the vicinity of the spectrometer probe to a fume stack, as a
precaution in case of tube breakage. The probe has no bottom
holes so that, in the unlikely event of a breakage, the sample
would be contained and the probe could at once be removed for
decontamination. We have used microcells for the tritium samples
because these are very strong and easily sealed and are readily
filled by syringe. Wilmad 100 µl and 470 µl cylindrical cells
fit inside standard 5 mm and 10 mm tubes respectively, being held
by means of a teflon chuck : insertion (and withdrawal) is by
means of a screwed rod (Figure 8). A small gauge is set up so
that the bulbs are mounted inside the n.m.r. tubes in the optimum
position with respect to the rf coil. The annular space between

the sealed cell and the n.m.r. tube is filled with inert liquid
(CCl_4) to prevent wobble, or with a deuteriated (or other) solvent
to provide in addition for field frequency locking if for some
reason the lock liquid cannot be used as sample solvent.   The
n.m.r. tubes are then capped and mounted in the probe spinner in
the normal way.   More recently we have favoured cells made from
3 mm precision tubing, sealed to short lengths of 5 mm tube to
facilitate closure with a serum cap.[83]   The tubes can then be
evacuated through a syringe needle (the vacuum is held for some
days), and subsequent filling by syringe, and sealing, are greatly
aided.   The sealed 3 mm sample tubes are mounted in 'nicked'
teflon rings and slid into 5 mm n.m.r. tubes : external liquid
can be added if necessary.   Alternatively, on special occasions
the bare sealed sample tube can be used in a 3 mm probe for maximum n.m.r. detection sensitivity (see Figure 8 for details of
mounting.

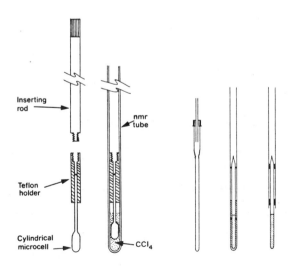

Figure 8.   Details of micro cells

The double containment and related precautions have made ^{3}H
n.m.r. spectroscopy a safe routine. No breakage or incident of
contamination has occurred during measurement of hundreds of
samples over a ten year period.

Of the conventional methods for the analysis of tritiated
compounds,[82] liquid scintillation counting is now highly refined
and automated. Provided essential precautions and necessary
standardisation are carried out, the method will quantitate the
tritium content with precision. The technique of autoradiography
readily locates tritiated compounds on paper chromatograms and
t.l.c. plates. Reverse dilution analysis serves for the positive
identification of specific compounds bearing tritium labelling,
whilst radiochromatography enables mixtures of labelled products
to be separated, and individual components to be identified by
mixed chromatography with known 'cold' compounds.

None of these methods will delineate the positions of
labelling or give the relative amounts of label in the various
positions in the molecules of a tritiated compound. Hitherto
when such information has been required, then step-wise chemical
degradation has been performed, with counting of the products at
each stage. However, such chemical reactions rarely proceed
quantitatively and may be accompanied by unpredictable side
reactions and unsuspected isotope exchange.[82,84] Moreover, the
approach can be very time consuming. The method of choice now is
tritium n.m.r. spectroscopy.[85,86,87] It is non-destructive and
rapid, and it reveals in a very direct manner the positions in a
tritiated compound which bear label, often stereochemical
features, and the relative amount of label at each site.

3.B. Nuclear magnetic properties of tritium

The tritium nucleus, the triton, has a spin $\underline{I}$ = 1/2 like the
proton but a higher magnetic moment and therefore a larger mag-
netogyric constant γ_t (<u>ca</u>. 4.5414 x 10^7 Hz T^{-1}) (see Section 1.C.).
The resonance frequency for a triton in a given magnetic field
will then be higher than for a proton by the ratio γ_t/γ_p
(<u>ca</u>. 1.06664) and the spectral dispersion slightly better. The
^{3}H n.m.r. frequency is 96 MHz at 2.11 Tesla for which the ^{1}H
n.m.r. frequency is 90 MHz.

The preceding ratio of magnetogyric constants is not directly measurable, the constants referring to bare nuclei but,
assuming that the ratio of the hydrogen isotopic screenings in
real compounds must be very close to unity, the experimentally
accessible Larmor ratio ω_T/ω_H should suffice. A good mean value
of 1.06663975 ± 3 x 10^{-8} was obtained[88] from measurements of the
triton and proton resonance frequencies at constant field made on
each of a wide variety of partially tritiated compounds.

The sensitivity of the triton to n.m.r. detection (at constant field) is somewhat higher than for the proton, being 1.21
(at 100% abundance).[7] This clearly facilitates the use of n.m.r.
methods for the investigation of tritium in tritiated compounds
where the isotope will normally be at rather low abundance, <u>e.g</u>.
10^{-1} - 10^{-3} % per site, corresponding to 30 mCi - 300 µCi of
radioactivity per site (1.11 GBq - 1.11 MBq). Use of a very
high field Fourier transform n.m.r. spectrometer will extend that
range downward by one order of magnitude.

Being a spin $\underline{I}$ = 1/2 nucleus and having a resonance frequency not dissimilar to that of the proton, the triton will
undergo relaxation in much the same way as has been described for
the proton (Section 1.C.), the upper spin state energy being exchanged with random fluctuating magnetic fields having the

correct frequency and phase, generated by chance by the molecular
'lattice' and its collection of nuclear spins. Hence triton
relaxation times should be similar to those of corresponding
protons. This is borne out by measurements of T_1 for the two
nuclei in partially tritiated compounds, under the same conditions
(Table 3). Line widths of the two nuclei are therefore comparable, and similar operating conditions in n.m.r. spectroscopy are
appropriate.

Table 3.   Experimental spin-lattice relaxation times for ^{3}H
and ^{1}H in the same sites.

Compound (position tritiated)	^{3}H	T_1/s ^{1}H	Correlation coefficients	
Water	0.29(6)	0.21(4)	0.999	0.997
Diethyl malonate (2)	0.63(7)	6.49(6)	0.991	0.998
Dimethyl sulphoxide	0.83(6)	0.66(8)	0.996	0.998
Benzimidazole (2)	1.58(6)	1.79(5)	0.996	0.998

3.C.   Characteristics of ^{3}H n.m.r. spectroscopy

It can be argued that ^{3}H n.m.r. chemical shifts should be
virtually the same as ^{1}H chemical shifts (and ^{2}H chemical
shifts).[88] This follows from the facts that the shielding of a
hydrogen nucleus in a compound in solution is very largely a
function of the local molecular environment and that replacement
of the hydrogen by an isotope, e.g. tritium or deuterium, will
not materially affect that local environment. Experimental
verification came from the Larmor ratio measurements (Section 3.B.)
made on a variety of partially monotritiated compounds at constant
field. The Larmor ratio was constant to 7 decimal places.[88]
Subsequent, more precise measurements indicate that there is a

very small dependence of hydrogen nuclear shielding upon bond
order but confirm that triton and proton chemical shifts are the
same to within ± 0.02 p.p.m.[89] This latter is then the magnitude
of the primary tritium isotope effect - hardly different from
normal routine n.m.r. error. The upfield shift in a tritium
resonance resulting from the + $\underline{I}$ effect of another tritium, as in
CHT_2 or CT_3 methyl groups, is $\underline{ca}$. -0.02 p.p.m. : the secondary
isotope effect is -0.01 p.p.m. for an additional vicinal triton.[89]
These shifts are just enough to be properly resolved at 96 MHz
(2.11 Tesla) and enable multiple labelling to be identified.

The effective identity of triton and proton (and deuteron)
chemical shifts means of course that the vast store of information
concerning proton chemical shifts is immediately available for
the interpretation of [3]H (and ^{2}H) n.m.r. spectra. A further consequence is that there is no essential need for an internal
tritiated reference for [3]H n.m.r. spectroscopy.[88,89] Our preferred technique is to use internal TMS or DSS in the normal way,
with a deuteriated solvent to provide the field-frequency lock
(a fluorine lock would be a good alternative). The position of
the TMS signal in the [1]H n.m.r. spectrum, $\underline{i.e.}$ its frequency, is
obtained from the spectrometer output after only a few pulses and
Fourier transformation. Multiplication by the Larmor ratio
(1.06663975) then gives a corresponding reference frequency for
the transformed [3]H n.m.r. spectrum. The ^{1}H transmitter and
matched probe are then changed for the ^{3}H accessories, and with
the same sample (and hence at exactly the same locked field), the
[3]H n.m.r. spectrum is acquired and eventually transformed. The
[3]H pulse point is known and the corresponding computer address.
Hence the calculated ^{3}H reference frequency can be assigned to
its correct address and the latter treated as the internal
reference point. The procedure is effectively the same as if

monotritiated TMS had been used as an internal reference. Recently we synthesised partially monotritiated TMS and confirmed the
validity of the ghost referencing procedure,[89] at the same time
revising the Larmor ratio slightly to $1.066639738 \pm 2 \times 10^{-9}$.

Because tritium is normally present in labelled compounds at
low abundance (0.1% per site or less), its presence is not seen in
a normal intensity proton n.m.r. spectrum, just as the presence
of natural ^{13}C (1% abundance) does not intrude usually in ^1H n.m.r.
spectra of organic compounds. The signals in a ^3H n.m.r. spectrum,
however, will show the expected coupling to adjacent protons. This
can be very useful in revealing the stereochemistry of the
labelled site. However the distribution of signal intensity
amongst the lines of multiplets will necessitate inconveniently
long acquisition times to build up an adequate signal-to-noise
ratio. Most frequently, therefore, tritium n.m.r. spectra are
acquired with concomitant proton noise decoupling so that only
single lines are recorded, one per chemically distinct site (as
in ^{13}C n.m.r. spectroscopy). The ^3H chemical shifts are thus
available by inspection.

Using samples of tritiated water, it was early shown that the
integrated ^3H n.m.r. signal intensity corresponded directly with
the radioactivity determined by counting.[85] Hence the tritium
analyses derived from undecoupled ^3H n.m.r. spectra are undoubtedly accurate to within normal n.m.r. limits ($\pm$ 2%). When ^3H
signal intensities are measured with simultaneous ^1H decoupling,
errors may arise from differential nuclear Overhauser effects
(n.O.e.s), but these are not large and can normally be ignored.[90]
Table 4 illustrates the fact that where ^3H n.m.r. analytical results can be compared with those from degradation and counting,
the agreement is generally very good.

A nuclear Overhauser effect is a change in signal strength

Table 4. ^{3}H Nuclear Overhauser effects (%) with relative ^{3}H
signal intensities (%) from (i) ^{1}H noise-decoupled and (ii)
n.o.e.-suppressed gated-decoupled spectra[a]

Compound and position	N.O.e. [b]	(i)	(ii)
L-[4,5-^{3}H]Leucine [c]			
5	40	77.3	77.0
4	40	18.9	18.9
impurity	29	3.7	4.1
[G-^{3}H]Toluene [d]			
Me	17	40.1	38.6
o	10	27.6	28.4
m + p	10	32.3	33.0
[G-^{3}H]Isopropylbenzene [d]			
Me	32	26.6	25.5
α-CH	22	10.5	10.8
ring	24	62.9	63.7
[1β,2β(n)-Testosterone [e]			
1β	0	35.7	35.2
2β	5	33.3	34.4
1α	0	11.9	12.3
2α	8	19.0	18.0

[a]Alternate blocks of 200 decay signals from the two acquisition
modes were accumulated, to minimise differential effects from
possible changes in spectrometer performance.
[b]Experimental values for the particular freshly-prepared
samples. [c]In D_2O. [d]Neat. [e]In $(CD_3)_2SO$.

as a result of a double irradiation experiment.[7,16] For a partially tritiated compound, the n.o.e. observed in the ^{3}H signal
intensities will be derived from the ^{1}H induced relaxation of the
tritons. Irradiation of the protons increases the population of
their upper spin state relative to the lower. To compensate, and
maintain an overall Boltzmann distribution between lower and upper
spin states for the whole of the system of interacting nuclei
(tritons and protons), a corresponding redistribution of triton
spins occurs to give a greater population in their lower spin
state. More rf energy is absorbed and the increase in triton signal strength results. The theoretical maximum n.o.e. for this
system is 47% $(\gamma_p/2\gamma_t)$[91] (as compared with 199% for ^{13}C on ^{1}H
irradiation). Errors in triton signal intensities due to ^{1}H
irradiation will then never be large and at the very worst would
be ± 10% but generally much less.[90] This is acceptable in comparison with the major effort and the many uncertainties involved
in conventional tritium analysis by chemical degradation and
counting. Detailed study has shown there is little variation in
n.o.e. between different tritons in a generally labelled compound
so that analysis of the tritium distribution by integration of the
^{3}H n.m.r. signals acquired with proton spin-decoupling is not
materially affected (see Table 5).[90] Nuclear Overhauser effects
can be fully eliminated, if required, by employing ^{1}H decoupling
with appropriate inverse gating (an n.o.e. suppression sequence
with delays of 10 x T_1 s between acquisitions)[92] but this increases
the spectrometer operating time by ca. 4 times, which is a cost
not often justifiable.[90]

   An analysis of nuclear spin-spin coupling between nuclei N
and N' via electron spin and electron orbital motion indicates[7]
that all contributions to the coupling constant $\underline{J}_{NN'}$ are proportional to the product of the magnetogyric constants, so that

Table 5. Distribution of tritium (i) by degradation and counting and (ii) by ^{3}H n.m.r.

Compound and position	Relative labelling	
	(i)	(ii)
[G-^{3}H]Phenylalanine		
side chain, 2	4	2
ring	71	74
[G-^{3}H]Benzopyrene		
6	8.6	11
[G-^{3}H]Valine		
α	10.5	12
β	77.7	74
γ	12	14
[1β,2β(n)-^{3}H]Testosterone		
1β + 2β	82-75	79
2α + 2β	50	51
[1α,2α(n)-^{3}H]Cholesterol		
1α + 1β	66-62	62.5
2α + 2β	40	37.5

Equation 3 should hold. It is thus possible to predict triton

$$J_{NN}/\gamma_N\gamma_N \ = \ J_{NN'}/\gamma_N\gamma_{N'} \ = \ J_{N'N'}/\gamma_{N'}\gamma_{N'} \ = \text{etc.} \quad \ldots 3$$

coupling constants to other nuclei, knowing the ratio of the appropriate magnetogyric constants. In particular, triton-proton coupling constants can be predicted from the corresponding proton-proton coupling constants and vice versa.[93,89] The relationships given in Equation 4 follow from Equation 3. Hence we may write Equation 5 using the best available experimental value for the

$$\underline{J}_{T,T} = \underline{J}_{H,T} \, (\gamma_T/\gamma_H) = \underline{J}_{H,H} \, (\gamma_T/\gamma_H)^2 \qquad \ldots 4$$

Larmor frequency ratio in place of the ratio of magnetogyric constants.[89]

$$\underline{J}_{T,T} = \underline{J}_{H,T} \times 1.06663974 = \underline{J}_{H,H} \times 1.06663974^2 \qquad \ldots 5$$

It has been a practice to calculate inaccessible proton-proton coupling constants (such as geminal $^2\underline{J}_{H,H}$ where the protons have the same chemical shift) from measurements of the deuteron-proton coupling constant (Equation 6) but this approach suffers

$$\underline{J}_{H,H} = \underline{J}_{H,D} \times 6.5144 \qquad \ldots 6$$

in precision from the small value of the observed splitting and uncertainties from possible line broadening, as well as the unfavourable numerical factor which magnifies those intrinsic errors. Much better practice is to use tritium substitution and calculation (Equation 5) from the observed tritium-proton coupling constant in the ^{3}H n.m.r. spectrum : the observation can be made as accurately as for measurement of a proton-proton coupling constant and the multiplying factor is not far from unity.[89,93]

3.D. Applications of ^{3}H n.m.r.

Uncertainty concerning the pattern of labelling in commercially available [G-^{3}H]phenylalanine [94,95] was readily resolved by ^{3}H n.m.r.[96] The aminoacid is labelled by being heated with tritiated water and a platinum catalyst at 135°C, and then labile tritium is back exchanged from the carboxyl and amino groups. The 96 MHz ^{3}H n.m.r. spectrum, acquired by the Fourier transform method and with ^{1}H decoupling (Figure 9), showed a single line from each labelled position, three at low field from the ring and three at high field from the side chain, indicating at once that

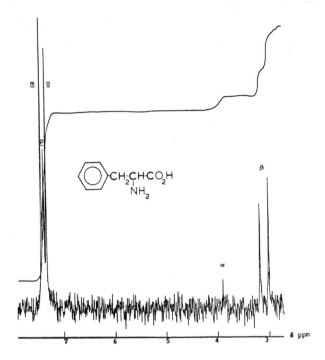

Figure 9. ^{3}H N.m.r. spectrum (^{1}H decoupled) of phenylalanine

the label was generally distributed. The assignments were made in

part from the ^{1}H n.m.r. spectrum and partly from specific labell-

ing of the ring, and the distribution of the label from the signal

intensities : there was little variation between four different

samples, 26% of the label being in the side chain and 74% in the

ring, with the ring uniformly labelled. The ^{3}H n.m.r. spectrum

gave directly the two chemical shifts of the non-equivalent

β-hydrogen atoms and (for the first time) the separate chemical

shifts of the ring hydrogens. The complete analysis took little

more than the spectral acquisition time of 8.5 hours for a 54 mCi

sample. With quadrature detection, this time is now halved. In

contrast, step-wise chemical degradation and counting occupied

1 man-month : the results were in very close agree-

ment, confirming in this early case the reliability

(and great convenience) of the ^{3}H n.m.r. method and

the absence of differential nuclear Overhauser

effects on the n.m.r. intensity ratios.

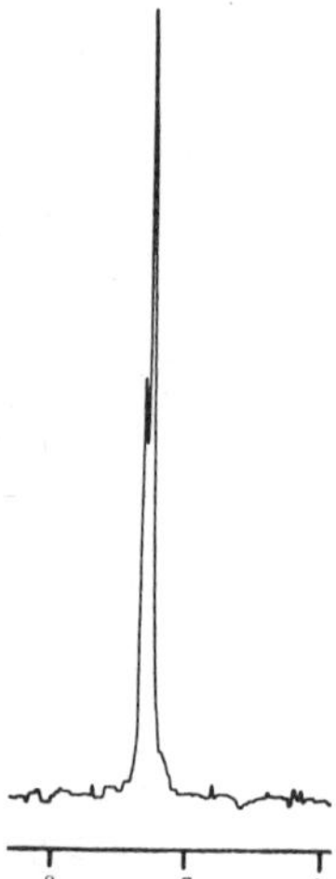

A specifically labelled L[2,4,6-^{3}H]phenyl-

alanine was prepared via catalysed dehalogenation

of a trihalogeno precursor with tritium diluted

with ordinary hydrogen gas. The ^{1}H decoupled ^{3}H

n.m.r. spectrum of the product (Figure 10) showed

only two singlets, at δ 7.35 and 7.39, with inten-

sities in the ratio of 2:1 in accord with expect-

ation. The higher field signal arises from tritium

at the 2,6-positions and the lower field signal

from that at the 4-position. Confirmation of the

Figure 10.

assignments came from the decoupled ^{3}H n.m.r. spectrum of

L[4-^{3}H]phenylalanine similarly prepared from the 4-halogeno com-

pound. This showed a single line at δ 7.39 from the p-tritium.

p-Fluoro[G-^{3}H]phenylalanine gave the well-resolved ^{3}H n.m.r.

spectrum shown in Figure 11, and again a full analysis of the

tritium distribution was derived at once from the signal inten-

sities.[96] About 21% of the total label was in the side chain,

with much more in the α-position than in phenylalanine, and the

rest of the label (79%) was equally distributed in the ring.

Knowing typical values for the ortho and meta proton-fluorine

coupling constants, possible values for the corresponding triton-

fluorine coupling constants can be calculated (Equation 3) and so

the assignments made with certainty as shown (Figure 11). In

comparison, specifically tritiated 4-fluoro[2-^{3}H]benzoic acid

showed only a 1:1 doublet at δ 8.04 in the ^{1}H decoupled n.m.r.

spectrum, with $J_{T,F}$ 5.9 Hz in accordance with the tritium being

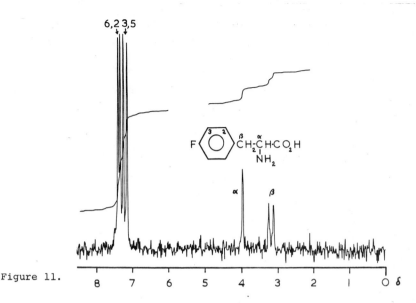

Figure 11.

entirely <u>meta</u> to the fluorine. The corresponding proton-fluorine coupling constant is 5.5 Hz.[97]

In none of the preceding examples was there any evidence for true multiple labelling. The triton signals in the ^{1}H decoupled spectra were all singlets. The exchange and other conditions employed were such as to yield products in which a small proportion of the molecules bore one tritium atom each. The observed ^{3}H n.m.r. spectra were then superpositions of the spectra of several different mono-labelled species. This of course is the normal situation in the n.m.r. spectroscopy of nuclei at low abundance.

The ^{3}H n.m.r. spectrum (^{1}H decoupled) of 4-amino[2,3-^{3}H] butyric acid, prepared from the corresponding alkene by catalytic addition of tritium gas (containing some ordinary hydrogen), showed evidence of true double labelling and so of the presence of the species 4-amino[2,3-^{3}H$_2$]butyric acid. The spectrum (Figure 12) shows an AB quartet (<u>J</u> 7.8 Hz) from the 3- and the 2-coupled

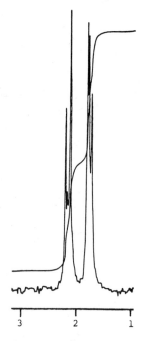

Figure 12. ^{3}H n.m.r.
spectrum (^{1}H decoupled)
of 4-amino[2,3-^{3}H]but-
yric acid

tritons, at δ 1.84 and 2.24 respectively.
Also visible are singlets at δ 1.85 and
2.24(5) from singly labelled molecules.
The secondary isotope effect observable
in the doubly labelled species is then
of the order of -0.01 to -0.005 p.p.m.

5-Hydroxy[G-^{3}H]tryptamine (29) as a
salt in D_2O solution gave a ^{3}H n.m.r.
spectrum (with ^{1}H decoupling) which
clearly indicated that the label was
fairly evenly distributed between the
four ring positions, with rather less in
the side chain (Figure 13). By observing the corresponding coupled ^{3}H n.m.r.
spectrum (Figure 14), a complete assignment of the resonances became possible,
as shown in Table 6. The singlet at
δ 7.31 must be from tritium in the lone
2-position. The fine doublet ($\underline{J}$ 1.9 Hz)
at δ 7.12 must be from the 4-position,
where the triton is coupled only to a

Table 6. ^{3}H N.m.r. results for the aromatic rings in
5-hydroxy[G-^{3}H]tryptamine (as the creatinine sulphate salt)
in D_2O containing DSS.

δ_T	Multiplicity ($J_{T,H}$/Hz)	Assignment	%
6.89	dd (9, 1.9)	6	22
7.12	d (1.9)	4	20
7.31	s	2	19
7.43	d (9)	7	22

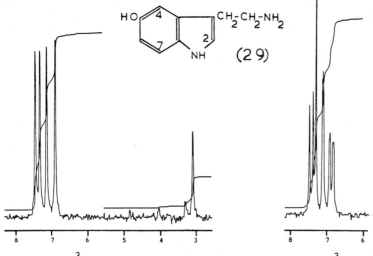

Figure 13. ³H N.m.r. spectrum of 5-hydroxy[G-³H]tryptamine with ¹H decoupling

Figure 14. ³H N.m.r. spectrum with ¹H coupling

hydrogen <u>meta</u> to it. The lowest field signal at δ 7.43, split as a wider doublet (J 9 Hz), must arise from a 7-triton, coupled <u>ortho</u> to hydrogen in the 6-position. In full agreement, the remaining tritium signal is a double doublet (J 9, 1.9 Hz) and so only assignable to the 6-position, which has hydrogen both <u>ortho</u> and <u>meta</u> to it.

Another example of the good resolution attainable in ³H n.m.r. spectra, as a result of the narrow line width and the satisfactory spectral dispersion, is provided in Figure 15. This shows the aryl region of the ¹H coupled ³H n.m.r. spectrum of [G-³H]nicotine (30)[98]. The assignments shown are completely confirmed by the first-order multiplets which arise from each triton as a result of coupling to surrounding protons. There is no double labelling and the spectrum is essentially a superposition of the ³H n.m.r. spectra of each monotritiated species present in the sample. The

integrated signal intensities
quantitate the distribution of
label directly. Incidentally,
assignments for the pyrrolidine
ring of nicotine have been derived from ^{1}H decoupled [2]H n.m.r.
spectra of some specifically
deuteriated derivatives.

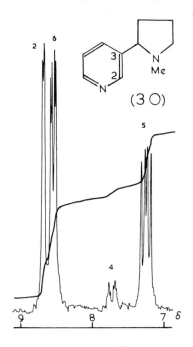

(30)

Figure 15.

When atropine (31) was
subjected to catalysed general
exchange conditions, much less
tritium was incorporated than
expected. The [3]H n.m.r. spectrum
(with ^{1}H decoupling) revealed that
label was confined largely to the
N-methyl group with lesser amounts
in the acyl group, as indicated
(Figure 16). Once again it is to be noted that the labelled
material comprises a mixture of singly labelled species, there
being no triton-triton coupling. Also, the chemical shifts of
the non-equivalent hydrogens of the acyl CH_2 group are available
by inspection, and there is an indication that the N-methyl group
comprises more than one tritiated species. The coupled [3]H n.m.r.
spectrum (Figure 17) confirmed the assignments in that each of the
three signals in the δ 4 region now appeared as the four-line X
part of an ABX spin coupled system. Two of the four lines from
the CTPh signal overlap the two lowest field lines of the adjacent
CHT.O group multiplet. In the δ 2 region can be seen a 1:2:1
triplet signal (J 12.8 Hz) centred at δ 2.23 from the CH_2T labelled
methyl group, a doublet (J 12.8 Hz) centred at δ 2.20 from a CHT_2
methyl group, and a weak singlet at 2.18, necessarily from a CT_3

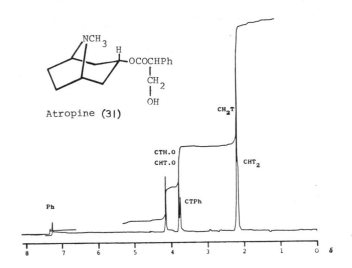

Figure 16. ^{3}H N.m.r. spectrum of [^{3}H]atropine (^{1}H decoupled)

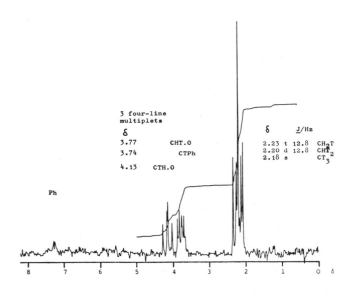

Figure 17. ^{3}H N.m.r. spectrum of [^{3}H]atropine showing coupling
to protons

methyl group. The upfield shifts, due to the isotope effect,
demonstrate the +I (inductive) effect of tritium with respect to
protium.

   In each of the three previous examples it would have been
an extremely difficult task to have delineated the labelling by
conventional degradation and counting. This was even more true
as regards [³H]vinblastine, an important anticancer drug with a
complex light sensitive structure (32) which is unstable except
in the protonated form. Labelling with tritium was successfully
achieved using [O-³H]trifluoroacetic acid at ambient temperature
for 2 hours and this was followed by removal of labile tritium
and purification by t.l.c. to give [³H]vinblastine at the high
specific activity of 8.2 Ci mmol^{-1} (303 GBq mmol^{-1}). Direct
comparison of the ³H and ¹H n.m.r. spectra of a 20 mCi sample at
once revealed that the label was confined to stable aromatic
positions (see Figure 18). The assignments in the indole ring
were made partly from the expectation that the multiplet signal
from H-14' would be relatively broadened by the adjacent

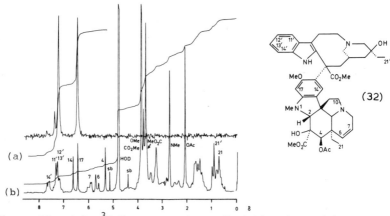

Figure 18.   (a) ³H N.m.r. spectrum of vinblastine sulphate
                   (20 mCi) in D₂O)

             (b) The ¹H n.m.r. spectrum of the same sample

quadrupolar nitrogen : the chemical shifts derived (and already

described) for protonated tryptamine lend strong support. Follow-

ing this non-destructive analysis of [^{3}H]vinblastine, the sample

was used for the successful development, by the radio-immunoassay

approach, of a high avidity antiserum, which can now be employed

in the routine monitoring of the extremely low doses of the drug

required for effective therapy of responsive cancers.[100] Hitherto

there was no suitably sensitive analytical method for determin-

ation of the drug in blood serum, which was a serious difficulty

in its therapeutic use. Moreover, the way is now open for studies

on the mode of action of vinblastine, employing tritium as a

tracer.

Much effort has been expended in attempting to achieve

specific tritium labelling in steroids for biochemical invest-

igations and in confirming the specificity of such labelling.[84]

This latter is now very easily achieved by ^{3}H n.m.r. spectroscopy,

which gives complete analyses directly, quickly, and non-destruct-

ively. Hitherto the chemical and biochemical methods of degrad-

ation (and counting) have been tedious and have not always given

consistent results, underlining the fundamental difficulty of

establishing the integrity of such chemical procedures. Figure 19

shows the ^{3}H n.m.r. spectrum, taken with ^{1}H decoupling, of

17α-hydroxy-11-deoxy[1,2(n)-^{3}H]corticosterone (33), prepared by

partial catalytic reduction of the corresponding 1,2,4,5-dien-3-

one with tritium gas. The doublets at δ 1.54 and 2.11 show there

is tritium in the 1α- and 2α-positions respectively. The split-

ting of the signals (J 4.2 Hz) arises from the coupling between

the axial 1α-triton and its equatorial 2α-triton partner in the

necessarily doubly-labelled molecules. That the reduction does

not proceed with complete stereoselectivity is shown by the

further two weak doublets at δ 1.90 and 2.35 (J 4.8 Hz),

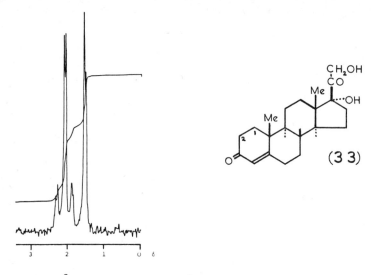

Figure 19.   [3]H N.m.r. spectrum (^{1}H decoupled) of the
[1,2(n)-^{3}H]corticosterone derivative (33)

respectively, from an equatorial 1β- and an axial 2β-triton.   The
signal intensities show there is an equal distribution of tritium
between the 1- and 2-positions, with 80% of the labelled material
bearing the label on the α-face and 20% on the β-face of the
molecules.   Similar detailed analyses of [1,2-^{3}H]testosterone
samples were readily achieved, and the assignments confirmed from
the splitting of the signals observed in the ^{1}H coupled [3]H n.m.r.
spectra.[84]

        The capability in [3]H n.m.r. spectroscopy of being able to

observe the coupling of tritons to neighbouring protons enables unexpected signals to be identified. A preparation of 5α-dihydro[4,5-³H] testosterone (35), by catalytic reduction of testosterone (34) with a mixture of tritium and ordinary hydrogen gas, gave a ^{1}H decoupled ^{3}H

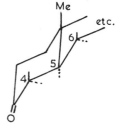

Figure 20. ^{3}H N.m.r. spectrum of 5α-dihydro [4,5-³H]testosterone (no decoupling)

Figure 21. Detail of stereochemistry

n.m.r. spectrum which showed <u>three</u> major signals, not two, and some minor ones. That the major signals at δ 1.40, 1.86, and 2.29 arise from tritium in the 5α-, 4α-, and 4β-positions, severally, was made clear by the ^{1}H coupled ^{3}H n.m.r. spectrum (Figure 20). This showed the highest field signal as a triplet of triplets, consistent with it originating from a 5α-triton coupled twice axially and twice equatorially to surrounding protons (Figure 21). The next signal was a doublet (<u>J</u> 15.6 Hz) showing signs of further splitting, in agreement with it being from the 4α-triton coupled geminally (to 4β-H) and equatorially-axially (to 5α-H). The third signal (δ2.29) was a sharp triplet (<u>J</u> 14.5 Hz), consistent with its assignment to a 4β-triton which would be coupled axially (to 5α-H) and geminally (to 4α-H).

The carcinogenic activity of polycyclic aromatic hydrocarbons, which occur widely in the environment, poses a continuing problem for which tracer studies with readily accessible tritiated derivatives are widely used. There has been an increasing need for more specific studies involving a knowledge of the distribution of tritium in these compounds, but there have been no adequate analytical approaches. [3]H N.m.r. spectroscopy readily solves the problem.[101] The narrow line width and the sufficient spectral dispersion ensure success where the corresponding use of ^{2}H labelling and [2]H n.m.r. analysis would fail because of the close proximity of the hydrogen chemical shifts for such compounds. Figure 22A shows a typical spectrum, that of [G-^{3}H]3-methylcholanthrene (36) labelled by catalysed exchange with tritiated water in acetic acid at 140°C. With ^{1}H decoupling most of the labelled positions give a separate [3]H n.m.r. line. These can be assigned by reference to the analysed [1]H n.m.r. spectrum measured at 220 MHz and so the tritium distribution can then be derived from the line intensities as indicated in Table 7. The results indicate that the aliphatic groups were the most heavily labelled, particularly the 3-methyl group, for which two chemical shifts were apparent. That the lines at δ 2.36 and 2.38 were from a CHT$_2$ and a CH$_2$T group was demonstrated in the undecoupled [3]H n.m.r. spectrum (Figure 22B) by a doublet (J 15.3 Hz) and a triplet signal (J 15.1 Hz) respectively. The decoupled ^{3}H spectrum (Figure 22A) gave the chemical shifts of the 1- and 2-methylene hydrogens directly, whereas of course these are not easily derived from the compact AA'BB' multiplet system in the [1]H n.m.r. spectrum

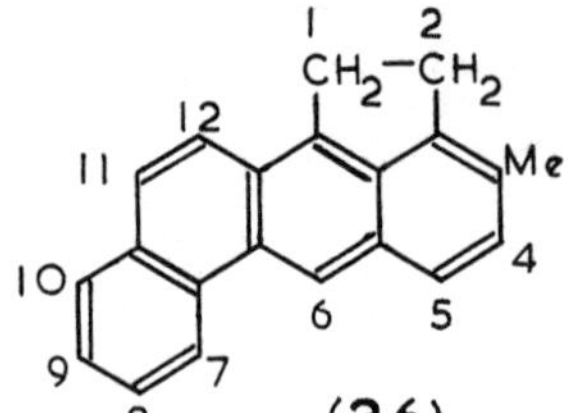

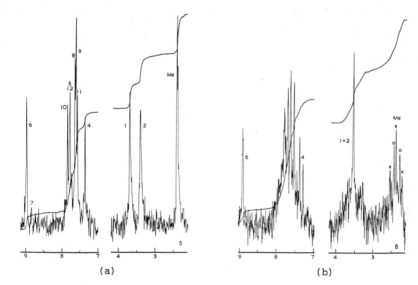

Figure 22. ^{3}H N.m.r. spectra of [G-^{3}H]3-methylcholanthrene
(a) with ^{1}H decoupling and (b) without

or from the coupled ^{3}H n.m.r. spectrum (Figure 22B). The aromatic
ring positions were reasonably uniformly labelled except for the
hindered 7-position which was only weakly labelled. The equally
hindered 6-position is activated and so had an almost full share
of label. Only the 5- and 12-positions had coincident chemical
shifts in perdeuteriodimethyl sulphoxide but the total signal
intensity suggested each position carried label. Separation of
such coincident signals can be achieved in principle by change of
solvent or judicious use of a shift reagent.

That the hydrogen chemical shifts in a multispin coupled
system are directly available by inspect-
ion from the ^{1}H decoupled ^{3}H n.m.r.
spectrum of a [G-^{3}H]labelled compound can
(37) materially assist the spectral analysis.
It was possible in this way and by

Table 7. Distribution of tritium in [G-^{3}H]3-methylcholanthrene
from the ^{3}H n.m.r. spectrum of a solution of d_6-DMSO measured at
96 MHz with proton noise decoupling.

Chemical shift δ	Assignment	Intensity %
2.36	3-CHT$_2$	} 20
2.38	3-CH$_2$T	
3.39	2-CHT	13
3.68	1-CHT	12
7.34	4	7.5
7.56	11	6
7.60	9	8
7.63	8	8
7.76	5,12	10.5
7.82	10	7.5
8.79	7	1
8.93	6	6

reference to related compounds to assign the chemical shifts in
phenanthridine (37)[98] although the complex ^{1}H n.m.r. spectrum had
not been analysed.

A first example of the potential usefulness of ^{3}H n.m.r. in
biosynthetical investigations employing tritium as a tracer was
the derivation by this means of the incorporation of [2-^{3}H]ace-
tate into penicillic acid (38) by *Penicillium cyclopium*.[31,102]
The biosynthesis of penicillic acid had been studied using
[^{14}C]acetate and [^{14}C]malonate as well as through ^{13}C labelling
and ^{13}C n.m.r. Hence the new approach could be compared in detail
with established methods. The earliest ^{14}C experiments had
established the polyketide nature of penicillic acid but not the
precise pathway through intermediate orsellinic acid (39),

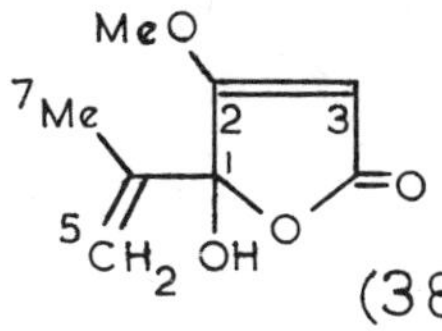

(38) 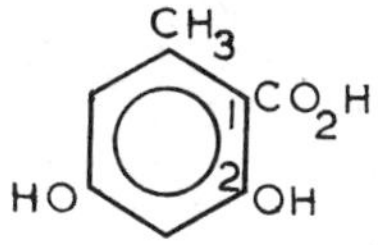   (39)

cleavage being feasible at either the 1,2- or the 4,5-bond [103,104]

(see Figure 23). A decision in favour of the 4,5-cleavage was

achieved only through the considerable effort of synthesising and

feeding [1-^{14}C]orsellinic acid.[104] In contrast, the feeding to

the mould of [2-^{3}H]acetate, and ^{3}H n.m.r. examination of the

resulting labelled penicillic acid, at once revealed the biosyn-

thetic pathway unambiguously : this necessarily involved 4,5-

cleavage of the intermediate orsellinic acid because tritium was

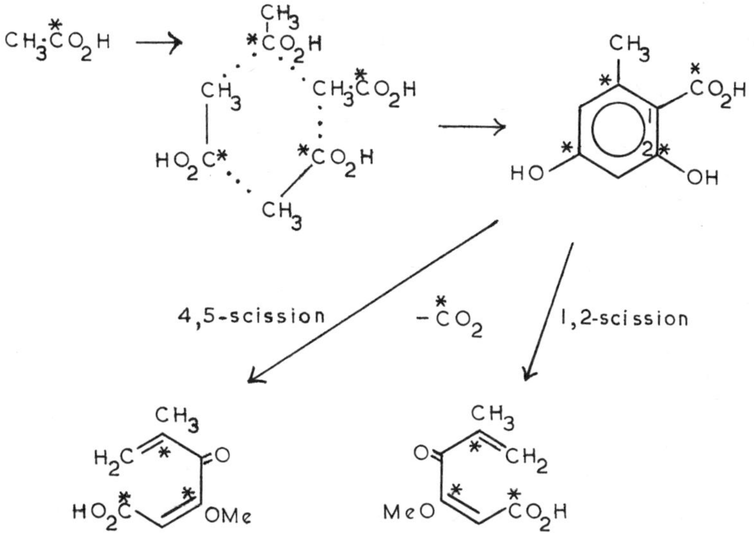

Figure 23. Biosynthesis of penicillic acid from [^{14}C]acetate.

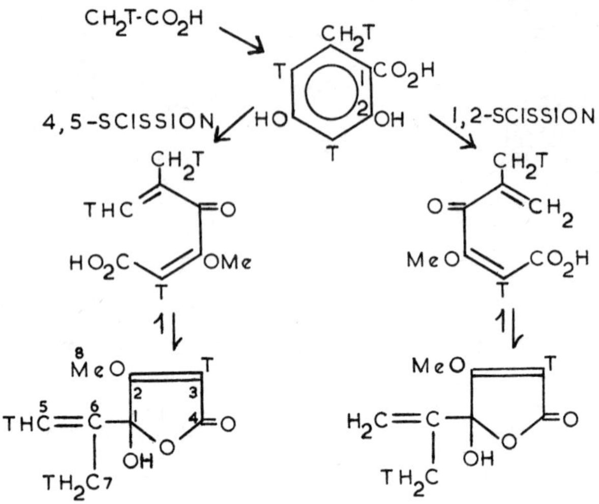

Figure 24. Biosynthesis of penicillic acid from [³H]acetate

incorporated into the vinylic methylene group of the penicillic

acid (Figure 24). The ³H n.m.r. spectrum (Figure 25A) showed

signals from the vinylic 5-tritons, one of which was very weak,

a singlet from the 3-triton and a triplet ($\underline{J}_{T,H}$ 15.5 Hz) from the

7-CH$_2$T methyl group. The assignments followed unambiguously from

the ¹H n.m.r. spectrum (see Figure 25B) of the same sample. The

strong, higher field 5-T signal was assigned to the triton <u>trans</u>

to the methyl group from the results of experiments with a shift

reagent. This orientation of the labelling at C-5 provided

hitherto unknown information about the stereochemistry of the

acetate incorporation. In further experiments, use of

[3,5-³H]orsellinic acid confirmed the role of orsellinate as

an advanced intermediate, whilst use of [2-³H]malonate

confirmed that malonate was a precursor and derived from acetate

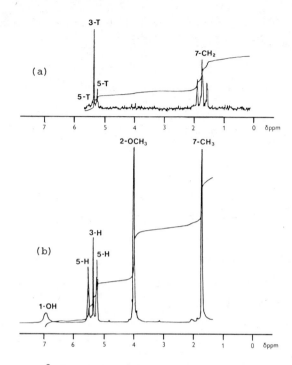

Figure 25. (a) ^{3}H N.m.r. spectrum of tritium labelled penicillic acid and (b) the ^{1}H n.m.r. spectrum

because the penicillic acid produced was labelled only in the 3- and 5-positions. The 7-methyl group was derived solely from starting acetate. The lower incorporation of label into the 3-position of penicillic acid relative to the 7-methyl group and into the 5-position relative to the 3-position clearly indicated the biosynthetic pathway as involving condensation of acetate with three malonate units to give rise on the enzyme surface to a tetraketide, followed by cyclisation to orsellinic acid and re-arrangement into penicillic acid. The preferential labelling in the 5-position indicated stereoselectivity in the last process. The various deductions[31] could be made with confidence because, as a result of detailed preliminary studies, it was possible to

employ conditions which both minimised exchange and maximised
the incorporation of label.

Important advantages thus emerged over the use of carbon
isotopes, in part because only isotopic hydrogen will distinguish
between sp^2 methylene sites. Equally only the use of isotopic
hydrogen will distinguish between possible stereospecific processes involving sp^3 prochiral methylene groups or involving
methyl groups (Section 2.C.).

This unique advantage of isotopic hydrogen was put to
excellent use in elucidating the stereochemistry of the cyclisation of 2,3-oxidoqualene (40) to cycloartenol (41),[105] a step in
the biosynthesis of sterols by photosynthetic organisms. The
cyclisation could conceivably proceed with either retention or
inversion of the configuration at the C-6 methyl group of the
precursor, as shown in Figure 26. Through detailed study of line
widths, long range couplings, and induced shifts, it was possible
to assign the cycloartenol ^{1}H resonances at δ 0.16 and 0.44 to
the exo and endo cyclopropylmethylene protons, respectively.
2,3-Oxidosqualene with a chirally labelled (R)-CHDT methyl group
(43) was synthesised from D-malic acid by an ingenious sequence
involving amongst others the steps depicted in Figure 27. In the
intermediate product (42), 30% of the molecules bore tritium in
the methyl group and all those also carried one deuterium and one
protium atom as shown by the ^{3}H n.m.r. spectra taken with and

(40)   (41)

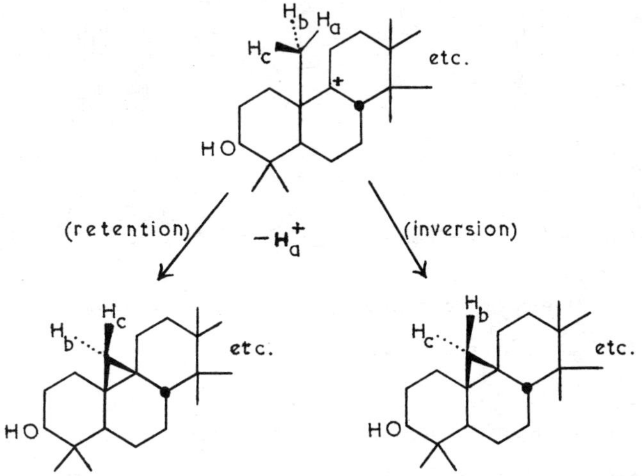

Figure 26. Formation of the cyclopropyl ring in cycloartenol

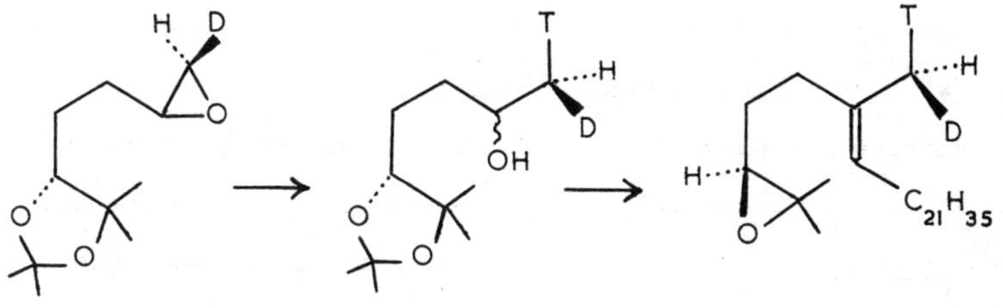

Figure 27. Part synthetic sequence for chirally methyl-labelled oxidosqualene

without ^{1}H decoupling (Figure 28). Incorporation of the labelled precursor oxidosqualene (43) by a microsomal fraction from Ochromonas malhamensis gave a 22% yield of labelled cycloartenol. This showed a major singlet at δ 0.168 in the ^{3}H n.m.r. spectrum

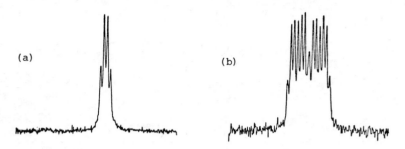

Figure 28.   (a) Proton decoupled ^{3}H n.m.r. spectrum of
compound (42) and (b) the coupled spectrum

both with and without ^{1}H decoupling, indicating the presence of
an exo cyclopropyl triton in molecules having an endo cyclopropyl
deuteron.   There was a lesser singlet at δ 0.456 assignable to an
endo cyclopropyl triton in molecules having an exo cyclopropyl
proton, as shown by the doublet splitting in the absence of the
^{1}H noise decoupling.   Thus the biological cyclisation had
occurred with retention of configuration (Figure 26).

In another biochemical study, ^{3}H n.m.r. was used [106] to
follow the 1,2-dehydrogenation of testosterone (44) by a fungus,
Cylindrocarpon radicicola. [1α,2α(n)-^{3}H]Testosterone (44), predominantly cis-labelled and in which the proportions of
[1α,2α-^{3}H$_2$] and [1β,2β-^{3}H$_2$] species were 5:1 [84] (Figure 29), was
incubated with the organism.   The product (45), in which the
original 17-hydroxy group had been oxidised, showed two lines in
the olefinic region of the ^{1}H decoupled ^{3}H n.m.r. spectrum
(Figure 30A), at δ 6.15 and 7.23 with intensities of 5:1, from
tritium at the 2- and the 1-position respectively.   In such
enones, the hydrogen adjacent to the carbonyl group has the higher field shift.   The undecoupled ^{3}H n.m.r. spectrum showed two
doublets, at the preceding origin positions with $J_{T,H}$ 12.6 Hz
and with exactly the same intensity ratio (Figure 30B).   Clearly

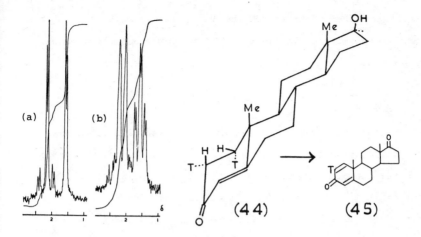

Figure 29. (a) ^{3}H N.m.r. spectrum of [1α,2α(n)-^{3}H]testosterone (44) with ^{1}H decoupling and (b) with full coupling

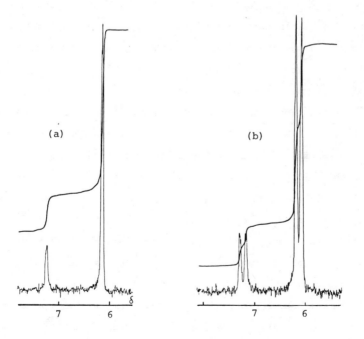

Figure 30. ^{3}H N.m.r. spectra of androsta-1,4-dien-3,17-dione (45), (a) with ^{1}H decoupling and (b) without

the fungal dehydrogenation of the A-ring of testosterone (44) had
occurred stereospecifically by 1α,2β-trans-diaxial elimination
to yield (45).

3.E. Summary

The foregoing examples and n.m.r. spectra will illustrate
the excellent properties of the triton as a magnetic nucleus for
high resolution studies. [3]H N.m.r. spectroscopy provides a most
convenient and non-destructive means for locating the nucleus in
tritium labelled compounds, for investigating the stereochemistry
of labelled sites, and for obtaining the relative distribution
when tritium is present at more than one site. These facts have
been abundantly clear for some years.[84,85,88,93,96] so the recent
statement in a leading Journal[37] that "When tritium is used as a
tracer, alone or in combination with C-14, positions and stereochemistry of labelling must be determined by extensive degradations" is rather surprising.

In quantitative [3]H n.m.r. work there is no general need to
use the rather time-consuming special n.O.e.-suppression sequence,
claimed as necessary.[110,111] The use of ^{1}H noise decoupling in
[3]H n.m.r. spectroscopy introduces little variation in n.O.e.
between the tritons in different positions in the molecules of
partially tritiated compounds having moderate specific radioactivity, as indeed can be concluded from a consideration of the
relaxation mechanism.[90] Hence quantitation of tritium distribution
from n.m.r. signal intensities is practically unaffected by [1]H
noise decoupling.

[3]H N.m.r. spectroscopy and use of tritium labelling appears
ideal for detailed study of exchange processes, of catalytic
hydrogenation, and the specificity of different catalysts.[98,107]
It can readily be applied to the solution of mechanistic

problems,[87,88,108] and for following biochemical transformations
as already indicated. A field of investigation hardly yet touched
upon, and for which ^{3}H n.m.r. could be uniquely appropriate, is
that of the self-radiolysis of tritiated compounds at high
specific radioactivity.[85,86] In addition, tritium has good
potential for use in solving purely n.m.r. problems because its
nuclear magnetic properties are similar to those of the proton and
yet it has a sufficiently distinct resonance frequency. Thus
chemical shifts[98,109] and specific coupling constants can readily
be determined,[89,93] where these are not accessible either readily
or at all from the ^{1}H n.m.r. spectrum. Other special applications
for the study of hydrogen bonding are emerging.[112]

4. Conclusions

It is hoped that the examples discussed in this brief review
will properly indicate the enormous potential that hydrogen
isotopic labelling has, when coupled with the use of the appropriate n.m.r. techniques, for the investigation of biochemical
transformations, reaction mechanisms, and catalysed exchange processes. Deuterium and tritium are complementary. Each has
particular advantages over the other for specific applications.
Thus when molecular motion is to be studied in macromolecules,
enzymes, or membranes, then deuterium labelling and ^{2}H n.m.r. can
provide the required data by virtue of the quadrupolar properties
of the deuteron. Where the highest spectral dispersion is essential, and full spin-coupling information is required, as in the
investigation of stereochemical problems, then use of tritium
labelling with ^{3}H n.m.r. provides the method of choice. If
problems are soluble through simple chemical shift and intensity
measurements and the spectral dispersion is unimportant, then
either deuterium or tritium will suffice. Whilst deuterium can

be used at high abundance without particular precautions, the use

of tritium at n.m.r. tracer levels (300 μCi - 30 mCi) does not pose

any special hazard provided the correct procedures and precautions

are taken. Indeed, the weak β-activity of tritium can provide an

added advantage in that tritiated compounds are readily detected,

e.g. on t.l.c. plates, and their purity is easily determinable by

sensitive radiochemical methods whose detail has long been

established.

5. References

[1] E.M. Purcell, H.C. Torrey, and R.V. Pound, Phys. Rev., 1946, 69, 37.

[2] F. Block, W.W. Hansen, and M.E. Packard, Phys. Rev., 1946, 69, 127.

[3] H.S. Gutowsky and C.J. Hoffman, J. Chem. Phys., 1951, 19, 1259.

[4] G. Lindström, Phys. Rev., 1950, 78, 817.

[5] H.A. Thomas, Phys. Rev., 1950, 80, 901.

[6] J.T. Arnold, S.S. Dharmatti, and M.E. Packard, J. Chem. Phys., 1951, 19, 507.

[7] J.A. Pople, W.G. Schneider, and H.J. Bernstein, "High Resolution Nuclear Magnetic Resonance", McGraw-Hill, New York, 1959.

[8] J.W. Emsley, J. Feeney, and L.H. Sutcliffe, "High Resolution Nuclear Magnetic Resonance Spectroscopy", Pergamon, Oxford, 1966, Vols. 1 and 2.

[9] L.M. Jackman and S. Sternhell, "Applications of Nuclear Magnetic Resonance Spectroscopy in Organic Chemistry", Pergamon, Oxford, 1969.

[10] A. Abragam, "The Principles of Nuclear Magnetism", Clarendon Press, Oxford, 1961.

[11] J.W. Akitt, "NMR and Chemistry", Chapman and Hall, London, 1973.

[12] A. Ault and G.O. Dudek, "NMR, An Introduction to Proton Nuclear Magnetic Resonance Spectroscopy", Holden-Day, San Francisco, 1976.

[13] J.R. Dyer, "Applications of Absorption Spectroscopy of Organic Compounds", Prentice-Hall, Englewood Cliffs, 1965, Chapter 4.

[14] I.Fleming and D.H. Williams, "Spectroscopic Methods in Organic Chemistry", McGraw-Hill, London, 1966, Chapter 4.

[15] W. Kemp, "Organic Spectroscopy", MacMillan, London, 1975, Chapter 3.

[16] R.M. Lynden-Bell and R.K. Harris, "Nuclear Magnetic Resonance Spectroscopy", Nelson, London 1969.

[17] D.W. Mathieson (ed), "Nuclear Magnetic Resonance for Organic Chemists", Academic Press, London 1967.

[18] W.W. Paudler, "Nuclear Magnetic Resonance", Allyn and Bacon, Boston, 1971.

[19] F. Scheinmann (ed), "An Introduction to Spectroscopic Methods for the Identification of Organic Compounds", Pergamon, Oxford, 1970, Vol. 1 and 1974, Vol. 2.

[20] B.P. Straughan and S. Walker (eds), "Spectroscopy", Chapman and Hall, London 1976, Chapter 2.

[21] N.S. Bhacca and D.H. Williams, "Applications of NMR Spectroscopy in Organic Chemistry", Holden-Day, San Francisco, 1964.

[22] N.S. Bhacca, L.F. Johnson, and J.N. Shoolery, "NMR Spectra Catalog", Varian Associates, Palo Alto, 1962, Vol. 1.

[23] N.S. Bhacca, D.P. Hollis, L.F. Johnson, and E.A. Pier, "NMR Spectra Catalog", Varian Associates, Palo Alto, 1963, Vol. 2.

[24] Sadtler Standard Spectra, Sadtler Research Laboratories Ltd, Philadelphia, 1972 _et seq._

[25] T. Cairns, "Spectroscopic Problems in Organic Chemistry", Heyden, London, 1964.

[26] T.C. Farrer and E.D. Becker, "Pulse and Fourier Transform NMR", Academic Press, New York, 1971.

[27] D. Shaw, "Fourier Transform NMR Spectroscopy", Elsevier, Amsterdam, 1976.

[28] P. Diehl and T. Leipert, _Helv. Chim. Acta_, 1964, _47_, 545 ; P. Diehl, in "Nuclear Magnetic Resonance Spectroscopy of Nuclei other than Protons", eds. T. Axenrod and G.A. Webb, Wiley-Interscience, New York, 1974, p. 275.

[29] H.H. Mantsch, H. Saito, and I.C.P. Smith, _Progr. N.M.R. Spectroscopy_, 1977, _11_, 211.

[30] Y. Sato and T. Oda, Tetrahedron Letters, 1976, 2695.

[31] J.A. Elvidge, D.K. Jaiswal, J.R. Jones, and R. Thomas, J.C.S. Perkin I, 1977, 1080.

[32] Y. Sato and T. Oda, J.C.S. Chem. Comm., 1977, 415.

[33] P.M. Dewick and D. Ward, J.C.S. Chem. Comm., 1977, 338.

[34] J.E. Nordlander, P.O. Owuor, and J.E. Haky, J. Amer. Chem.Soc., 1979, 101, 1288.

[35] D.E. Cane and R.H. Levin, J. Amer. Chem. Soc., 1975, 97, 1282.

[36] D.E. Cane and G.G.S. King, Tetrahedron Letters, 1976, 4737.

[37] D.E. Cane and S.L. Buchwald, J. Amer. Chem. Soc., 1977, 99, 6132.

[38] A.J. Birch and S.F. Hussain, J. Chem. Soc. C, 1969, 1473.

[39] G. Snatzke, A.F. Thomas, and G. Ohloff, Helv. Chim. Acta, 1969, 52, 1253.

[40] D.E. Cane and P.P.N. Murthy, J. Amer. Chem. Soc., 1977, 99, 8327.

[41] D. Arigoni, Ciba Foundation Symp. Biosynth. Terpenes and Sterols, 1959, 1958, 239.

[42] J.J. Britt and D. Arigoni, Proc. Chem. Soc., 1958, 224.

[43] A.J. Birch and H. Smith, Ciba Foundation Symp. Biosynth. Terpenes and Sterols, 1959, 1958, 259.

[44] A.J. Birch, R.W. Richards, H. Smith, A. Harris, and W.B. Whalley, Tetrahedron, 1959, 7, 241.

[45] B. Dockerill and J.R. Hanson, J.C.S. Perkin I, 1977, 324.

[46] W.B. Whalley, B. Green, D. Arigoni, J.J. Britt, and C. Djerassi, J. Amer. Chem. Soc., 1959, 81, 5520.

[47] C. Djerassi, B. Green, W.B. Whalley, and G.G. DeGrazia, J. Chem. Soc. C, 1966, 624.

[48] M. Sheves, E. Berman, Y. Mazur, and Z.V.I. Zaretskii, J. Amer. Chem. Soc., 1979, 101, 1882.

[49] B.W. Bycroft, C.M. Wels, K. Corbett, and D.A. Lowe, J.C.S. Chem. Comm., 1975, 123.

[50] H.H. Mantsch, H. Saito, L.C. Leitch, and I.C.P. Smith, J. Amer. Chem. Soc., 1974, 96, 258.

[51] F.A.L. Anet and A.R.J. Bourn, J. Amer. Chem. Soc., 1965, 87, 5250.

[52] G.C. Levy and G.L. Nelson, J. Amer. Chem. Soc., 1972, 94, 4897.

[53] C. Brevard and J.M. Lehn, J. Amer. Chem. Soc., 1970, 92, 4987.

[54] J.A. Glasel, S. Hendler, and P.R. Srinivasan, Proc. Nat. Acad. Sci. USA, 1968, 60, 1038.

[55] J.A. Glasel, V.J. Hruby, J.F. McKelvy, and A.F. Spatola, J. Mol. Biol., 1973, 79, 555.

[56] A.P. Zens, P.T. Fogle, T.A. Bryson, R.B. Dunlap, R.R. Fisher, and P.D. Ellis, J. Amer. Chem. Soc., 1976, 98, 3760.

[57] J. Andrasko and S. Forsén, Chemica Scripta, 1974, 6, 163.

[58] H.G. Hertz and M.D. Zeidler, Ber. Bunsengesellshaft Phys. Chem., 1964, 68, 821.

[59] H.G. Hertz, Ber. Bunsengesellschaft Phys. Chem., 1967, 71, 979.

[60] G. Engel and H.G. Hertz, Ber. Bunsengesellshaft Phys. Chem., 1968, 72, 808.

[61] J. Granot, A.M. Achlama, and D. Fiat, J. Chem. Phys., 1974, 61, 3043.

[62] J.A. Glasel, Nature, 1968, 220, 1124.

[63] J.A. Glasel, J. Amer. Chem. Soc., 1970, 92, 375.

[64] D.E. Woessner and B.S. Snowden, J. Colloid Interface Sci., 1970, 34, 290.

[65] G.E. Ellis and K.J. Packer, Biopolymers, 1976, 15, 813.

[66] F.W. Cope, Biophys. J., 1969, 9, 303.

[67] M.M. Civan and M. Shporer, Biophys. J., 1975, 15, 299.

[68] A. Seelig and J. Seelig, Biochemistry, 1974, 13, 4839.

[69] Z. Luz, R.C. Hewitt, and S. Meiboom, J. Chem. Phys., 1974, 61, 1758.

[70] B. Deloche, J. Charvolin, L. Liebert, and L. Strzelecki, J. Phys., 1975, 36C-1, 21.

[71] J.W. Emsley, J.C. Lindon, and G.R. Luckhurst, Mol. Phys., 1975, 30, 1913.

[72] J.J. Visintainer, E. Block, R.Y. Dong, and E. Tomchuk, Canad. J. Phys., 1975, 53, 1483.

[73] G.W. Stockton, C.F. Polnaszek, A.P. Tullock, F. Husan, and I.C.P. Smith, Biochemistry, 1976, 15, 954.

[74] A. Seelig and W. Niederberger, Biochemistry, 1974, 13, 1585.

[75] J. Charvolin, P. Manneville, and B. Deloche, Chem. Phys. Letters, 1973, 23, 345.

[76] E. Oldfield, D. Chapman, and W. Derbyshire, Chem. Phys. Lipids, 1972, 9, 69.

[77] G.W. Stockton, K.G. Johnson, K.W. Butler, C.F. Polnaszek, R. Cyr, and I.C.P. Smith, Biochim. Biophys. Acta, 1975, 401, 535.

[78] G. Arvidson, G. Lindblom, and T. Drakenberg, FEBS Letters, 1975, 54, 249.

[79] J.H. Davies, K.R.Jeffrey, M. Bloom, M.I. Valic, and T.P. Higgs, Chem. Phys. Letters, 1976, 42, 390.

[80] P.A.J. Gorin, Canad. J. Chem., 1974, 52, 458 ; P.A.J. Gorin and M. Mazurek, ibid, 1975, 53, 1212 ; cf. G.L. Lebel, J.D. Lapoza, B.G. Sayer, and R.A. Bell, Anal. Chem., 1971, 43, 1500.

[81] U. Sankawa, H. Shimada, T. Sato, T. Kinoshita, and K. Yamaski, Tetrahedron Letters, 1977, 483.

[82] E.A. Evans, "Tritium and its Compounds" (2nd edn.), Butterworths, London, 1974.

[83] J.M.A. Al-Rawi and J.P. Bloxsidge, Org. Magnetic Resonance, 1977, 10, 261.

[84] J.M.A. Al-Rawi, J.P. Bloxsidge, J.A. Elvidge, J.R. Jones, V.E.M. Chambers, V.M.A. Chambers, and E.A. Evans, Steroids, 1976, 28, 359.

[85] J.P. Bloxsidge, J.A. Elvidge, J.R. Jones, and E.A. Evans, Org. Magnetic Resonance, 1971, 3, 127.

[86] J.A. Elvidge, J.R. Jones, V.M.A. Chambers, and E.A. Evans, "Tritium Nuclear Magnetic Resonance Spectroscopy", Chapter 1 in "Isotopes in Organic Chemistry" (ed. E. Buncell and C.C. Lee), Elsevier, Amsterdam, 1978, Vol. 4.

[87] V.M.A. Chambers, E.A. Evans, J.A. Elvidge, and J.R. Jones, "Tritium Nuclear Magnetic Resonance Spectroscopy", Review 19, The Radiochemical Centre, Amersham, 1978.

[88] J.M.A. Al-Rawi, J.P. Bloxsidge, C. O'Brien, D.E. Caddy, J.A. Elvidge, J.R. Jones, and E.A. Evans, J.C.S. Perkin II, 1974, 1635.

[89] J.P. Bloxsidge, J.A. Elvidge, J.R. Jones, R.B. Mane, and M. Saljoughian, Org. Magnetic Resonance, 1979, 12, 574.

[90] J.P. Bloxsidge, J.A. Elvidge, J.R. Jones, R.B. Mane, and E.A. Evans, J. Chem. Research (S), 1977, 258.

[91] J.H. Noggle and R.E. Schirmer, "The Nuclear Overhauser Effect", Academic Press, New York, 1971.

[92] D. Canet, J. Magnetic Resonance, 1976, 23, 361 ; R.K. Harris and R.H. Newman, ibid, 1976, 24, 449.

[93] J.M.A. Al-Rawi, J.A. Elvidge, J.R. Jones, and E.A. Evans, J.C.S. Perkin II, 1975, 449.

[94] R.B. Herbert and I.T. Nicolson, J. Label. Compounds, 1974, 9, 567.

[95] M.C. Clifford, E.A. Evans, A.E. Kilner, and D.C. Warrell, J. Label. Compounds, 1975, 11, 435.

[96] J.M.A. Al-Rawi, J.A. Elvidge, J.R. Jones, V.M.A. Chambers, and E.A. Evans, J. Label. Compounds and Radiopharmaceuticals, 1976, 12, 265.

[97] Sadtler Standard Spectra, 6757 M.

[98] J.A. Elvidge, J.R. Jones, R.B. Mane, and J.M.A. Al-Rawi, J.C.S. Perkin II, 1979, 386.

[99] J.P. Bloxsidge, J.A. Elvidge, J.R. Jones, R.B. Mane, V.M.A. Chambers, E.A. Evans, and D. Greenslade, J. Chem. Research (S), 1977, 42.

[100] J.D. Teale, J.M. Clough, and V. Marks, Brit. J. Clin. Pharmacol., 1977, 4, 169.

[101] J.M.A. Al-Rawi, J.P. Bloxsidge, J.A. Elvidge, J.R. Jones, V.M.A. Chambers, and E.A. Evans, J. Label. Compounds and Radiopharmaceuticals, 1976, 12, 293.

[102] J.M.A. Al-Rawi, J.A. Elvidge, D.K. Jaiswal, J.R. Jones, and R. Thomas, J.C.S. Chem. Comm., 1974, 220.

[103] A.J. Birch, G.E. Blance, and H. Smith, J. Chem. Soc., 1958, 4582.

[104] K. Mosbach, Acta. Chem. Scand., 1960, 14, 457.

[105] L.J. Altman, C.Y. Han, A. Bertolino, G. Handy, D. Laungani, W. Muller, S. Schwartz, D. Shauker, W.H. de Wolf, and F. Yang, J. Amer. Chem. Soc., 1978, 100, 3235.

[106] J.M.A. Al-Rawi, J.A. Elvidge, R. Thomas, and B.J. Wright, J.C.S. Chem. Comm., 1974, 1031.

[107] J.A. Elvidge, J.R. Jones, M.A. Long, and R.B. Mane, Tetrahedron Letters, 1977, 4349.

[108] J.A. Elvidge, J.R. Jones, R.B. Mane and M. Saljoughian, J.C.S. Perkin I, 1978, 1191.

[109] J.A. Elvidge, J.R. Jones, R.B. Mane, V.M.A. Chambers, E.A. Evans and D.C. Warrell, J. Label. Compounds and Radiopharmaceuticals, 1978, 15, 141.

[110] L.J. Altman and N. Silberman, Steroids, 1977, 29, 557.

[111] L.J. Altman and N. Silberman, Analyt. Biochem., 1977, 79, 302.

[112] L.J. Altman, D. Laungani, G. Gunnarsson, H. Wennerström, and S. Forsén, J. Amer. Chem. Soc., 1978, 100, 8264.

Mass Spectrometric Methods of Isotope Analysis

D.S. Millington   and   W.D. Unsworth

VG-Micromass Ltd., 3, Tudor Road, Altrincham, Cheshire.

Introduction

The first and undoubtedly the most notable achievement of
the field of positive ion analysis, pioneered by Thomson in the
early nineteen hundreds, was the discovery of stable isotopes.
Subsequently, it was also realised that the chemical properties
of an element are determined by its atomic number and not by
atomic weight.

Thomson began experimenting in 1910 with his "positive ray
analyser", a device in which a collimated beam of ions was
passed through a combined magnetic and electric field.[1] The ions
were deflected in two orthogonal directions and recorded on a
photographic plate as a series of parabolic curved images,
separated from each other according to their mass to charge
ratios.

During analysis of a sample of neon, in 1912, the expected
line corresponding to mass 20 plus another corresponding to mass
22 and not attributable to any known element were observed.[2]
Thomson's suggestion that he had in fact identified a second
stable isotope of neon was subsequently confirmed by Aston, who
designed and built new instruments, called mass spectrographs,
for the purpose of systematically identifying the isotopic
components of the known elements.[3] By 1924, over fifty had been
investigated and the mass and abundance figures, when used to
calculate atomic weights, were in good agreement with the
chemically determined values. Dempster also built an instrument

at around this time which employed an electrical recording method,[4] the principle of which is employed in most of today's commercially produced mass spectrometers.

The groundwork for the main techniques in mass spectrometry was thus established by these early pioneers, at least twenty years before the appearance of the first commercial product.

An entire field in mass spectrometry, accounting for a large proportion of the output of commercial instruments today, has evolved from the basic ability of a mass spectrometer to separate and determine abundance ratios of the isotopes. This is referred to as "accurate isotope ratio mass spectrometry" and two types of instrument have been specially developed for use in this field. The stable isotope enrichment spectrometer, designed to handle low molecular weight gaseous samples, is usually dedicated to the analysis of one pair of isotopes and can achieve the best precision for isotope abundance ratios of any type of instrument. The thermal ionisation spectrometer handles involatile samples and gives abundance ratios for a wide range of the heavier elements. These instruments and some of their applications, which involve many interesting and important branches of science, are discussed in Part I of this chapter.

By far the largest area of application of mass spectrometers during the past twenty years has been in the fields of chemistry and biochemistry. The development of techniques to analyse the wide variety of naturally occurring and synthetic organic compounds by mass spectrometry is still continuing at a rapid pace. The most significant of these is the on-line coupling of gas chromatography with mass spectrometry,[5] which was achieved in the mid-1960s and has provided the chemical sciences with their most powerful analytical tool. Later, the emergence of quantitative techniques in mass spectrometry and combined gas chromatography - mass spectrometry has established the powerful technique of stable isotope dilution as the primary method for such assay procedures[6] The need to measure isotope abundance ratios in molecules of biological interest has stimulated the provision of many new instrumental techniques on scanning organic mass spectrometers. The most recent developments in mass spectrometry include improvements in selectivity of the detection method, such as the use of increased resolution, in order to enhance the accuracy and reliability of quantitative measurements at low sample concentration. These factors are discussed in Part II of this chapter.

As this chapter will make clear, the application of stable isotopes in both inorganic and organic mass spectrometry techniques is rapidly increasing, a fact which emphasises the interrelationship of mass spectrometry and isotopes since the days of Thomson and Aston.

Part I.   Isotope Ratio Mass Spectrometry

The Notation.   The field of application for accurate isotope ratio mass spectrometry has broadened considerably in the past few years owing mainly to improvements in the instrumentation, enabling the measurement of very small differences in relative isotope abundance with a high degree of confidence. For most applications, absolute abundance ratios are not required and in fact are very difficult to determine accurately for a number of reasons. Isotope abundance ratios are therefore usually expressed as relative differences with respect to an internationally accepted arbitrary standard, since such differences can be determined with high precision and known accuracy. A ratio difference is defined by the " δ value" calculated as follows:

$$\delta = \frac{\text{ratio for sample - ratio for standard}}{\text{ratio for standard}} \times 1000$$

These δ values, expressed as parts per mil (‰) are usually small positive or negative numbers that are easy to memorise and are additive to a first approximation over a wide range of values.

A sample having a $\delta^{13}C$ value of ← 10 means that the $^{13}C/^{12}C$ ratio is higher by 1% than that of the standard. State-of-the-art instrumentation now permits δ value measurement to precisions of better than 0.02‰ as determined by replicate analysis. The isotope ratio instruments are therefore highly specialised for their particular applications and the emphasis in design has been placed on stability and reproducibility.

General Description of the Instruments

The magnetic sector[7] spectrometer is the type predominantly used for accurate isotope ratio measurement. The principal components of the instrument are shown in Figure 1. The ions are produced by either low energy electron bombardment or surface ionisation in the 'ion source' and given a high kinetic energy corresponding to the accelerating potential V in the direction of the mass analyser. Molecules ionised in this manner have a range of internal energies and some undergo fragmentation, in a manner dependent on the molecular structure. The accelerated ion beam experiences a magnetic field B at right angles to its direction of travel, causing dispersion into several beams whose trajectories in the magnetic field are determined by both the momentum and charge carried by the ions. The mass per unit charge is related to the magnetic field (B) and the accelerating voltage (V) by the equation

$$\frac{m}{e} = \frac{B^2 r^2}{2V}$$

Figure 1

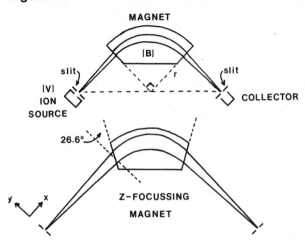

where r is the radius of curvature, which is fixed by the position of the collector.

The magnetic field also provides direction focussing of the ion beam and its geometry determines the position of the source (or object) slit and the collector (or image resolving) slit (Figure 1). A recent development on commercial instruments is the provision of magnets with "extended" geometry to provide z-direction focussing (i.e. in the plane of the magnetic field) in addition to focussing in the plane perpendicular to the magnetic field[25] This involves shaping the magnet polepieces at a special angle relative to the direction of the ion beam as illustrated in Figure 1 and it enhances the sensitivity by a factor of 3-4.

A normal mass spectrum is obtained by varying B so as to scan the separate ion beams across the collector. In isotope ratio spectrometers, the objective is to measure the relative intensity of two or more beams adjacent to or at least within a few mass units of each other. Instead of scanning, therefore, a multiple collector system is used continuously to monitor the

salient ion currents or a single detector is shared between

them by alternately switching the magnetic field through

preselected values in a cyclic manner.

The ion currents are preferably detected by Faraday

collectors, which have the most stable and linear response over

a wide dynamic range of any device for measuring small currents.

High ohmic resistors are used to provide a measurable voltage

from the very small charges, typically 10^{-9} to 10^{-15} coulomb

per second, collected from the ion beams. The signals are then

amplified and recorded.

The ion source, flight path and collector are housed in a

vacuum system which requires efficient pumping, since the

performance of isotope ratio instruments is directly related to

the pressure in the analyser system (<u>vide infra</u>), which should

be less than 10^{-8} mbar.

Resolution and Abundance Sensitivity

Before describing isotope ratio instruments in more detail,

the definition and significance of these important performance

parameters are explained.

<u>Resolution</u>.   Resolution is normally defined as the value of

$M/\Delta M$ where an ion beam at mass M is separated from another of

equal intensity at mass M $+$ ΔM such that, for pseudo-

Gaussian peak shapes, there is an overlap between the two peaks

of 10 per cent relative to the peak height.

This situation is represented in Figure 2A. At mass 500,

therefore, the resolution is 500 if M $=$ 1 a.m.u. For the

magnetic sector instruments used in isotope ratio measurements,

resolution is dependent mainly on the size of the magnet.

Stable isotope enrichment spectrometers typically employ magnetic

Figure 2

201

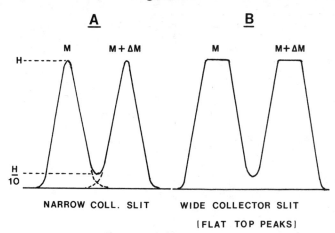

Resolution $= \dfrac{M}{\Delta M}$

radii of about 6 cm, having a resolution at full transmission of around 250-300.

In thermal ionisation instruments the magnet radius is generally considerably larger, 30 cm for example, with resolution capabilities of 2000 - 3000.

Generally, these instruments are operated at lower than optimum resolution, in the "flat-topped peak" mode shown in Figure 2B, which minimises errors caused by uncertainty in defining the peak maxima, especially in the peak jumping technique. Abundance sensitivity. Abundance sensitivity is in effect a specialised definition of resolution, of great importance to the isotope ratio mass spectroscopist as a measure of the ability to detect a small peak next to a very large one. It is expressed as the ratio of the height of the "tail" at the mass of interest relative to the height of the large peak at the adjacent mass, which may be 1×10^{-5} or 10 ppm, as in Figure 3. These low-level tails may constitute an undesirable signal at the mass of interest, thus limiting the detectability of small peaks. In

Figure 3

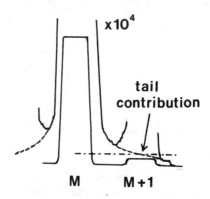

M   M+1

Abundance Sensitivity ≈ 10ppm

order to increase the detectability of small signals the
abundance sensitivity must obviously be as low as possible: in
other words the tailing of large peaks must be minimised. This
is achieved mainly by maintaining an ultra-high vacuum
(pressures of 10^{-8} mbar and below) in the analyser, where ions
are scattered by collisions with neutral molecules and with the
walls of the analyser. In the latter type of collision, the
scattered ions can be prevented from reaching the collector by
special attention to the shape and position of the analyser tube
walls and by the use of baffles. With these improvements, the
abundance sensitivity will be in the ppm range at mass 200-300,
for a mass difference of 1 a.m.u. at a resolution of 500 with
flat-topped peaks.

When needed, further improvements in abundance sensitivity
are achieved by using energy filters or double stage instruments
which employ an additional magnetic or electric sector [8-10] but
these are highly specialised, large and expensive instruments.

Isotope ratio mass spectrometers are conveniently

subdivided into two major groups, those employing electron
ionisation and used to analyse gaseous samples, and those using
surface or thermal ionisation for analysis of solid samples.
These two instrument types are now described in more detail.

Stable Isotope Enrichment Spectrometers

Instruments specially designed to analyse the low molecular
weight 'organic' elements in gaseous form use the multiple
collector principle[12] for high precision isotope ratio
measurement and are generally referred to as stable isotope
enrichment (S.I.E.) spectrometers. The spatial separation of
adjacent mass ions at the focal plane of a magnetic sector
instrument diminishes with increasing mass and the multiple
collector arrangement is feasible up to about mass 70 with a 6cm
radius magnet. A larger radius magnet is required for analysis
of higher mass gases, such as SF_6 or UF_6. The typical 6cm radius
magnetic instruments are used to measure ratios such as $^2H/^1H$
(as H_2), $^{13}C/^{12}C$ (as CO_2), $^{18}O/^{16}O$ (as CO_2 or H_2O), $^{15}N/^{14}N$
(as N_2) and $^{34}S/^{32}S$ (as SO_2).

Sample introduction. In S.I.E. spectrometers, a double inlet
system is used in which one gas reservoir is dedicated to the
reference and the other is used to admit the samples. The gas
pressures in the two reservoirs are independently adjustable (by
changing the volume) so that the major isotope peak gives the
same ion current for both reference and sample gases, an
important feature for reproducibility. Each reservoir is
connected to the ion source via a capillary line about 1m long x
0.15mm int. diam. with a restriction at the source end.

A system of magnetically operated valves is employed for
rapid switching between the sample and reference gases without

significant mixing, because a single measurement involves the
sequential isotope ratio determination of sample and reference
several times, usually 6-10, in order to determine the precision
of the measurement expressed as a standard deviation.

Ionisation method.   The sample gas molecules are ionised by
electron bombardment inside a small chamber. Although not very
efficient (only 1 in 10^4 molecules approximately are converted
into positive ions), electron ionisation sources[13] are simple,
very stable and reliable for long periods of time. For the
particular problem of hydrogen/deterium analysis, an "open"
source is used to minimise a side-reaction that produces
interfering $(^1H_3)^+$ ions by ion-molecule $(^1H_2^+ + {}^1H_2)$ collisions.
The probability of this reaction increases as the square of the
partial pressure of hydrogen.

Ion detection.   The essential features of double and triple
collector systems are shown in Figure 4. Most systems employ
the double collector (Figure 4A). The principles of the isotope
ratio technique as they apply to the determination of $\delta^{18}O$
values for CO_2 are now outlined.

The magnet current is adjusted to enable the ion beams of
mass 44 and 45 to pass into the major collector and that of mass
46 enters the minor collector. Secondary electrons emitted from
the collectors are repelled back to them by electron "suppressor"
plates (Figure 4). This removes spurious effects causing peak
shape distortion, especially at the top and near the baseline,
which would otherwise reduce the precision of the ratio
measurement.

The ratio actually measured is shown in the following
equation:

$$\text{Measured ratio} = \frac{{}^{12}C^{18}O^{16}O + {}^{13}C^{17}O^{16}O + {}^{12}C^{17}O^{17}O}{{}^{12}C^{16}O^{16}O + {}^{13}C^{16}O^{16}O + {}^{12}C^{16}O^{17}O}$$

The desired ratio $= \dfrac{^{12}C^{18}O\ ^{16}O}{^{12}C^{16}O\ ^{16}O}$ and is obtained after

Figure 4

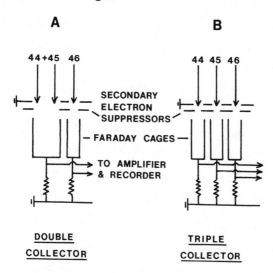

A

B

44+45 46

44 45 46

SECONDARY
ELECTRON
SUPPRESSORS

— FARADAY CAGES —

TO AMPLIFIER
& RECORDER

DOUBLE
COLLECTOR

TRIPLE
COLLECTOR

correction for the other isobaric ion abundances.

With the triple collector arrangement (Figure 4B) the separate ion beams of mass 44, 45 and 46 are collected individually and it is possible then to determine $^{13}C/^{12}C$ and $^{18}O/^{16}O$ ratios simultaneously.

Because the ratios are not determined absolutely but expressed as δ values relative to a standard, the effects of background signals due to residual analyser gas cancel out. However, each ratio requires correction for neighbouring peak tail contribution, sample mixing effects in the inlet system and other factors associated with the sample preparation.

The quality of the resistors and amplifiers used in the detection circuit obviously affects the ultimate precision of isotope ratio determinations and further improvements are obtained by use of fully automated sample introduction and measurement systems that have recently become available

commercially. The necessary correction factors are automatically

applied using an on-line computer system which also determines

the standard deviation for the δ value determination. Because

a single measurement may take up to 20 min, automatic loading

and measurement of up to 20 samples without operator intervention

has been made available on some commercial systems and this

increases the precision still further. (The best achievable

with current technology is around 0.017‰). Thus the errors

involved in the preparation and handling of samples for isotope

ratio measurement are usually much greater than those inherent in

the measurement technique.[13] The overall precision in isotope

ratio determinations is typically around 0.1‰.

Applications of S.I.E. Spectrometers

S.I.E. spectrometers can detect very small differences in

isotope abundance ratio of the 'organic' elements and are thus

able to measure natural variations arising for example from

differences in the biological origin of substances containing

the elements in question. The geochemical and environmental

applications of such measurements are far-ranging. In addition,

the study of metabolic processes in vivo using stable isotope

labelled precursors as tracers is assuming great importance in

the medical field, where the use of radioactive isotopes is

unpopular owing to the potential health hazard. Some of the

specific applications are now briefly reviewed.

Pollution monitoring. The fate of man-made CO_2 is the subject

of much discussion owing to the geological importance of

increasing CO_2 concentration in the atmosphere, particularly

with regard to the "greenhouse effect".[14] The distribution of

man-made CO_2 is studied by $^{13}C/^{12}C$ ratio determination because

fossil carbon has about 18% less ^{13}C than indigenous atmospheric

CO_2 owing to accumulated isotope effects during photosynthesis.

Therefore, the increase of CO_2 due to the burning of fossil fuels

is gradually depleting the ^{13}C content of atmospheric CO_2.

The production of wood cellulose from glucose involves

relatively few biochemical steps and is therefore regarded as a

suitable indicator of the atmospheric CO_2 at the time of formation.

Thus $\delta^{13}C$ values of CO_2 from combustion of wood cellulose are

being studied in order to gain knowledge of the global

distribution of fossil fuel-derived CO_2.[15]

Similarly, the extent of sulphur pollution can be estimated

owing to the difference in $^{34}S/^{32}S$ ratio between natural sulphur

compounds and industrial H_2S and SO_2[16, 17]. The industrial

pollutants are significantly enriched in ^{34}S.

Monitoring of food substitutes. Adulteration of foodstuffs by

synthetic or less expensive equivalents is a problem of common

concern where large scale manufacture of foods takes place.

Classical chemical methods are inapplicable owing to the chemical

similarity of such additives to the natural or authentic material,

but the small variations in isotope composition of organic matter

resulting from differences in origin can be measured by S.I.E.

spectometry. Natural vanillin, for example, has a $\delta^{13}C$ value

of 19.9 - 20.8 whereas that of synthetic vanillin is between 26.9

and 28.7[23].

The undeclared addition of cane sugar to maple syrup at

levels down to 10% can be detected owing to a difference in $\delta^{13}C$

of around 12‰.[24]

Geochemistry. Stable isotope geochemistry has recently been

reviewed[30]. As one example, the origin and evolution mechanisms

of petroleum have been studied by S.I.E. spectrometry[18]. Results

have shown that only a small part of organic matter is converted

into petroleum, that the process occurs at low tmperatures (not

more than $200^\circ C$) and that the source of the paraffin-naphthene
fraction is mainly fatty acids and that of the aromatic
hydrocarbon fraction is mainly lignin-type compounds.

Medical applications. S.I.E. spectrometry has great potential
for tracer experiments in this field owing to its high accuracy
and applicability to stable isotope labelling. The factors
involved in tracer experiments with ^{13}C-labelled drugs have
been reviewed[20] and it has been shown that an enrichment of
50 ppm in the ^{13}C content of an extract can safely be detected.
For example, a single dose of aspirin (300 mg) spiked with 0.1%
[13]C-labelled drug is sufficient measurably to perturb the $\delta^{13}C$
value of a 24 hr. pooled urine (acidic fraction). Only
microgram quantities of the pooled sample are required for the
analysis.

[13]C-labelled substrates chosen for their possession of a
target bond whose cleavage liberates a [13]CO_2-containing
functional group can be used as diagnostic indicators of
pathological conditions by measuring differences in the rate or
extent of metabolism to [13]CO_2.[19]

A simple technique for the classically very difficult task
of estimating the tissue protein turnover rate in humans based on
$\delta^{15}N$ measurements following ingestion of a single dose of
[15]N-labelled glycine has been described.[21] The same research
group have also developed a method for precise measurement of the
total body water content, requiring a 1-2g dose of D_2O and a 5 μl
sample of any body fluid.[22]

Analysis of nuclear fuel and waste. Special mass spectrometers
have been designed for the specific purpose of analysis of
uranium as UF_6 gas, including, for the first time in accurate
isotope ratio spectrometry, a quadrupole mass spectrometer which
has proved suitable for field studies.[33]

Thermal Ionisation Mass Spectrometers

The thermal ionisation (T.I.) instrument is designed to measure isotope abundance ratios of non-volatile elements by combustion at a specific temperature on a filament. The overall design is much the same as for the S.I.E. spectrometer (Figure 1). The magnetic sector is usually considerably larger and the double collector is replaced by a single detector, usually a Faraday collector or, for small samples, a Daly type scintillation detector.

The ionisation process. When an atom or molecule is evaporated from a heated surface, a fraction of the atoms escape as positive ions. This is explained on the basis that there is a probability that the surface has a greater affinity for an outer electron orbital of an escaping atom than for the atom itself. The ionisation efficiency in this process is expressed by Langmuir's equation:[25]

$$\frac{N^+}{N^o} = \exp \left[\frac{e\ (W - I)}{kT} \right]$$

where N^+/N^o is the fraction of positive ions to neutral atoms, e is the electronic charge, W is the work function of the surface and I is the ionisation potential (in volts), k is Boltzmann's constant and T is the temperature in K.

The metal surface should have a high melting point and a high work function, such as rhenium or tungsten[26]. The most efficient ionisation takes place when I is slightly larger than W, as with the alkali-metals, when efficiencies approaching 100% are feasible. A few elements with very high ionisation potential are not amenable.

The value of I - W is specific for each element and means

that it is possible to adjust the temperature so that only one
element is ionised in the presence of a mixture. The introduction
of multiple filaments[26] such as the popular triple filament shown
in Figure 5 was a major improvement in the technique. The sample
and reference standard are applied in microgram amounts to the

Figure 5

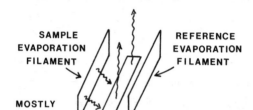

POSITIVE IONS

SAMPLE
EVAPORATION
FILAMENT

REFERENCE
EVAPORATION
FILAMENT

MOSTLY
NEUTRALS

IONISING
FILAMENT

Triple filament for T I source

outer filaments, usually as an inorganic salt layer. One of them
is heated until the element or its oxide is vaporised and the
third filament is maintained at the correct temperature for
thermal ionisation. This enables the evaporation rate and
ionisation efficiency to be carefully controlled independently
during the isotope ratio measurement, which usually takes several
minutes, until the sample is exhausted. The reference ratio is
then determined in order to calculate the value. T.I. sources
are less stable and much more prone to systematic errors than
those used in S.I.E. spectrometry.[13] Reproducibility in the
sampling and mass spectrometry techniques is vital to the success
of this type of analysis and automation has been introduced into
the technique. Several samples can be loaded and analysed without
operator intervention in the most modern systems.

Applications of T.I. Spectrometry

A comprehensive account of the range of applications of T.I. instruments is given in a recent review article[13] therefore only a brief summary will be presented here.

Atomic weights and nuclear constants. Accurate isotope ratio measurements are being used to re-calculate the atomic weights of multi-isotopic elements. The new values are accepted by I.U.P.A.C. as standard owing to the high precision of their measurement compared with the chemically determined values[28].

The half-lives of radioactive elements such as ^{137}Cs ($\sim$ 30y) and ^{241}Pu ($\sim$ 15y) are determined with better accuracy than by any other technique using a stable isotope ratio ($^{135}Cs/^{133}Cs$ or $^{240}Pu/^{239}Pu$) as an internal standard. For elements with longer half-lives isotope dilution is combined with ion counting techniques for high precision measurements.

Age determination. An expanding field of T.I. mass spectrometry is accurate age determination on microgram amounts of sample by studying isotope decay systems such as $^{87}Rb \rightarrow ^{87}Sr$, $^{147}Sm \rightarrow ^{143}Nd$ and $^{235}U \rightarrow ^{207}Pb$. These applications have been reviewed[29]. The age of the universe was recently estimated to within an accuracy of 1% by the Rb/Sr method[31].

Pollution monitoring. An ingenious experiment to determine unequivocally the contribution of lead from automobile exhausts to the overall lead pollution in the environment is being undertaken in Italy[32]. Lead from a particular source having a characteristic low value for the ratio $^{206}Pb/^{207}Pb$ was introduced into gasoline in various regions of Italy in 1974. Since then, the depletion of ^{206}Pb in various environmental samples including airborne particulates and human blood has been followed by accurate relative lead isotope ratio determinations.

The results of the survey up to June 1979 are soon to be announced[33].

Part II.   Organic Mass Spectrometric Methods

Most organic mass spectrometers are used for routine chemical analysis. Chemists and biochemists are generally in favour of having at their disposal a choice of sample introduction systems and of ionisation techniques in order to handle the various types of organic compound encountered. The emphasis in design of organic mass spectrometers is therefore towards versatility, especially with regard to ease of operation and changeover between the different operating modes. Most instruments manufactured at the present time are equipped as standard with electron ionisation (E.I.) and chemical ionisation (C.I.) sources, provide for analysis of both positive and negative ions, have gas chromatograph connections for both packed and capillary columns and a direct insertion probe, plus, of course, the almost mandatory datasystem.

This diversity of instrumentation plus the fact that studies involving stable isotopes are very common in all organic applications of mass spectrometry make a complete coverage of the field beyond the scope of this chapter. Attention has been focussed on recent instrumental developments that have been applied in quantitative analyses using stable isotope dilution. This is prefaced by a short review of the general applications of organic mass spectrometry in isotope analysis and indicates the limitations of the techniques.

Determination of the Isotopic Composition of Molecules by Accurate Mass Measurement

The most important contribution of the mass spectrometer in the identification of compounds of unknown structure is the

determination of molecular formulae. Just as each element is
characterised by the masses and relative abundance of its
isotopes, so a molecule is characterised in the same manner,
since it is only a composite of elements with known isotopic
mass and abundance parameters. The elemental composition of a
molecule is best determined by accurate mass measurement of the
lowest mass ion in the isotope cluster corresponding to the
intact ionised molecule. This ion is the sum of the major
isotopes of the constituent organic elements and is referred to
as the "molecular ion". The number of possible formulae at a
given mass reduces dramatically as the precision in the
determination of accurate mass increases, and the best precision
is achieved using double focussing mass spectrometers[34]. Double
focussing is achieved by addition of an 'electric sector'

Figure 6

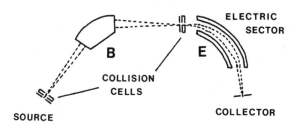

'Reversed geometry' double focussing

(Figure 6) consisting of a pair of curved plates to which an
electrostatic field (E) is applied. The magnetic field B and
electric field E together compensate for velocity spread in the
ion beam which blurs the image on a single focussing instrument.
The ion beam is said to be focussed in both energy and direction,
hence the term double focussing. The addition of an electric

sector usually has a dramatic effect on the ultimate resolution,
typically increasing it by a factor of 10-50 times for the same
size magnet. A precision of $\pm$ 1 or 2 ppm in mass measurement is
readily attainable by the technique of peak matching[35] at a
resolution of 10000 and this is usually sufficient for
unequivocal determination of the elemental composition with the
help of tables or, more likely, computer programs. High
resolution is also required to separate multiplet peaks
occurring at the same nominal mass such as the first isotope
peak of any organic compound, which is composed of ^{13}C-containing
and ^2H-containing ions (vide infra).

Computer techniques can achieve similar precision to peak
matching but these are more tedious and prone to systematic
errors. A recent development in computer technology is the
commercial realisation[36] of a technique for accurate mass
measurement at low resolution[37] based on a small number of
internal reference ions and an accurately pre-calibrated mass
scale. The technique is attractive because it applies to all
the popular ionisation modes (E.I, C.I, and F.D.), utilises the
full sensitivity of the mass spectrometer and is one of the
only practical methods for accurate mass determination of
compounds eluting from a G.C. column, especially if it is a
capillary-type column[38].

Typical errors are in the order of 10 ppm for double
focussing spectrometers and 20 ppm for single focussing machines
with reproducible scan law characteristics. In most cases this
precision is sufficient for elemental composition, especially
for typical low molecular weight compounds analysed by G.C-M.S.

Precision of Measurement of Relative Isotopic Abundance
Organic mass spectrometers are not designed to achieve high
precision isotope ratio measurements, but quite good results are

obtained with magnetic sector instruments by simulating the

techniques used in "accurate isotope ratio mass spectrometry"

described in the foregoing section. If the pure compound of

interest is introduced by direct probe or reservoir inlet system

and the magnet cycled slowly across the molecular ion region,

with the instrument tuned to give flat-topped peaks (Figure 2)

with good resolution, the relative abundance of the isotopes can

be determined to a precision of within 0.1%. Slightly better

precision is possible when a Faraday collector is substituted for

the conventional electron multiplier and some instruments are in

fact provided with both types of detector. When the sample is

introduced <u>via</u> a G.C. column, such precision is unattainable

under the dynamic conditions of the experiment. The best

technique in this case is to employ selected ion monitoring by

accelerating voltage switching at constant magnetic field, with

flat-topped peak conditions, and integrating the signal areas for

each ion corresponding to the G.C. peak profile using a computer.

The precision achieved by this method for small quantities of

$C_{12} Cl_{10}$ injected onto the G.C. column is shown in Table 1.   The

precision of around 1% for the major isotope ratios at the 300 pg

Table 1. Reproducibility of isotope ratio determination
by M.I.D. for small amounts of $C_{12} Cl_{10}$ analysed by GC/MS
on a 16 cm single focussing spectrometer. Figures in
brackets are the coefficients of variation.

Amount injected on GC column	Ratio $\frac{496}{498}$	Ratio $\frac{500}{498}$
300 pg	0.700($\pm$ 0.5%)	0.841 ($\pm$ 0.8%)
20 pg	0.688($\pm$ 4.4%)	0.854 ($\pm$ 5.4%)

level did not improve significantly when larger amounts of sample

were analysed, but ratios for the less abundant isotopes were

improved.

When the isotope abundance ratios are determined from

individual mass spectra obtained under normal full scan conditions,

the precision is related to scan speed and sample quantity (owing

to statistical effects) and is typically not better than 10% for
any type of spectrometer. Averaging the data from several scans,
say during elution of a GC peak, will improve this figure. The
precision required in isotope analysis of organic ions depends on
the application. Some applications are described in the following
section. It should be noted that organic quadrupole mass
analysers are generally unsuitable for high precision isotope
ratio measurement owing to poor peak shape and mass discrimination.
Elemental composition by isotope ratio measurement. It has almost
been forgotten that isotope ratios determined for a molecular ion
group can be used to determine elemental composition. Tables
have been published for this purpose and even now one of the
leading mass spectrometry laboratories is working on a
sophisticated computer program to determine possible molecular
formulae from isotope ratio data[39]. It is fair to say that
accurate mass determination is a faster and more reliable
technique, but when such techniques are not available, isotope
ratio methods can be useful. One of the main disadvantages is
that the relative differences in isotope abundance for different
possible combinations of elements diminish with increasing mass,
and of course the number of possible combinations also increases
with mass. Consequently, the precision required in the ratio
measurements must also be increased with mass increase and there
is a practical limit set by the instrumentation. In addition,
natural isotope abundance variations of the elements, particularly
for carbon, become more significant at higher masses. Very often,
a combination of isotope mass and ratio values, both determined
with moderate precision, plus application of the rules of valency
and common sense will suffice for unequivocal elemental
composition determination of ions up to mass 500. For example,
the two formulae $C_{11} H_{12} N_4$ and $C_{10} H_{16} O_4$ differ in mass by only

1.5 mmu (7.5 ppm) but they can easily be distinguished on the basis that the ratio for the first and second isotopes (M + 1 : M + 2) is high for $C_{11} H_{12} N_4$ (15.85) and low for $C_{10} H_{16} O_4$ (8.17).

The presence of elements other than C, H, O and N is often indicated by a distinctive isotope pattern, such as Cl, Br, S and unfortunately chemists and biochemists tend only to notice the isotope pattern and make use of it for diagnostic purposes in these special cases.

Tracer Experiments with Isotope Labelling

Tracer experiments requiring measurement of incorporation of an isotopically labelled substrate into specific metabolites using organic mass spectrometers are difficult and mostly impracticable owing to limitations in the precision of isotope ratio measurement (vide supra), unless incorporation of the label is in the percent range. However, the contribution to the first isotope peak of a molecular ion due to natural ^2H is relatively small compared with that of ^{13}C, and when these two contributions can actually be resolved from one another ^2H enrichment due to incorporation can be measured at lower levels of incorporation. Unfortunately, the resolution required to separate this doublet is beyond the scope of all but the most powerful (and large) commercial instruments and corresponds to M/0.0029, where M is the mass of the molecular ion. For example, the resolution needed at mass 100 is over 30,000 and at mass 300 is about 100,000.

This technique has been applied in conjunction with Fourier-transform NMR to determine the position and extent of incorporation of ^{13}C-labelled and ^2H-labelled ethanol into steroid metabolites, with the triple goals of investigating the contributions of different metabolic pathways, precursor compartmentation and isotope effects.[40]

An account of the criteria for estimating the degree of
labelling in ions and solutions to various problems in isotope
labelling has been published.[41]

Other Applications of Isotope Labelling in Metabolic Studies

Isotope labelling is used with mass spectrometric detection
for a variety of reasons in metabolic studies, particularly
involving drugs, where precision in isotope ratio measurement is
not required. To identify drug metabolites in body fluids, for
example, a mixture of unlabelled and judiciously labelled precursor
is used to confer a characteristic isotope pattern that is easily
recognised by mass spectrometry. More specifically, differences
in the metabolic rate of enantiomers can be studied by labelling
one of them and looking at differences in isotope ratio of the
metabolites using selected ion monitoring.[42] The various uses of
isotope labelling in metabolic studies are the main reasons for
the rapidly expanding application of mass spectrometry in the
fields of pharmacology, biochemistry and medicine.[43, 44] The use
of isotope dilution in assays by mass spectrometry is discussed as
a separate topic later in the text.

Isotope Labelling in Organic Mass Spectrometry

Specific isotope labelling is employed routinely in the
investigation of all processes occurring in the ion source of
a mass spectrometer. It is a mandatory tool for the elucidation
of mechanisms of fragmentation of ions, although most of this work
is of little practical value to analytical or research chemists
and biochemists.

In CI sources, it is sometimes not clear whether an observed
ion at the highest mass is a protonated species or a higher adduct
ion. For example, when ammonia is used as reactant gas, an
observed ion may be due to MH^+, $[M + NH_4]^+$ or neither. Repeating
the analysis with deuterium-labelled ammonia will clarify the

situation.

Isotope labelling with mass spectral analysis is widely
used as an aid in structure determination. Classic examples are
found in the area of peptide sequencing[45]

Quantitative Mass Spectrometry by Isotope Dilution

The principles of the isotope dilution technique as they
apply to quantitative measurement of the concentration of a
particular known compound in a complex biological or chemical
environment are as follows. A known amount of the compound,
specifically labelled to provide an increase in the molecular ion
mass of at least 2 units, is added to the sample which contains
an unknown amount of the compound under investigation. After
allowing time for equilibration with the biological matrix, the
labelled and unlabelled molecules will be mixed in a definite
ratio, which is assumed to be unaltered significantly by any
subsequent handling or chemical modification of the sample.

The compound is then extracted from the biological sample,
purified by chromatography and, if necessary, chemically
converted to a form suitable for analysis by mass spectrometry.
The spectrometer determines the ratio of signals for a particular
doublet (or group of doublets) corresponding to a characteristic
ion (or group of ions) occurring in both the labelled and non-
labelled compounds' mass spectra. The absolute concentration of
the compound in the sample is obtained from a "standard curve",
constructed previously by similar experiments utilising a range
of known amounts of the unlabelled compound, mixed always with
the same amount of the labelled compound, which is referred to
as an internal standard. This technique is the most accurate of
the known methods for quantitative analysis of organic compounds,
being as it is the only 'absolute' method available. From the
mass spectrometric point of view, the requirements of the

instrument are apparently very straightforward. The instrument
is tuned to monitor a few specific ion masses while the sample is
being introduced into the ion source. For most work of this type,
a simple, compact, low-cost instrument is all that is needed and
these requirements are adequately met by the quadrupole mass
spectrometer.[46] Further advantages of the quadrupoles are their
ease of operation through automatic control of all the instrument's
functions using microprocessor or computer logic, and their
ability to detect both negative and positive ions simultaneously.[47]
In addition, they are able to jump rapidly between the various
characteristic ions of interest in any part of the mass scale with
minimal dead-time. There is probably little difference in
performance between the quadrupole and magnetic sector instruments
operating at low resolution, but double focussing sector
instruments have distinct advantages over both when low detection
limits are sought, as described further on in the chapter.

Quantification from the direct insertion probe. The direct probe
is a valid means of introduction of samples that would decompose
on GC columns. The sample to internal standard ratio is determined
by an adaptation of the integrated ion current technique.[48] A good
example of its use is in the extraction and quantitative
measurement of trace elements in tissue samples by forming
metal-tetraphenylporphin complexes.[49] A good account of the
practical aspects of the direct probe technique has been recently
published.[50]

Quantification using the gas chromatograph. The on-line coupling
of gas chromatography and mass spectrometry was undoubtedly one of
the greatest breakthroughs in modern analytical science. It is
applied in every branch of organic chemistry as the supreme
qualitative analytical tool.[51] Quantitative G.C.-M.S. developed
from the construction of multiple peak monitoring systems on

magnetic instruments based on accelerating voltage switching.[52]

The detector is tuned at the lowest mass ion of interest M_1 and

the other specific ions of interest, M_2, M_3 ... etc. are located

by alternately setting the accelerating voltage to the values

$M_1/M_2 V$, $M_1/M_3 V$ ... etc. where V is the initial value. Such

voltage switching is now carried out by fast digital logic

circuitry under computer control. Typically, at least two

characteristic ions per component are required for positive

identification and their intensity ratios compared with the

corresponding ions in the labelled internal standard should be

consistent, otherwise chemical interference must be suspected and

steps taken to eliminate the source of interference, for example

by further purification of the sample before G.C.-M.S.

Limit of Detection by G.C.-M.S. Assays using Isotope Dilution

Paradoxically, although mass spectrometers can detect

picogram amounts of pure compounds by G.C.-M.S., such detection

levels are very rarely achieved in the analysis of biological

samples. This is mainly due to interference on the chromatogram

traces from spurious isobaric ions from other compounds entering

the ion source via the G.C. column.

The detection limit is therefore determined by the selectivity

available with the mass spectrometer rather than by the absolute

sensitivity.[53] Problems with interference are encountered when

analysis at the sub-ppm level is required. At the ppb level and

below it is usually necessary to increase selectivity, since

further sample clean-up steps before G.C.-M.S. become undesirable

owing to excessive losses in material.

Methods of increasing selectivity. Selectivity in G.C.-M.S.

assays is achieved by limiting the number of interfering signals

reaching the detector. This can be achieved in a number of ways.

Chemical ionisation[54] reduces fragmentation but does not always

give enhanced sensitivity compared with electron ionisation.
However, a biologically interesting compound derivatised
specially to introduce a strongly electronegative group may
ionise preferentially under electron capture conditions in a
C.I. source and the molecular anion may be detected with very
high selectivity compared with other biological material, thus
affording a means of improving the detection limit.[55]

Field ionisation (F.I.)[56] is a technique that normally
yields only molecular ions and therefore has very high
selectivity. Recent improvements in the design of multipoint
F.I. sources[57] have improved the sensitivity of the technique so
markedly that subnanogram level quantification of the drug
imipramine, using a $^2H_7$-labelled analogue as the internal
standard, has recently been achieved by G.C.-F.I.M.S.[58]

More possibilities for increasing the selectivity of E.I.
methods are available with double focussing mass spectrometers
(Figure 6). By increasing the resolution and monitoring the
exact mass of an ion of interest, a reduced window in the mass

Figure 7

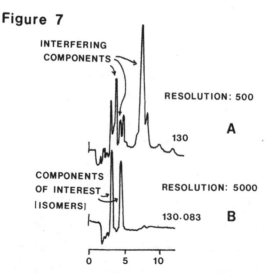

scale is open to the detector as a result of narrowing the
resolving slits and the spectrometer can then discriminate
against interfering isobaric ions (Figure 7). The inherent
advantage of this approach is a great improvement in specificity
compared with low resolution operation.

It is advisable to monitor more than one characteristic ion
even at high resolution, but existing hardware on many high
resolution instruments will allow single ion detection only. In
this case, the molecular ion is the best signal to monitor because
it is unlikely that another compound will give rise to a fragment
ion with the same elemental composition, especially as it would
be an "odd-electron" species and these are much less common in
mass spectra than "even-electron" ions. With high resolution
S.I.D. in general, only isomers cause significant problems of
interference and it is thus important to ensure that the GC
column employed is capable of resolving any such isomers as may
occur in the biological sample.[59]

Metastable ions are those which undergo fragmentation between
the ion source and detector of a mass spectrometer. Recent
developments applicable to double focussing mass spectrometers
have increased the potential applications of metastable peaks in
the analytical and biomedical fields, in particular the forced
decomposition of ions at specific points between the ion source
and collector by the use of collision gas (collisional) activation,
CA,[60] and new methods for detection of metastable ion products
from transitions occurring in the field-free regions of the MS
without interference from normal ions.[61]

Most unimolecular decompositions of metastable ions occur in
the field-free region between the ion source and first analyser
(Figure 6). Typically about 1-5% of the ions leaving the source
fragment in this region and their products are normally

undetected. However, if the MS is tuned to transmit a
particular parent ion and the magnetic (B) and electric (E)
fields are scanned simultaneously downwards so that the ratio
B/E is constant throughout the scan, a pure metastable daughter
ion spectrum is obtained and the normal mass spectral ions are
defocussed.[62]

Figure 8

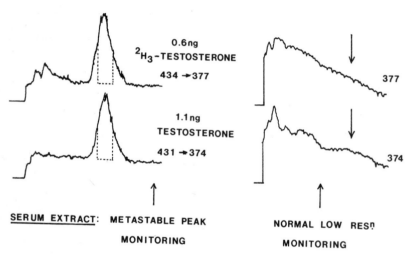

SERUM EXTRACT: METASTABLE PEAK MONITORING   NORMAL LOW RES. MONITORING

When the principles of selected ion monitoring are applied
to metastable peaks instead of normal mass spectral peaks, there
is a large gain in specificity, as with the high resolution
method.[63]

Selected ion monitoring traces obtained using this new
technique to quantify testosterone in human blood extracts[64] to
which $^2H_3$-labelled testosterone was added as the internal
standard, are compared with the conventional M.I.D. traces in
Figure 8.

Exciting possibilities for mixture analysis, even from the
direct probe, are afforded by the technique of mass analysed

ion kinetic energy spectrometry (M.I.K.E.S.),[70, 71]which examines
metastable ion products formed in the inter analyser region of a
'reverse geometry' mass spectrometer (Figure 6).

Quantification using Field Desorption

Those compounds not amenable to analysis by any E.I. or C.I.
technique may be ionised by Field Desorption (F.D.)[56]. Recently,
several examples have been published using this technique to
quantify compounds inaccessible to other ionisation techniques
by stable isotope dilution. These include choline and acetyl
choline in rat brain tissue[65]

The precision of isotope ratio measurement, and the
sensitivity of the technique, are improved by enhancement of the
signal to noise ratio using a multichannel analyser to integrate
the signals. Figure 9 shows the result of accumulating 1000 scans

Figure 9

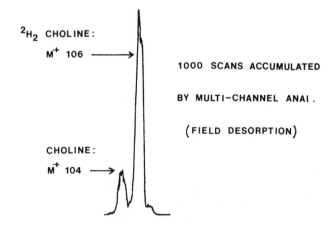

$^2H_2$ CHOLINE:

M^+ 106 ⟶

1000 SCANS ACCUMULATED

BY MULTI-CHANNEL ANAL.

(FIELD DESORPTION)

CHOLINE:

$\ddot{M}^+$ 104 ⟶

over the molecular ion region during F.D. of a mixture of choline

and $^2H_2$-choline, achieved with a commercially available

F.D.-quadrupole system equipped with a multichannel analyser

facility[66]. Resolution was sacrificed for better sensitivity.

Scope of HPLC-MS in Quantification

Finally, the commercial development of versatile interfaces

for high performance liquid chromatography (H.P.L.C.) and mass

spectrometry[67,68] has opened up another new field in chemical

analysis. Preliminary data[68,69] have suggested that

quantitative analysis using stable isotope dilution is feasible

without the problems of derivatisation encountered in G.C.-M.S.

Figure 10 shows the first commercial interface suitable for

use with magnetic sector instruments.

Figure 10   HPLC -MS   INTERFACE
(SCHEMATIC)

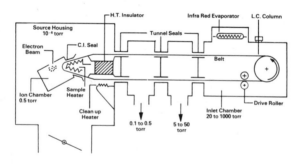

Acknowledgement.   The authors wish to express their thanks to

Dr. Martin Elliot, VG-Isotopes Ltd., for helpful discussions

during the preparation of this manuscript.

References

1.   J.J. Thomson, Phil. Mag., 1911, 21, 225

2.   J.J. Thomson, "Rays of Positive Electricity", Longmans,
        Green and Co., London, 1913.

3.   F.W. Aston, "Isotopes," Edward Arnold & Co., London, 1924.

4.   A.J. Dempster, Phys. Rev., 1918, 11, 316.

5.   W.H. McFadden, "Techniques of Combined Gas Chromatography/
        Mass Spectrometry," Wiley and Sons, New York, 1973.

6.   B.J. Millard, "Quantitative Mass Spectrometry," Heyden,
        London, 1978, p. 135.

7.   J.H. Beynon, "Mass Spectrometry and its Applications to
        Organic Chemistry," Elsevier, London, 1960,
        Chapter 1, p.8.

8.   N.J. Freeman and N.R. Daly, J. Sci. Instrum., 1967, 44, 956.

9.   F.A. White, F.M. Rourke and J.C. Sheffield, Appl. Spectrosc.
        1958, 12, 46.

10. R.P. Morgan, J.H. Beynon, R.H. Bateman and B.N. Green,
        Int. J. Mass Spectrom Ion Phys., 1978, 28, 171.

11. A.O. Nier, Rev. Sci. Instr., 1947, 18, 398.

12. H.A. Sreaus, Phys. Rev. 1941, 59, 430.

13. P. de Bièvre, "Advances in Mass Spectrometry," Heyden & Son,
        London, 1978, Vol. 7A, 395.

14. F.J. Möller, J. Geophys. Res., 1963, 68, 3877

15. H,D, Freyer and L. Wiesberg, Naturwissen., 1973, 11, 517

16. H.R. Krouse, Nature, 1977, 265, 45

17. J.O. Nriagu, "Isotope Ratios as Pollutant Source and
        Behaviour Indices," I.A.E.A. Vienna, 1975.

18. A.P. Vinogradov and E.M. Galimov, Geochem. Int., 1970, 7,
        217.

19. H. Heinrich and W.D. Lehmann (Univ. Hosp. Hamburg),
        personal communication.

20. G E. von Unruh, D.J. Hauber, D.A. Schoeller and J.M. Hayes,
     Biomed. Mass Spectrom., 1974, 1, 345.

21. D. Halliday, J.S. Garrow, and P.A. Rodgers, Proc. Nutr.
     Soc., 1974, 33, 37A.

22. D. Halliday and A.G. Miller, Biomed. Mass Spectrom., 1977,
     4, 82.

23. J. Bricout, C. Fontes, R. Letolle, A. Mariotti and
     L. Merlsiat, "Isotope Ratios as Pollutant Source and
     Behaviour Indices", publ. I.A.E.A. Vienna, 1975,
     p. 359.

24. L.W. Doner and J.W. White, Science, 1977, 197, 891.

25. I. Langmuir and K.H. Kindon, Proc. Roy. Soc. (London), 1925,
     107, 661.

26. F.A. White, in "Mass Spectrometry in Science and Technology",
     1968, p. 62.

27. W.G. Cross, Rev. Sci. Instr., 1951, 22, 717.

28. P. de Bièvre, Z. Anal. Chem., 1973, 264, 365.

29. S.E. Church, Rev. Geophys. Space Phys., 1975, 13, 98

30. H.P. Taylor, Jr., Rev. Geophys. Space Phys., 1975, 13, 102.

31. W. Compston, J.J. Foster and C.M. Gray, The Moon, 1975, 14,
     445.

32. F. Magi, S. Facchetti and P. Garibaldi, International
     Atomic Energy Agency - SM-191/27, 1975.

33. S. Facchetti, 8th International Symposium on Mass
     Spectrometry, Oslo, Aug. 1979. (Personal
     communication).

34. J.H. Beynon in "Advances in Mass Spectrometry", Pergamon,
     London, 1959, p. 328.

35. A.O. Nier, in "Nuclear Masses and Their Determination",
     Pergamon, Oxford, 1957, p. 185.

36. P. Powers, P.H. D'Arcy, J.C. Bill and M.H. Wallington,
    Proceedings of the 26th Annual Conf. on Mass
    Spectrometry and Allied Topics, St. Louis, 1978,
    p. 480.

37. W.F. Haddon and H.C. Lukens, Proceedings of the 22nd Annual
    Conf. on Mass Spectrometry and Allied Topics,
    Philadelphia, 1974, p. 436.

38. K. Hall, D.S. Millington and G. Shackleton, VG-Micromass
    technical literature, Insight No. 8, 1978.

39. I.K. Mun, R. Venkataraghavan and F.W. McLafferty,
    Analyt. Chem., 1977, 49, 1723.

40. D.M. Wilson, A.L. Burlingame, D. Hazelby, S. Evans,
    T. Cronholm and J. Sjövall, Proceedings of the 21st.
    Annual Conference on Mass Spectrometry and Allied
    Topics, p. 357.

41. Yu. N. Sukharev and Yu. S. Nekrasov, Org. Mass Spectrom.,
    1976, 11, 1232.

42. H. Miyazaki and H. Abuki, Chem. Pharm. Bull. Japan, 1976,
    24, 2572.

43. B.J. Millard, in "Mass Spectrometry, vol. 5", The Chemical
    Society, 1979, p. 186.

44. See also "Proceedings of the Symposium on Stable Isotopes:
    Applications in Pharmacology, Toxicology and Clinical
    Research", MacMillan Press, London, 1978.

45. See for example H.R. Morris, D.H. Williams and R.P. Ambler,
    Biochem. J. 1971, 125, 189, also R.E. Lovins, Prac.
    Spectrosc., 1979, 3, 19.

46. P.A. Dawson, "Quadrupole Mass Spectrometry and its
    Applications", Elsevier, Amsterdam, 1976.

47. D.F. Hunt, G.C. Stafford, F.C. Crow and J.W. Russell,
    Analyt. Chem., 1976, 48, 2098.

48. A.A. Boulton and J.R. Majer, J. Chromatography, 1970, 48, 322.

49. K.-S. Hui, B.A. Davis and A.A. Bolton, Neurochem. Res., 1977,
        2, 495.

50. "Quantitative Mass Spectrometry", Ed. B.J. Millard, Heyden,
        London, 1978, p. 91.

51. W.H. McFadden, "Techniques of Combined Gas Chromatography/
        Mass Spectrometry", Wiley-Interscience, New York, 1973.

52. C.C. Sweeley, W.H. Elliot, I. Fries and R. Rhyhage, Anal.
        Chem., 1966, 38, 1549.

53. D.S. Millington, J. Reprod. Fert., 1977, 51, 303.

54. F.H. Field and B.O. Munsen, Accounts Chem. Res., 1968, 1, 42.

55. D.F. Hunt, in "High Performance Mass Spectrometry: Chemical
        Applications" (A.C.S. Symposium Series, No. 70).
        Washington, D.C., American Chemical Society, 1978, 150.

56. H.D. Beckey, "Principles of Field Ionisation and Field
        Desorption Mass Spectrometry", Pergamon, Oxford, 1977.

57. J.H. McReynolds and M. Anbar, Int. J. Mass Spectrom. Ion.
        Phys., 1977, 24, 37.

58. H. d'A. Heck, N.W. Flynn, S.E. Buttrill, R.L. Dyer and
        M. Anbar, Biomed. Mass Spectrom., 1978, 5, 251.

59. D.S. Millington, J. Steroid Biochem., 1975, 6, 239.

60. F.W. McLafferty, et. al., J. Amer. Chem. Soc., 1975, 97, 2298.

61. J.H. Beynon and R.K. Boyd, "Adv. in Mass Spectrometry",
        Heyden, London, 1978, 7B, 1115.

62. D.S. Millington and J.A. Smith, Org. Mass Spectrom., 1977, 7,
        264.

63. S.J. Gaskell and D.S. Millington, "Quantitative Mass
        Spectrometry in Life Sciences. II", Elsevier, Amsterdam,
        1978, 135.

64. S.J. Gaskell and D.S. Millington, Proceedings of the 27th
        Ann. Conf. on Mass Spectrom. and Allied Topics,

Seattle, 1979, to be published.

65. W.D. Lehmann, H.-R. Schulter and N. Schröder, <u>Biomed. Mass</u>
    <u>Spectrom.</u>, 1978, <u>5</u>, 591.

66. W.D. Lehmann, C. Smith, T. Russell and D.S. Millington,
    "Insight No. 8.", VG.-technical literature, 1979.

67. W. McFadden, H.L. Schwartz, and S. Evans, <u>J. Chromatog.</u>, <u>122</u>,
    389.

68. D.A. Yorke, P. Burns, D.A. Baty and B.N. Green, Proceedings
    of the 27th Ann. Conf. on Mass Spectrom. All. Topics,
    Seattle, 1979, to be published.

69. D.E. Games, personal communication.

Applications of Isotopes in Drug Metabolism

D.R. Hawkins

Huntingdon Research Centre,
Huntingdon, Cambs. PE18 6ES

In the broadest sense metabolism studies are concerned with the behaviour and fate of a drug in a biological system. The term drug may be interpreted as a pharmaceutical product, pesticide, food additive or contaminant or an industrial chemical, while the biological system may be anything from a specific enzyme preparation from mammalian tissue, bacteria, animals and man to soil and plants. In this presentation the discussion will be confined primarily to studies with pharmaceutical products in animals and man, although the general principles are applicable to the other situations.

Metabolism studies are now an integral part of the safety evaluation and clinical development of new drugs and are also important in more fundamental studies on the mode of action and in producing leads to the development of safer and more efficacious drugs. These studies are usually carried out in the same animal species which are used in toxicity studies and also in man, to provide a species comparison. As a first stage basic information is required on mammalian metabolism to provide data which may be helpful in interpreting the results from toxicology studies and also in assessing the relevance of extrapolating toxicity data in animals to the potential toxicity in man. These studies are designed to provide information on absorption, tissue distribution, biotransformation and excretion and retention of the drug and/or its metabolites. Data on the concentrations and kinetics of the drug in plasma are also important as this may reflect the presence of the drug at the sites of action and determine the onset and duration of the pharmacological action and/or toxic effects. Distribution studies indicate which tissues are exposed to the drug and its metabolites and also whether there is any specific accumulation in a particular tissue. Identification of metabolites will give an insight into the metabolic pathways and perhaps through the discovery of pharmacologically active or toxic metabolites lead to the design of a better drug.

Radioisotope labels can provide a very sensitive and specific method of locating and measuring particular compounds and, with the advent and development of liquid scintillation counting as a routine technique for measuring radioactivity, radiolabelling has become the preferred method for studying the metabolic fate of foreign compounds in biological systems[1].

Besides radioisotopes, stable isotopes, particularly ^{2}H, ^{13}C, ^{15}N and ^{18}O, have found increasing use as labels in drug metabolism[22a]. While stable isotopes do not replace radioisotopes, since they cannot readily allow the same type of information to be obtained, their unique properties can be exploited for certain aspects of metabolism studies. Thus stable isotopes are available as a complementary tool to radioisotopes, which is demonstrated by the increasing combined use of both isotopes in the same study.

Important Isotopes

Radioisotopes. The most important radioisotopes used in drug metabolism are the low energy β-emitters shown in Table 1. The three important characteristics are the particle emission energy, half-life and specific activity. The half-life is the time required for the radioactivity to decline to one half the original amount at any given time and the specific activity for an isotope is the amount of radioactivity per milliatom. The unit of radioactivity is the Curie (Ci) which is that amount of radioactivity representing 2.22×10^{12} disintegrations per minute. The SI unit has been defined as the Becquerel (Bq) which is equivalent to one disintegration per second (1 Curie = 3.7×10^{10} Bq). Carbon-14 and tritium are by far the most commonly used isotopes. Carbon-14, a moderate energy β-emitter, has a very long half-life of 5730 years and compounds can be obtained with specific activities of 50-60 mCi/mmol, which is approaching 100 per cent incorporation of ^{14}C. Higher molar specific

activities can be achieved if a compound is labelled at more than one position.

Tritium is a weaker β-emitter and consequently is less easily detected and

measured than carbon-14. However, the half-life of 12.35 years is long

enough to cause no problems with experiments and the high specific activities

obtainable sometimes make this the isotope of choice. Sulphur-35 is

comparable to ^{14}C in terms of specific activity and β-energy but the short

half-life can be inconvenient for certain experiments. Other less commonly

used isotopes are ^{36}Cl and the phosphorus isotopes ^{33}P and ^{32}P. Of the

phosphorus isotopes, ^{33}P has some advantages over the less expensive and more

commonly used ^{32}P. Phosphorus-33 has a slightly longer half-life and, being

less energetic than ^{32}P is less hazardous to handle.

Table 1

Isotopes used in drug metabolism studies

Isotope	Half-life	Maximum energy of β-particles	Maximum specific activity	
Carbon-14	5730 years	0.155 MeV	62 mCi/matom	2.29 x 10^9 Bq
Tritium	12.3 years	0.018 MeV	29 Ci/matom	7.40 x 10^{11} Bq
Sulphur-35	87.4 days	0.167 MeV	14488 Ci/matom	5.37 x 10^{14} Bq
Chlorine-36	3.07 x 10^5 years	0.709 MeV	1.2 mCi/matom	4.44 x 10^7 Bq
Phosphorus-32	14.3 days	1.71 MeV	9120 Ci/matom	3.37 x 10^{14} Bq
Phosphorus-33	25.2 days	0.25 MeV	5200 Ci/matom	1.92 x 10^{14} Bq

Stable isotopes. Naturally occurring stable isotopes of carbon, hydrogen,

nitrogen and oxygen, the most common elements of organic compounds, have been

known for more than 40 years. The most abundant stable isotopes of these

elements are deuterium (^{2}H), carbon-13, nitrogen-15 and oxygen-18 (Table 2).

While the corresponding radioactive isotopes, ^{3}H and ^{14}C, are available, long-

lived radioactive isotopes of nitrogen and oxygen are not known. All the
isotopes are commercially available in a variety of forms with isotopic
purities of greater than 90% in many cases. The use of a particular isotope
will be governed by the nature and purpose of the study.

Table 2

Abundance of stable isotopes

Isotope	Natural abundance (%)
Carbon-12	98.89
Carbon-13	1.11
Nitrogen-14	99.63
Nitrogen-15	0.37
Oxygen-16	99.76
Oxygen-17	0.04
Oxygen-18	0.2
Hydrogen (^{1}H)	99.985
Deuterium (^{2}H)	0.0156

Detection and Measurement of Isotopes

The most commonly used techniques for the detection and measurement of
isotopically labelled compounds used in drug metabolism are shown in Table 3.
The two most important techniques used for radioactive and stable isotopes
are liquid scintillation counting[3] and mass spectrometry (m.s.)[4,5]. These two
techniques exploit the fundamental properties of each type of isotope,
providing analytical methodology which optimises both specificity and
sensitivity. Besides liquid scintillation counting radioactivity can be
detected and measured using autoradiography, involving exposure to X-ray film
and gas flow proportional counters which are commonly incorporated into

equipment such as thin layer scanners and in gas chromatographic detectors. These and other techniques are described in detail in a useful text[6]. Nuclear magnetic resonance spectroscopy (n.m.r.) is becoming more useful both for radioisotopes (tritium[7] where it can be used to locate precise positions of labelling in organic compounds and for stable isotopes where ^{13}C-n.m.r.[8] has a potential for the detection of ^{13}C-labelled compounds as well as being useful in structural determination. Two less well used techniques for stable isotopes are infra red and the microwave plasma detector[9,10], both of which are dependent upon the different spectroscopic properties associated with bonds involving natural and isotopic atoms.

Table 3

Techniques for the analysis of isotopically
labelled compounds

Radioactive isotopes	Stable isotopes
Liquid scintillation counting	Mass spectrometry
Autoradiography	NMR
Gas flow proportional counter	Infra red
NMR	Microwave plasma detector

Synthesis of Labelled Compounds

The position of labelling is of the utmost importance in drug metabolism to provide the information required and to ensure that meaningful results are obtained. In most cases the objective is to incorporate a label which will remain associated with the parent drug and all the pharmacologically and toxicologically important metabolites. On occasions this may require labelling in more than one position if, as a result of metabolism, the molecule can be cleaved into two biologically important components. There may be instances whereby labels are situated in other positions either to provide information about the mechanisms of biotransformation or to enable an assessment of the importance of a particular biotransformation.

Radioisotopes. A consideration which may affect the choice of radioisotope is the specific activity required. This is determined by the weight of compound in the proposed dose and the minimum amount of radioactivity per dose necessary to give the required sensitivity. The lowest dose level used in animal and human studies is likely to be near to the proposed human therapeutic dose. The amounts of radioactivity administered in animal studies are usually about 5-20 μCi to a rat and about 50-100 μCi to a dog. Carbon-14 would usually be the isotope of choice but if the dose level is very low it may be necessary to resort to the use of tritium in order to obtain a sufficiently high specific activity. Carbon-14 has the advantage that it can be incorporated into the carbon skeleton of organic compounds, for example an aromatic ring or aliphatic chain, and consequently is more readily situated in a metabolically stable position. Tritium can be labile and frequently at least some of the tritium is lost from the compound resulting in the formation of tritiated water. This loss of tritium may occur either by simple exchange procedures, either before or after metabolism at adjacent positions, or as a result of metabolism (e.g. hydroxylation) at the position of attachment of tritium. The final decision on the choice of isotope and the positions of labelling will result from a consideration of the feasibility of synthetic routes and also possibly the economic factors.

The most common primary sources of ^{14}C, ^{3}H and ^{35}S are shown in Table 4 together with some examples of their use in synthetic reactions. Since many of the data derived from drug metabolism studies involve measurements of radioactivity, the radiochemical purity of the compounds used is of prime importance. The specificity of the analyses cannot be assured unless the administered radioactivity is associated with the compound under investigation and not impurities.

Table 4

Sources and use of radioisotopes in synthesis

Isotope and source	Synthetic use

Carbon-14

$Ba^{14}CO_3$
$(^{14}CO_2)$

Carbonation of organometallic derivatives to give ^{14}C-carboxylic acids
$$\longrightarrow -^{14}CO_2H \longrightarrow -^{14}CO_2CH_3$$
$$\longrightarrow -^{14}CH_2OH \longrightarrow -^{14}COCl$$
$$\longrightarrow -^{14}CHO$$

$Na^{14}CN$

Nucleophilic displacements of halides to give ^{14}C-nitriles
Sandmeyer reaction for aromatic ^{14}C-nitriles
$$-^{14}CN \longrightarrow -^{14}CH_2NH_2$$
$$\longrightarrow -^{14}CO_2H$$

$^{14}CH_3I$

Methylation of alcohols, amines, phenols

Tritium

^3H_2O

General tritium exchange

$^3H_2$

Exchange, reduction of halides and un-saturated compounds

$LiAl^3H_4/$
NaB^3H_4

Reduction

     ketone $\longrightarrow$ alcohols
     nitriles $\longrightarrow$ amines
     esters $\longrightarrow$ alcohols

C^3H_3I

Methylation of alcohols, amines, phenols

Sulphur-35

$P_2^{35}S_5$

Carbonyl $\longrightarrow$ thione

$^{35}SO_2$

Sulphates from alcohols, phenols

$H_2^{35}SO_4$

Sulphonation

$Na_2^{35}S$

Preparation of thiols

The chemical purity is also important but may not necessarily be the same as the radiochemical purity. In practice radiochemical purities of greater than 98% should be attained, although in certain cases this may not be possible.

The radiolabelled drugs (1—5) were required for use in metabolism studies
and the positions of labelling which were selected are indicated. The antibacterial agent nifuroxazide (1) contains two groups where a radiolabel
could be situated, the nitrofurfuryl and p-hydroxybenzoyl groups. Although
it would have been easier to label the benzoyl group this part of the molecule
could be released as p-hydroxybenzoic acid in vivo, which is much less important,
toxicologically, than the nitrofurfuryl group[11]. Amitriptyline N-oxide (2) is
a new tricyclic antidepressant and since metabolic degradation of the side-
chain is a possibility, radiolabelling the dibenzocycloheptane ring was
desirable. Although the label was incorporated at one of the synthetic stages,
involving carbonation with $^{14}CO_2$, due to the symmetry of the subsequent
intermediates carbon-14 was ultimately distributed over three positions[12].
Labelling the benzoyl group or ethyl side-chain in the anorectic agent (3) would
be unsuitable due to possible ester hydrolysis releasing benzoic acid and
oxidative N-dealkylation producing acetate, respectively. A suitable label
was built into the molecule at the benzylic carbon, since if metabolic
degradation of the side-chain occurred it would usually terminate at this
carbon[13]. The tranquilliser haloperidol is one of a series of drugs
containing a p-fluorobenzoylpropionyl group. Metabolism studies have been
conducted in animals and man using a specific tritium-labelled compound (4)
since the readily available m-fluoro-tritiobenzene could be used for several
drugs. However, animal studies have shown that the major biotransformation
involves N-dealkylation with the subsequent formation of metabolites such as
p-fluoropropionic acid. In these studies no information was obtained about the
biological fate of the substituted piperidine and for this a second labelled
form would be necessary. Drugs which are natural products and derived from
fermentation processes or isolated from plant material frequently have very
complex structures and synthesis of the ^{14}C-drug may not be feasible. In
these cases it may be necessary to resort to tritium-labelling and if possible
it is preferable to devise a method for introducing the label at a specific

position. The alkaloid vincamine (5) has been synthesised with a tritium
label at the carbon adjacent to nitrogen. The label was incorporated by
catalytic reduction of a quinolizinium salt using tritium gas[15]. The radiolabel was shown to be metabolically stable in animal experiments.

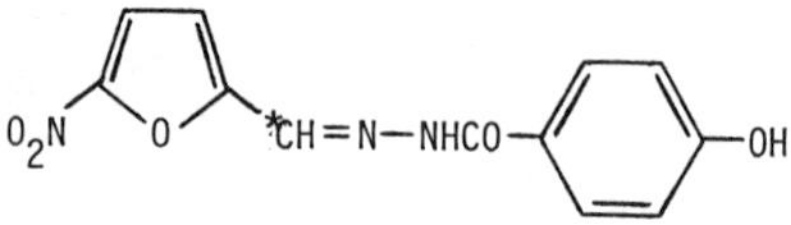

(1)

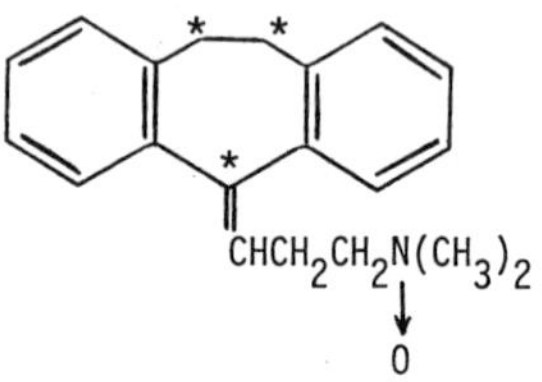

(2)

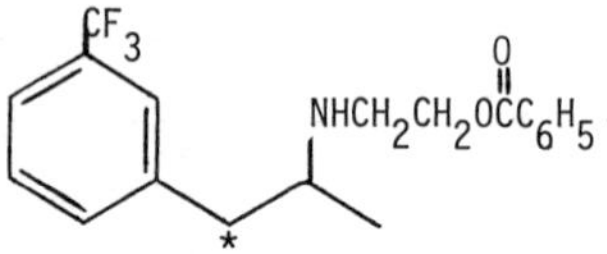

(3)

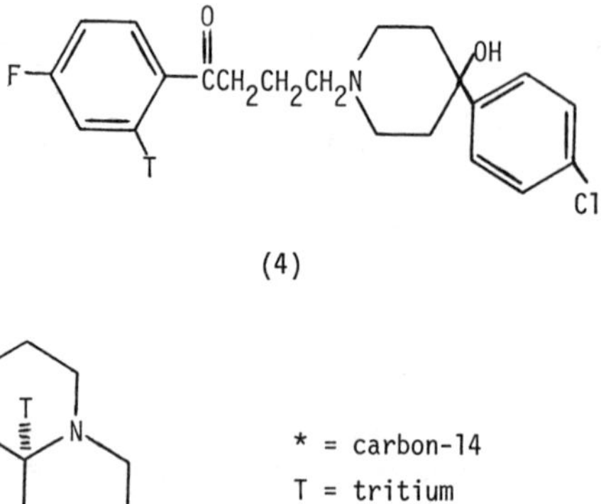

(4)

* = carbon-14
T = tritium

(5)

Stable isotopes. The consideration and methodology applied to the synthesis of
stable isotope labelled compounds is similar to that for radioisotopes. In
most cases compounds are specifically labelled and isotope incorporations are
usually high (> 90%), although isotope ratios may be adjusted prior to use by

dilution with non-labelled or even radiolabelled drug. Some of the primary
isotope sources are the same as radioisotopes namely, $Ba^{13}CO_3$, $Na^{13}CN$, $^{13}CH_3I$
and D_2O and deuterated complex metal hydrides. In addition oxygen-18 is
available as $H_2^{18}O$, $^{18}O_2$, $H_2S^{18}O_4$ and nitrogen-15 as $H^{15}NO_3$, ^{15}N-urea and
^{15}N-aminoacids. These and other commercially available compounds enable most
drugs to be synthesised with incorporation of stable isotopes.

Radioisotope Studies in Animals

Excretion studies. Animal studies would normally be carried out in a rodent
species (rat or mouse) and the dog or a non-human primate, since these are
the species which would be used in toxicology studies. The compound would be
administered by the route intended during normal usage and/or the route used
in toxicology studies, if this is different. In all the preliminary experiments
the total radioactivity concentrations in the biological samples are measured
in order to establish the pattern of excretion, tissue distribution and plasma
kinetics of the drug and its metabolites. Studies are usually divided into
different phases, but in a typical excretion study urine and faeces would
be collected separately at intervals after dosing the radiolabelled drug and
blood samples may also be taken at frequent intervals. The objective is to
obtain a near quantitative recovery of the administered radioactivity and in
most cases collection of excreta for about 5 days is sufficient. If
radioactivity is eliminated in the expired air either as a volatile metabolite
or $^{14}CO_2$ it would be necessary to trap out this radioactivity to obtain a
quantitative recovery.

Reference to studies which have been carried out with a series of cinnamoyl-
piperazines (6 and 7) illustrate some typical animal experiments[16,17]. Due
to the possible in vivo hydrolysis of the cinnamoylamide group, studies with

(6, R = OCH_2CH_3)

(7, R = N⟨⟩)

Δ and * denote positions
of two alternative
[14]C-labels

the ethyl ester (6) were conducted using two radiolabelled forms. The two compounds synthesised, contained carbon-14 in the piperazine ring and in the β-carbon of the trimethoxycinnamoyl group respectively. Experiments using each of these compounds ensured that if amide hydrolysis occurred information on the fate of each of the two groups would be obtained. However, animal experiments showed that the patterns of absorption, distribution and excretion of radioactivity and the chromatographic profile of radioactive components excreted were similar for both labelled forms indicating that amide hydrolysis was not a significant pathway of metabolism. For subsequent studies on the pyrrolidine derivative (7) location of carbon-14 in the cinnamyl α-carbon was considered satisfactory. Experiments in rat and dog showed that during 5 days after administration of an oral dose the rat excreted 37% and 58% and the dog 33% and 69% of the administered dose in the urine and faeces respectively[17]. In each case more than 70% of the dose was excreted during the first 24 h and less than 1% during the fifth day. Plasma concentrations of radioactivity reached a maximum value of 0.25% dose ml^{-1} in the rat at 30 min and 0.011% dose ml^{-1} in the dog at 90 min. These concentrations then declined with a half-life of about 90 min and 2.5 h respectively. This information immediately enables some assessment of the extent of absorption and the rate of excretion of the drug in each species. At the same time those samples which contain appreciable amounts of radioactivity can be selected for further analysis such as metabolite investigations.

Plasma samples containing sufficient concentrations of radioactivity can be analysed, after solvent extraction, by chromatographic techniques such as thin layer chromatography (t.l.c.) to determine the proportions of the parent drug and metabolites by measuring the amounts of radioactivity associated with each separated component. In this way it was shown that the compound (7) represented more than 60% of the total radioactivity in the peak plasma samples of each species. A concentration-time profile of the total radioactivity and parent drug in the rat is shown in Figure 1.

Figure 1

Mean plasma concentrations of radioactivity (●) and unchanged drug (▲) after oral administration of the ^{14}C-cinnamoylpiperazine   (7)

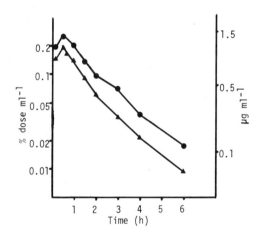

The limit of sensitivity for measurement of radioactivity depends primarily upon the specific activity of the compound but also on other factors such as background levels, counting efficiencies and counting times.   If a rat of body-weight about 200 g is administered 10 μCi of a ^{14}C-drug it is possible to measure with ease less than 0.01% dose in a 24 h urine sample and to measure a concentration of about 10^{-5}% dose ml^{-1} of plasma.   The sensitivity in terms of mass is obviously dependent upon the specific activity.   Some examples of

the approximate detection limits for different doses of a drug with molecular
weight 300 are shown in Table 5. It can be seen that for doses of 0.1 and
0.05 mg the specific activity must be increased to 30 and 60 mCi/mmol
respectively in order to achieve the same % dose limit of detection. For a

Table 5

The approximate limits of detection of radioactivity
in plasma for different doses of a ^{14}C-drug (m.wt. 300)
administered to a 200 g rat

Dose administered		Specific activity	Limit of detection	
mg	µCi	mCi/mmol	% dose ml^{-1}	ng ml^{-1}
1	10	3	10^{-5}	10
0.1	10	30	10^{-5}	1
0.05	10	60	10^{-5}	0.5
0.01	2	60	5×10^{-5}	0.5
0.01	10	300	10^{-5}	0.05

0.01 mg dose the specific activity is greater than the theoretical maximum for
one carbon-14 and to maintain the same detection limit an alternative labelling,
such as tritium, would be required. The chemical detection in the picogram to
nanogram range compares favourably with the most sensitive physico-chemical
methods available.

Tissue distribution studies. The tissue distribution of radioactivity can be
studied by two techniques, one involving sacrifice of the animal and dissection
of the tissues for the measurement of radioactivity and the other a more
qualitative technique, called whole body autoradiography. For the measurement
of radioactivity in tissues, samples of homogenised tissue or in some cases
whole small tissues are combusted. The combustion products, containing radioactivity as $^{14}CO_2$ and $^{3}H_2O$ etc., are absorbed in a scintillator for measurement.

By taking tissues from animals sacrificed at different times after dosing a profile of the uptake and rate of elimination of radioactivity, representing the drug and its metabolites, in tissues can be established.

Whole body autoradiography, being qualitative and at the best semi-quantitative, is complementary to the dissection technique. The animal, usually the rat, is sacrificed and the frozen carcass mounted on a microtome stage in a cryostat[18]. The carcass is cut longitudinally to one or more levels to expose the major tissues and organs. Sections of the whole animal, 20 microns thick, are cut with the microtome and after freeze drying are exposed to X-ray film for several weeks. The film is then developed to give an autoradiograph, with dark areas corresponding to the presence of radioactivity. By reference to the animal section it is possible to locate all the important tissues and organs. A semi-quantitative picture of the absorption, distribution and excretion of the drug is obtained, which will indicate if there is any unusual localisation or retention of radioactivity in a particular tissue. This technique has the advantage that it includes more tissues than it is practicable to dissect out and measure, particularly the smaller tissues. In addition it will reveal any regional distribution within a tissue such as in the areas of the brain, kidney medulla and cortex. Studies with pregnant animals will provide information on the extent of placental transfer of radioactivity and on the distribution within the foetus.

Some autoradiographs obtained from sections of pregnant rats administered oral doses of the diuretic and anti-hypertensive drug etozolin (8) are shown in Figures 2 and 3[19]. These animals were administered seven consecutive daily oral doses during day 10 to 17 of gestation in order to detect any accumulation of radioactivity in the foetus. At 3 h high concentrations of radioactivity can be seen particularly in the gastrointestinal tract, liver, kidneys and urinary tract and blood (Figure 2). There were also high concentrations in the placenta, but only a low level in the foetus, where the radioactivity was generally

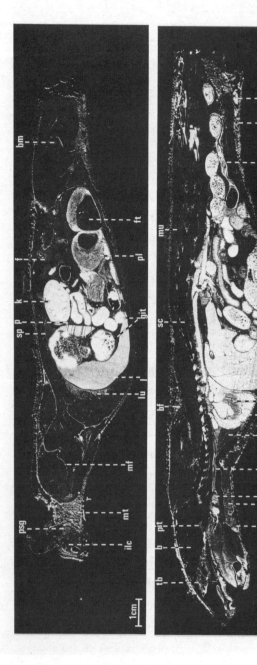

Figure 2: Whole-body autoradiograph of a rat killed 3 h after the last of seven daily oral doses of ^{14}C-etozolin. Key to abbreviations: b: brain; bf: brown fat; bl: blood; bm: bone marrow; f: fat; fr: fur; ft: foetus; git: gastrointestinal tract; ilc: inferior lachrymal gland; k: kidney; l: liver; ln: lymph node; lu: lung; mf: muscle fascia (epimysia); mt: mammary tissue; mu: muscle; my: myocardium; oe: oesophagus; p: pancreas; pg: preputial gland; pit: pituitary; pl: placenta; psg: parotid salivary gland; sc: spinal cord; sp: spleen; ssg: submaxillary salivary gland; th: thymus; ty: thyroid; tb: turbinate (nasal); ut: uterine tissue.

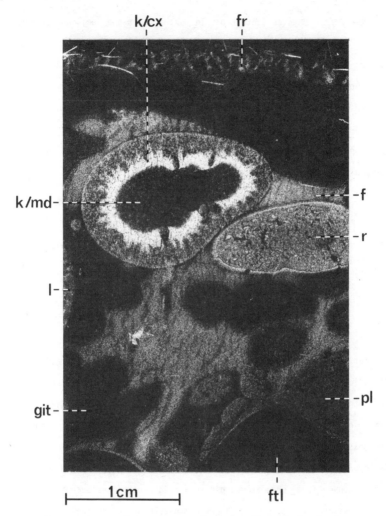

Figure 3:   Enlargement of part of a whole-body
autoradiograph of a rat killed 2 days after the
last of seven daily oral doses of ^{14}C-etozolin
showing the kidney and peripheral tissues. Key to
abbreviations: f: fat; fr: fur; ftl: foetal liver;
git: gastrointestinal tract; k/cx: kidney (cortex);
k/md: kidney (medulla); l: liver; pl: placenta;
r: rectum.

distributed. At 2 days concentrations in most tissues had declined but were
still high in the kidneys. An enlargement of this section of the autoradiograph
(Figure 3) showed that there was an uneven distribution within the kidney, and
that most of the radioactivity was associated with the boundary between the
cortex and medulla. This study showed that the placenta appeared to be a
barrier to the transfer of radioactivity to the foetus and indeed toxicological
studies confirmed that there were no toxic effects in pregnant rats and their
foetuses after oral administration.

(8)

Investigation of metabolites. Following the measurement of total radioactivity
in biological samples the second phase of investigations involves further
analysis to determine the nature of the radioactivity. The objective is to
obtain an optimum separation of the radioactivity into individual components
and to measure the proportions of these components in terms of radioactivity.
Suitable organic extracts of urine, plasma and tissues are prepared and subjected to conventional chromatographic separation techniques such as thin
layer chromatography (t.l.c.), high performance liquid chromatography (h.p.l.c.)
and gas chromatography (g.c.). Radioactive components separated by these
techniques can be detected and measured by a variety of techniques such as
autoradiography and thin layer scanning (t.l.c.), flow cell radioactivity
detector or off-line liquid scintillation counting of eluant (h.p.l.c.) and
gas flow proportional counting (g.c.)[6]. By selection of the most appropriate
technique and chromatographic conditions a metabolic profile for different
species can be generated. This enables an assessment of whether the drug is

Figure 4

Thin layer radiochromatograms of extracts of
urine samples obtained after oral doses of
^{14}C-67332 (9) to rat and man

A. Rat urine

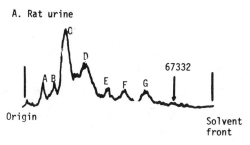

B. Human urine

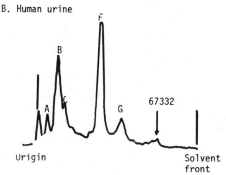

extensively metabolised and whether metabolites corresponding to authentic
reference compounds are present. However, chromatographic data alone are insufficient to be used as proof of the presence and identity of a metabolite. Any
species differences or apparent similarities will also be readily distinguished.
A typical thin layer radiochromatogram of two urine extracts is shown in Figure 4.

F⟨benzene⟩—C(=O)—CH₂CH₂CH₂N⟨piperazine * *⟩NCO₂—⟨cyclohexane⟩

$(* = {}^{14}C)$

(9)

These urine samples were obtained after oral administration of the ^{14}C-drug

67332 (9) to rat and man[22]. After t.l.c. of the urine extracts the radio-

chromatograms were produced using a thin layer scanner. Although similar

metabolites were formed by both species the quantitative differences were

readily apparent.

Radioisotope Studies in Man

The study of drug metabolism in man provides invaluable information to assist

in the process of safety evaluation and clinical drug development. In order to

obtain information similar to that from animal studies it is necessary to

administer the radiolabelled compound. After careful consideration of the

nature of the study by appropriate experts and ethical committees it is often

possible to obtain approval to carry out studies involving administration of a

radiolabelled drug to a small number of selected healthy human volunteers.

Besides the usual clinical approval of the study by ethical committees an

additional important consideration concerns estimation of the radiological

exposure of whole body and target tissues after administration of a given

radioactive dose. For this purpose the data on the concentrations and half-

lives of radioactivity in animal tissues, usually two species, after administra-

tion of the same labelled drug at the same or similar dose levels are used to

calculate the likely radiation dose to man. For many compounds, after

inclusion of appropriate safety factors, approval in the U.K. may be obtained

from the Isotope Panel of the Department of Health and Social Security to

administer a single dose of a radiolabelled drug to a limited number of healthy

volunteers.   Commonly used doses are about 50 μCi of ^{14}C or 100 μCi of ^{3}H,

although in some cases it may be necessary to reduce these doses if it has

been found in animal studies that there is uptake of high concentrations of

radioactivity into specific tissues from where it is only slowly eliminated.

There are some drugs, particularly basic compounds, that bind to the pigment

melanin, resulting in high concentrations in the eyes. While this may not have

any toxicological significance, the extrapolated radiological exposure in man
may require a greater restriction on the amount of radioactivity which can be
administered.

Although a 50 μCi dose of a ^{14}C-drug to a 60-70 kg man is relatively a much
smaller dose than is commonly used in animal studies, much valuable information
can be obtained from a study in only a few human subjects. The results
obtained from metabolism studies with a new antidepressant drug, amitriptyline
N-oxide (10) illustrate the value of such experiments[20,21]. Approval was
obtained to administer   an oral dose of 50 μCi (25 mg) of the ^{14}C-drug
to four healthy human volunteers. It was found that 84-91% of the administered
radioactivity was excreted in the urine during 9 days and 3 to 7% in the faeces
over 6 days[20]. A large proportion (mean of 47% dose) was excreted in the urine

* = carbon-14

(10)

during the first 6 h but the rate of urinary excretion was subsequently much
slower and about 1% was measured in the 24-h urine samples collected during
the ninth day. Peak concentrations of radioactivity in plasma occurred at 1 h
in three subjects, representing 0.37 μg equiv.ml^{-1}. Analysis of the plasma samples
by t.l.c. of methanol extracts showed that the unchanged drug was a major
component. The proportions of this compound accounted for about 75% of the
total radioactivity in the 1 h samples, corresponding to about 0.26 μg ml^{-1},
and concentrations were below the limits of detection at 24 h (< 0.04 μg ml^{-1}).
Extracts of individual urine samples were also analysed by t.l.c. and the

proportions of metabolites measured. By comparison with authentic reference

compounds and mass spectrometry of isolated compounds the major metabolites

were identified. The rates of excretion of these metabolites were calculated.

The major component was the unchanged drug but the 10-hydroxylated metabolite

(11) was identified as a major component in the first 6 h samples. Other

metabolites were 10-hydroxyamitriptyline (12) and 10-hydroxynortriptyline (13)

resulting from successive reduction of the N-oxide function to the tertiary

amine and N-demethylation. A biotransformation pathway of the drug in man can

be proposed as shown (Scheme 1). A comparison of the metabolism in rat, dog

and man showed that biotransformation was much more extensive in animals and

only 1% and 10% of the dose was excreted unchanged in 0-12 h urine by rats

and dogs respectively[21], compared to about 30% in man. Animal urine samples

contained a complex mixture of metabolites with up to about twenty components

resolved by t.l.c.

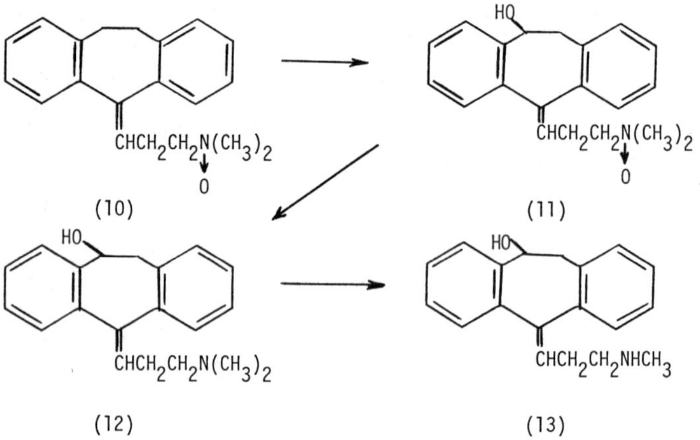

Scheme 1

The postulated biotransformation of amitriptyline
N-oxide in man

With the information obtained from these studies methodology can be developed
to allow further human studies on the drugs' kinetics and metabolism using
non-radioactive drug doses and formulated products. The analysis of plasma
samples in human studies can sometimes suffer from a lack of sensitivity due to
the relatively low specific activities used. In the study outlined above the
peak concentration of radioactivity represented about 1500 dpm ml^{-1}, but if much
lower levels are encountered chromatographic separation and measurement of
metabolites in terms of radioactivity is not possible. However, measurement
of total radioactivity can provide useful information particularly if combined
with analysis of the same samples for the parent drug by a non-radioactive
analytical method (e.g. g.c. or h.p.l.c.). These results will give the proportion of the parent drug and indicate whether metabolites are major components
circulating in plasma.

Use and problems associated with labile radiolabels. Normally the objectives
of metabolism studies require the use of radiolabels in metabolically stable
positions, but in some instances, to answer specific questions, this may not be
the case. Some of the major biotransformation processes in vivo include O- and
N-dealkylation with the formation of metabolites such as phenols or alcohols
and secondary or primary amines respectively. This is an oxidative process and
the alkyl group is lost as an aldehyde, such as formaldehyde, which would be
largely eliminated in the expired air as $^{14}CO_2$. If the alkyl group is labelled
with carbon-14 measurement of $^{14}CO_2$ eliminated in the expired air would give a
measure of the rate and extent of this specific biotransformation.

The complete collection of $^{14}CO_2$ from a small animal like a rat is relatively
straightforward since the animal can be conveniently housed in a closed system
and the total $^{14}CO_2$ trapped. It is more difficult for larger animals and man
and in these cases it may only be feasible to measure the rate of elimination
at certain intervals throughout the experiment. In studies with N-^{14}C-methyl-
nortriptyline (14) in rats about 23% of an intraperitoneal dose was eliminated

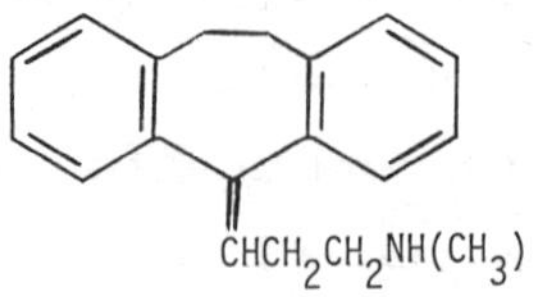

CHCH$_2$CH$_2$NH(CH$_3$)

(14)

as $^{14}CO_2$ during 8 h by male rats[23]. The initial rate of formation was slower
after an oral dose although the total amount was similar. This was attributed
to the absorption of the compound from the intestinal tract which occurred over
a period of several hours. Prior administration of the drug iproniazid, an
enzyme inhibitor, reduced the formation of $^{14}CO_2$ to about 12% demonstrating
that N-demethylation had been reduced. This can be a useful method of studying
drug interactions of this type. Studies on the metabolism of the antidepressant
drug viloxazine (15) in animals using the 1-^{14}C-ethoxy compound enabled a
species comparison of the extent of O-dealkylation[24]. 70% and 30% of an oral
dose to rats and mice was eliminated as $^{14}CO_2$ in the expired air, but since
the doses were quantitatively recovered in the urine and faeces of the dog and
rhesus monkey this could only be a minor pathway in these species. Use of an
alternative labelled form enabled confirmation that the phenol (16) was a major
metabolite in the rat.

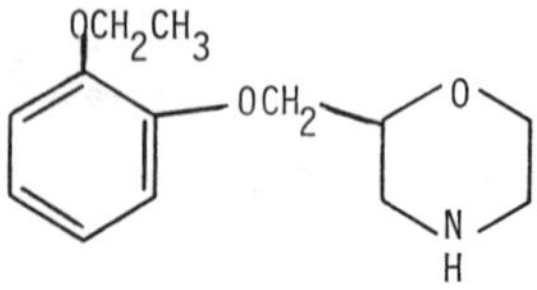

(15) 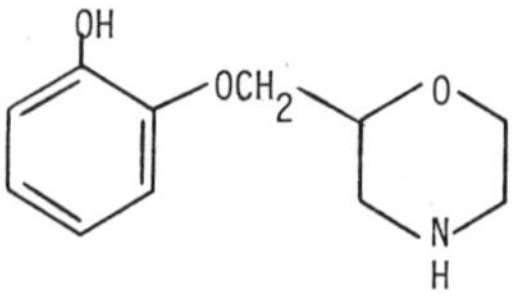   (16)

Non-metabolite residues. While it is relatively easy to measure small amounts
of radioactivity in biological samples some care should be given to the interpretation of the results since the detection of radioactivity may not necessarily
reflect the presence of the drug and/or its metabolites. Although the radiolabel may be derived from the drug it can be converted into a non-metabolite
residue which is of no relevance to the study of the drug's metabolism[25]. A
commonly encountered problem of this type occurs with tritium-labelled
compounds. The most economical way of labelling a compound is by general
tritium exchange whereby tritium may be incorporated at several positions in
the molecule. Although labile tritium may be removed before use, in vivo
loss of tritium after administration of the compound to animals or man may
occur either by simple exchange processes or as a result of metabolism (i.e.
hydroxylation) at carbons to which tritium is attached. Tritium can, of course,
also be lost from specifically labelled compounds as a result of metabolism.
The tritium released in vivo generates tritiated water which equilibrates with
body water and would be included in total radioactivity measurements on plasma,
tissues and urine. To avoid this interference tritiated water should be
removed by freeze-drying the samples before measurement of non-volatile (drug-
related) radioactivity.

After administration of the neuroleptic drug (17), which had been generally
labelled with tritium, to rats and dogs it was found that about 35-45% of the
radioactivity administered to rats and 26% of that administered to dogs was
released in vivo as tritiated water. The effect of the formation of tritiated
water is particularly noticeable in measurements of plasma radioactivity. Due

(17)

to the generally longer half-life of tritiated water in the body (3.5 days in

rats) relative to the drug and its metabolites the proportion of tritiated water

increases with time (see Figure 5). Even with compounds where loss of tritium

is low the tritiated water produced can give appreciable interference. The

relative proportions of tritiated water in a plasma sample obviously depend

upon the concentrations of the drug and its metabolites and the greatest

impact would be apparent for compounds giving low plasma concentrations and

with short half-lives.

Figure 5

Plasma concentrations of total and non-volatile
radioactivity after oral administration of $[$G-^{3}H$]$-
labelled drug (17) to rats

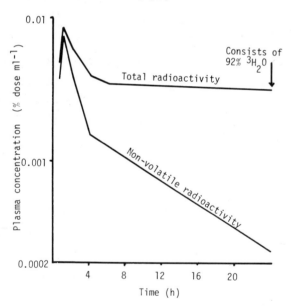

Another type of problem can occur with ^{14}C-labelled compounds when metabolic

degradation occurs resulting in the formation of small carbon compounds which

can be incorporated into biosynthetic pathways. The in vivo formation of

radiolabelled acetate, succinate and oxalate will result in the formation of radioactive residues in tissues, which are not drug-related. After administration of ^{14}C-labelled bicarbonate, acetate and succinate to rats most of the radioactivity is eliminated in the expired air as $^{14}CO_2$ and some radioactivity is incorporated into liver glycogen (bicarbonate and succinate) and glycogen, fatty acids and cholesterol (acetate)[26]. When the labelled drug viloxazine (18) was administered to dogs residual levels of radioactivity with a long half-life were measured in blood and tissues[27]. Similar levels were not detected when the O-ethyl labelled drug was used. Urea isolated from urine

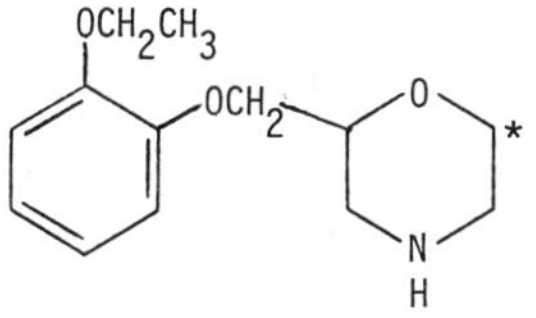

$$(* = {}^{14}C)$$

(18)

was shown to be labelled with carbon-14. Some of the major metabolites identified were formed through biotransformation in the oxazine ring and it was suggested that cleavage of this heterocyclic ring resulted in the formation of ^{14}C-glyoxylic acid (19) (Scheme 2), which was efficiently incorporated into biosynthetic pathways.

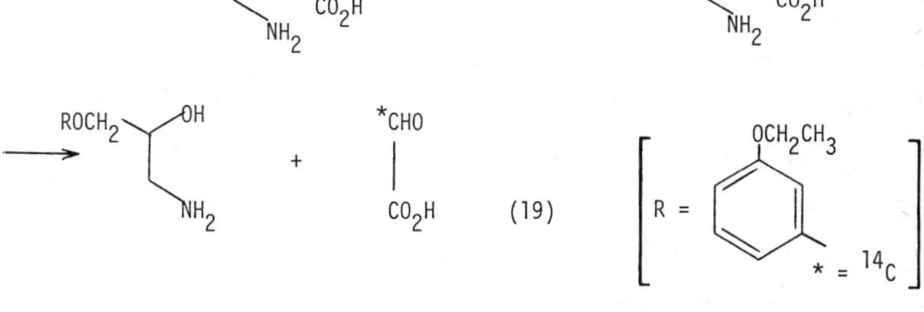

Scheme 2. Postulated formation of glyoxylic acid from ^{14}C-viloxazine

Stable Isotopes

The use of stable isotopes as opposed to radioisotopes in drug metabolism has
a great appeal, particularly for human studies, due to the absence of potential
radiological hazard. However, one of the greatest single drawbacks to the use
of stable isotopes is the lack of instrumentation for the routine detection
and assay of these isotopes in biological samples. In absolute terms the
sensitivity of methods for detecting stable isotopes is not as good as those
for detecting radioactive isotopes. However, there are some properties of
stable isotopes which enable information to be obtained which would not be
available by the use of the corresponding radioactive isotope. Stable isotopes
are thus more likely to complement radioisotopes and this is shown by the
increasing combined use of both isotopes in many studies, whereby each provides
information peculiar to its own properties.

In general, only limited studies using radiolabelled compounds are performed
in man, providing there is no reason to suspect a significant radiological
exposure to any body tissue or organ. Such a situation might occur with a
drug having a long half-life or one which is localised in a particular tissue.
An indication that a drug has these characteristics would be obtained from
animal studies and in certain instances it may be deemed inadvisable to carry
out any radiotracer studies in man. The study of compounds for use in women
and children poses a different ethical problem and there is almost universal
reluctance to use radioactivity in these subjects. In these situations studies
using stable isotope compounds could well provide much more information than
using the non-labelled drug.

In recent years the major impact of stable isotopes in drug metabolism has
been in the detection and identification of metabolites, mechanistic studies
concerned with investigations into the pathways of metabolism and indirectly
in the development of sensitive and specific methods for the analysis of drugs
and metabolites in biological fluids[2,2a].

Detection and identification of metabolites. The main physico-chemical
technique for the identification of metabolites is mass spectrometry due to
the small amounts of material available (ng to μg range) and the relatively
impure state of many of these samples. It is very seldom that a metabolite
sample has a high chemical purity and it is usually contaminated with endogenous
biological components, and other impurities from solvents, silica gel etc.
Hence in many cases a metabolite represents only a relatively small proportion
of the total sample. It becomes obvious that this poses problems in recognising
the contribution of the metabolite in a recorded spectrum. In some instances
this problem is alleviated when a drug and its metabolites have a very
characteristic fragmentation pattern, but otherwise it is more difficult, except
when naturally-occurring isotopes of atoms such as chlorine or bromine are
present. The natural abundance isotope ratios $^{35}Cl/^{37}Cl$ (ratio 3:1) and $^{79}Br/^{81}Br$
(ratio 1:1) mean that compounds possessing one (or more) of these atoms will
exhibit characteristic doublets (or multiplets) with a separation of two atomic
mass units (a.m.u.) in their molecular ions and any fragments containing these
atoms. The presence of these doublets helps in the recognition of metabolites,
as endogenous halogen-containing compounds are seldom encountered. For compounds
not containing these natural-abundance isotopes, artificial isotope doublets
can be produced by using a mixture of unlabelled and stable isotope labelled
drugs. It is important that the isotope label is situated in a metabolically
stable position and also in a position such that the major fragments in the
mass spectrum also contain the label.

^{13}C-labels can be incorporated at the same positions as a ^{14}C-label and thus
the same synthetic route can be used for both compounds. The drug 4-morpholino-
2-piperazinothieno$\sqsubset$3,2-d$\sqsupset$pyrimidine (20) was synthesised with ^{13}C and ^{14}C
at position 2 in the pyrimidine ring, for use in animal metabolism studies.
The doubly labelled compound offers the advantages of radioactivity for
obtaining quantitative kinetic data, and for the separation and isolation of
metabolites, while the stable isotope helps in the detection and structural

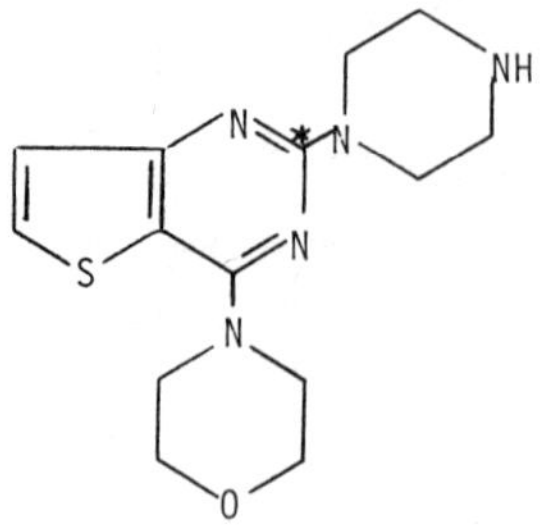

(20) 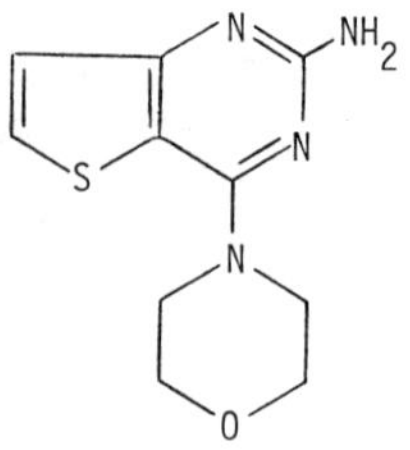   (21)

identification of isolated metabolites. Material containing equal amounts of the ^{12}C and ^{13}C compounds with a tracer amount of ^{14}C was used in the studies and thus metabolites could be distinguished by M, M+1 doublets in the mass spectra. Metabolites were isolated by two-dimensional t.l.c. of urine or urine extracts, and removal of radioactive areas after autoradiography. The components eluted from the silica gel were examined, in turn, by m.s. Spectra were obtained by slowly increasing the probe temperature, until peaks with the expected isotope distribution were observed. Metabolite spectra were obtained almost free of impurities and by this procedure eight compounds were identified, resulting from biotransformation of the piperazine or morpholine rings, one of which was the amine (21).

Deuterium-labelling may be used in a similar way and by the judicious choice of label positions it may not only aid in the detection of metabolites but can also provide confirmation of structural assignments. McMahon et al. [28] have used three labelled forms of propoxyphene including the $^2H_7$ form (22) and the $^2H_2$ form (23). A common fragment in the electron impact mass spectra of propoxyphene and related compounds is the benzyl ion (C_7H_7, m/e 91) and the $^2H_7$-labelled compound gives a corresponding fragment (C_7D_7) at m/e 98. In human studies, subjects were dosed initially with a 1:1 mixture of the

```
   D    D
    \  /      OCOCH2CH3
     []         |
D--[  ]--CD2C — CHCH2N(CH3)2
     []      |    |
    /  \     Ph  CH3
   D    D
```

```
  /==\              OCOCH2CH3
 |    |               |
 |    |--CH2C — CHCH2NCH3
  \==/     |    |    |
          Ph   CH3  CD2H
```

(22)   (23)

unlabelled and $^2H_7$-labelled compounds and urine extracts examined by g.c.-m.s. Peaks containing m/e 91/98 doublets were considered to represent propoxyphene metabolites. Metabolites resulting from ester hydrolysis, N-demethylation and aromatic hydroxylation were assigned from an examination of the mass spectra. To investigate metabolites resulting from N-demethylation the undiluted $^2H_2$-compound was administered. Assuming that deuterated and unlabelled N-methyl groups are metabolically indistinguishable, loss of one methyl group from $^2H_2$-propoxyphene results in metabolites consisting of a 1:1 mixture of unlabelled and ^2H-labelled forms and a characteristic isotope pattern. It was also shown that hydroxylation had occurred in the benzyl group rather than the 2-phenyl group by the observation that one deuterium atom was lost when the metabolite was formed from the $^2H_7$-compound.

Specific deuterium-labelling can help to assign unambiguous structures to metabolites which would otherwise require the use of n.m.r. or synthesis of authentic reference compounds. Aromatic hydroxylation is a common biotrans-formation but m.s. alone will seldom provide information on the position of hydroxylation. Use of deuterated ellipticine (24) has assisted in the solution of this type of problem[29]. A metabolite isolated from rats administered a mixture of ^{14}C-ellipticine and 7,9-dideuteroellipticine gave a mass spectrum showing an M,M+1 doublet for the molecular ion indicating that one deuterium atom had been lost. The spectrum was identical to that of 9-hydroxyellipticine but the possibility existed that the 7-hydroxy metabolite would give the same

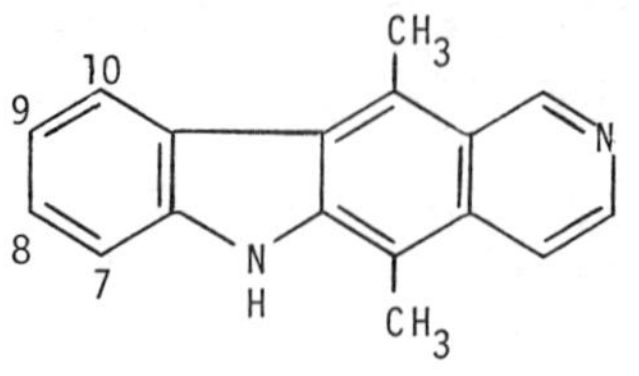

(24)

spectrum. However, the same metabolite derived from specifically labelled 9-deuteroellipticine contained no deuterium, unequivocally establishing the position of hydroxylation.

The in vitro and in vivo metabolism of N-n-butylbarbitone (25), a short-acting anaesthetic, has been studied using the ^{14}C- and ^{15}N-labelled drugs in which the ^{14}N/^{15}N isotope ratio was adjusted to about 1:1[30]. Metabolites were examined by g.c.-m.s. and detected by the appearance of M,M+1 doublets and identified as 2-hydroxy-, 3-hydroxy- and 2,3-dihydroxy-butyl derivatives (26).

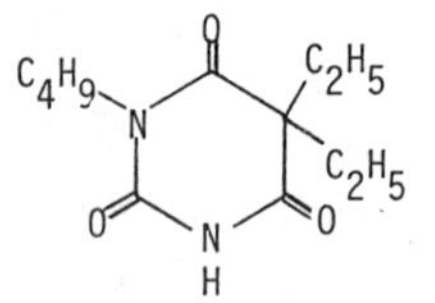

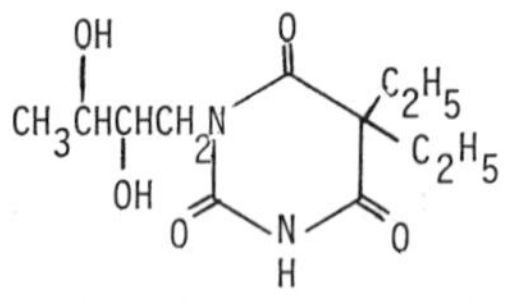

(25)   (26)

It is often necessary to derivatise metabolites prior to mass-spectral examination and methylation is frequently used for carboxylic acids, alcohols, phenols and amines. However, methylation may also re-methylate positions at which biotransformation (O- and N-demethylation) has occurred. This may cause problems in the structural elucidation but can be overcome by the use of deuterated derivatising agents, and deuterated diazomethane can be conveniently prepared for this purpose. The mass spectrum of mepyramine (27) shows a characteristic methoxybenzyl ion at m/e 121 and examination of a methylated urine extract showed the presence of a component corresponding to mepyramine[31]. However, the deuteromethylated extract contained the same component giving the fragment ion m/e 124, demonstrating that, in fact, urine contained the O-demethylated metabolite. Similarly, the structure was assigned to a further metabolite (28) due to the presence of m/e 154, corresponding to a methoxy-deuteromethoxybenzyl substituent, rather than m/e 157 from a dideuteromethyl derivative. In a similar way a comparison of the mass spectra of a metabolite derivatised with unlabelled and deuterated reagents will immediately indicate how many functional groups of a particular type are present. This is most useful when a spectrum of the underivatised metabolite cannot be obtained and may aid in subsequent structural assignments.

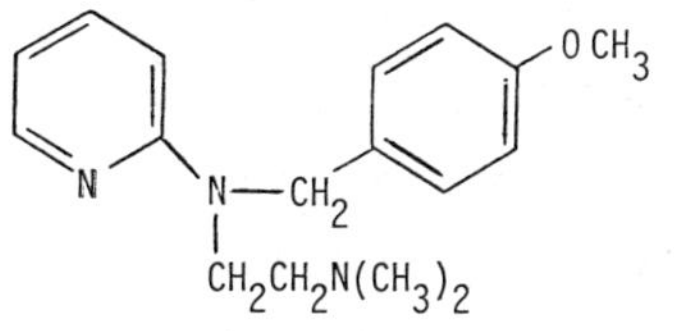

(27)

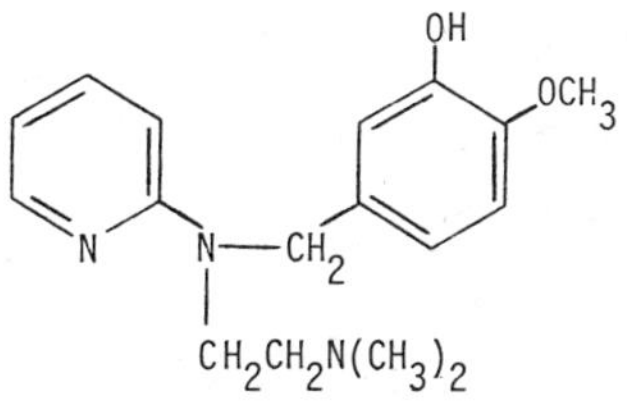

(28)

Measurement of drug and metabolite concentrations. In drug development there is an increasing demand for highly sensitive and specific methods for measuring drugs and their metabolites in biological fluids. Since there is a trend for drugs to become more potent, resulting in lower therapeutic doses and lower physiological concentrations, there is an increasing use of mass spectrometry as the basis of analytical methods. In this case the mass spectrometer is used in the selective-ion monitoring mode when it behaves as a very sensitive and specific detector.

Internal standards are invariably used in these methods, to compensate for
losses occurring during derivatisation, extraction and adsorption on g.c.
columns and to compensate for variability in the sensitivity of the mass
spectrometer during analysis of samples at different times. They can also act
as carriers to reduce the loss of compounds by adsorption on glassware and
columns during the manipulations in isolation and analysis. Consequently, an
ideal internal standard is one which behaves identically to the compound being
measured. An isotopically labelled form of the drug meets many of these
requirements and is also capable of being separately detected using m.s. The
use of such standards has been discussed in a recent review[32].

The labelled internal standard selected should be one which provides a molecular
ion and/or characteristic fragment ions which are separated usually by at
least 2 a.m.u. from the parent drug. It is preferable that these ions occur at
high mass since they are less likely to be affected by background interference,
either due to column bleed or biological impurities, although the greater the
relative intensity of the ion the greater the sensitivity of measurement. The
standard should also have the highest isotopic purity possible, since any of
the unlabelled compound in the standard will appear as a background and limit
the sensitivity. This is particularly important where the standard is being
used as a carrier and is present in as much as one hundredfold excess. The
basis of the method is one of isotope dilution, in which the internal
standard is added, in excess, to the biological fluid containing the drug or
metabolite. A calibration curve is constructed by measurement of the ratio of
the abundance of the two ions, corresponding to the unlabelled and labelled
components. Using this technique sensitivites in the 10^{-12}g (picogram) range
are often obtainable, which is equivalent to the sensitivity obtained with a
g.c. electron-capture detector.

Studies with the potent analgesic pethidine (29, R = H)have shown that besides
the parent drug, human plasma contains a pharmacologically active metabolite,

norpethidine[33]. For an investigation of the pharmacokinetics of these

components a g.c.-m.s. analytical method was developed. A deuterium-labelled

derivative of each compound was synthesised, $^2H_4$-pethidine (29, R = D) and

$^2H_5$-norpethidine (30). The molecular ions of pethidine (m/e 247) and $^2H_4$-

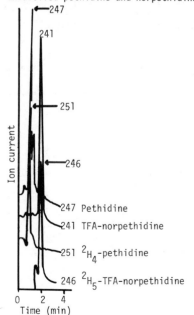

(29)   (30)

Figure 6

G.c.-m.s. selected ion chromatogram of human
plasma extract containing deuterated and
unlabelled pethidine and norpethidine

247 Pethidine

241 TFA-norpethidine

251 $^2H_4$-pethidine

246 $^2H_5$-TFA-norpethidine

pethidine were monitored. The trifluoroacetyl derivative of norpethidine was
used to improve the chromatographic characteristics and a fragment ion (M-88)
was monitored. Since this fragment ion would only have contained two of the
deuterium atoms from a piperidine-labelled derivative this was considered un-
satisfactory and the $^2H_5$-derivative was used instead, monitoring the ions
241 and 246. A representative chromatogram of a plasma extract containing the
two compounds and their internal standards is shown in Figure 6. The
sensitivity of this method was about 25 ng/ml for pethidine and 5 ng/ml for
norpethidine.

Mechanistic studies. Stable-isotope labelled drugs have been used to obtain
information on mode of action and specifically on the importance of certain
metabolic pathways on the extent and duration of pharmacological activity and
in some cases toxicity. The large mass difference between hydrogen and
deuterium leads to a lower zero-point energy of the C-D bond resulting in a
greater energy necessary for bond breaking. In metabolic processes where C-H
bond breaking is rate-determining substitution of deuterium for hydrogen leads
to isotope effects and a decrease in the rate constants.

A study of deuterium isotope effects in metabolism has been made using N-
substituted amphetamines (31) in man[34]. These experiments were designed to

show the importance of C-H bond breaking in the two major biotransformation
processes, deamination and N-dealkylation. Approximately equal amounts of
unlabelled and labelled deuterium forms were administered and the deuterium
ratio of the unchanged drug, isolated from urine, was measured by m.s. The
proportion of deuterium in isopropylamphetamine (31, R = $CH(CH_3)_2$) isolated
from urine after administration of a mixture containing the deuterated analogue
(31, R = $CD(CH_3)_2$) increased from 48% to 90% indicating a much slower rate of
N-dealkylation of the deuterated analogue. The relative proportions of unchanged drug and the metabolites of deamination and dealkylation represented
10, 45 and 45% in the urine. It is interesting that the isotope effect was
only observed for the d-isomer, while the l-isomer, which is excreted mostly
unchanged, showed no change in the hydrogen/deuterium ratio. The biological
half-lives of isopropylamphetamine and the deuterated analogue were 2 and 4 h
respectively. This result is consistent with the generally accepted mechanism
of N-dealkylation involving α-carbon hydroxylation as an intermediate step.
The magnitude of deuterium isotope effects varies considerably and in some
cases deuterium substitution may result in a change in the metabolic pathways
and either an increase in the importance of minor pathways or the formation of
new metabolites.

N-Desmethyldiazepam (32), an active metabolite of diazepam, is converted by
hydroxylation at C-3 into oxazepam (33), which is also a potent anticonvulsant
in mice and accumulates in brain. Marcucci et al.[35] have evaluated the effect

(32)   (33)

of deuterium substitution at C-3 on metabolism and pharmacology. After

intravenous administration of dideuterated N-desmethyldiazepam to mice, brain

concentrations of the drug were higher and concentrations of oxazepam were

lower than those obtained from the unlabelled compounds. These results were

paralleled by a reduction in the duration of anticonvulsant activity from 20

to 5 h.

A change in metabolism may often result in a change in the pharmacological

activity or toxicity and these mechanistic studies can help to provide inform-

ation on the processes which are important in determining this activity.

Structural modification, such as substitution at a position which deuterium-

labelling has indicated may be important to the activity of a drug, can lead

to the development of safer and more efficacious drugs. Even therapeutic use

of deuterium-labelled drugs could be advantageous and an antibacterial

combination drug containing 3-fluoro-$\underline{D}$-alanine-2-^{2}H has been used for this

purpose. 3-Fluoro-$\underline{D}$-alanine operates as an inhibitor of $\underline{D}$-alanine synthesis

by bacteria; however, the compound is oxidatively metabolised with the release of

fluoride, a potentially toxic component. The use of a deuterated derivative

reduces the oxidative metabolism and the extent of liberation of fluoride.

For drugs where the activity is due to the formation of specific metabolites,

studies with the deuterium-labelled drugs can assist in distinguishing which

metabolic pathway is responsible for the activity. Studies with the anti-cancer

drug cyclophosphamide (34) showed that the 4-^{2}H$_2$ compound produced an isotope

effect in the formation of the 4-keto metabolite (35) but there was no isotope

effect associated with the anti-tumour activity[36]. However, deuterium

substitution at C-5 caused a marked effect on the subsequent β-elimination

reaction and a decrease in the anti-tumour activity. Thus formation of

acrolein (36) and phosphoramide mustard (37) is rate-determining for the anti-

tumour activity and their formation appears to be responsible for the action of

the drug.

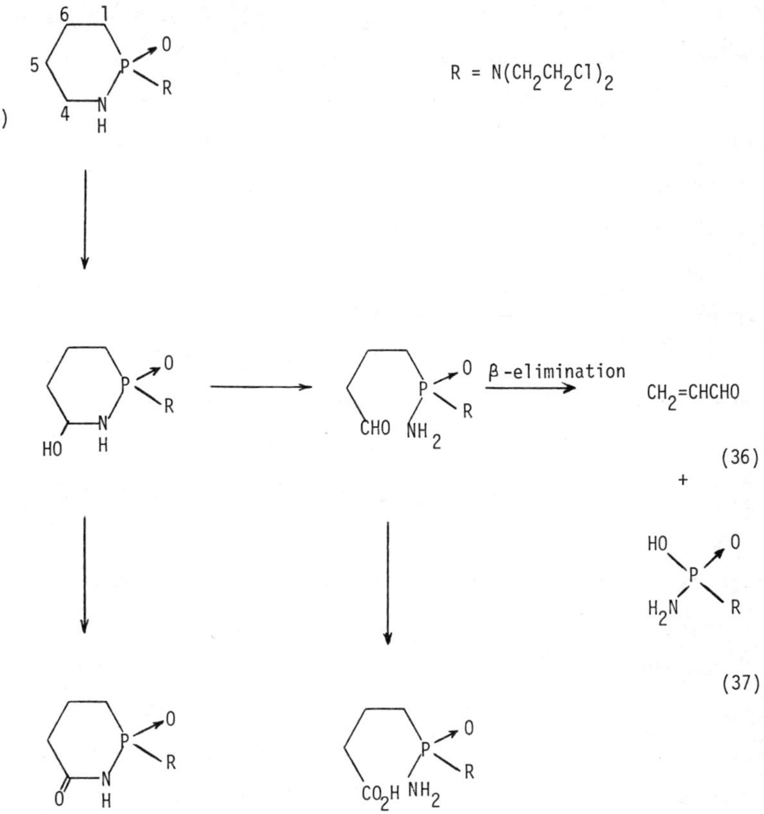

$R = N(CH_2CH_2Cl)_2$

(34)

β-elimination

$CH_2=CHCHO$

(36)

+

(37)

(35)   (38)

Often when a drug possesses one or more chiral centres biological activity resides
largely or even exclusively in one enantiomer or diastereoisomer. These
differences in activity may well be reflected in the metabolism. The relative
proportions of enantiomers of a drug and its metabolites in plasma and urine
following administration of the racemate may be determined by m.s. when one
enantiomer is isotopically labelled.

Cyclophosphamide (34) is asymmetric at the phosphorus and the metabolism of

the enantiomers in mice has been studied using the $^2H_4(+)$ and $^2H_4(-)$cyclo-

phosphamide[37]. Approximately equal amounts of $^2H_0(-)/^2H_4(+)$ and $^2H_4(+)/^2H_0(-)$

isomers were administered separately to mice and the enantiomer ratios for the

parent compound and the two metabolites measured by comparing the relative

intensities of the $[M-CH_2Cl]^+$ ions in the mass spectra. The results showed

that while there was only a small difference in the enantiomeric composition

of the parent drug, there was a depletion in the enantiomer of the 4-keto

metabolite (35) from (−)-cyclophosphamide (34) and a depletion in the enantiomer

of carboxyphosphamide (38) from (+)-cyclophosphamide.

Use of NMR for metabolite identification. The main technique which has been

used for the detection and measurement of stable isotope labelled drugs is

m.s. However, with the developments in instrumentation for ^{13}C-n.m.r. this

technique holds great potential for supplementing m.s. in metabolic studies

using stable isotopes. As carbon-13 is present in a natural abundance of 1.1%,

^{13}C-n.m.r. spectra can be obtained on unlabelled compounds. The information

from this technique can be very useful in elucidating metabolite structure as it

can provide a unique register of oxidative or reductive biotransformation occur-

ring at a particular carbon atom. This technique has been used to identify

metabolites of the anticonvulsant drug (39, R = CH_3)[38] and quinidine (40, R = H)[39].

(39)   (40)

M.s. indicated metabolites resulting from oxidation in the aromatic and
quinuclidine rings respectively and, since the carbon-13 resonances of these
groups in the parent drugs could be assigned, a shift in the resonances of the
hydroxylated and adjacent carbons enabled unambiguous assigment of the
structures as (39, R = CH_2OH) and (40, R = OH).

These spectra were obtained using samples of about 2 mg of the purified
metabolites but the possibilities of using ^{13}C-enriched compounds is very
attractive, since incorporation of ^{13}C of 90 atom % would increase the
sensitivity by almost two orders of magnitude. There is also the advantage
that spectra of such labelled compounds can be recorded under conditions
where only the signals from isotopically enriched carbons are observed and
thus it should be possible to obtain spectra on metabolites without the need
for extensive purification from endogenous compounds encountered in biological
samples. For these studies ^{13}C-labels should be incorporated at strategic
positions in the molecule either at or adjacent to positions of biotransformation. The tricyclic antidepressant drug amitriptyline (41, R_1 = R_2 = CH_3) has
been synthesised with ^{13}C-labels at positions 10(11), which would indicate
metabolites resulting from oxidation at these positions, and at the γ-carbon
of the side-chain, which would provide a sensitive indicator for biotransforma-
tions occurring at the adjacent nitrogen[12]. The ^{13}C-n.m.r. spectrum of a crude
extract of control rat urine to which 1 mg of each of $^{13}C_2$-labelled amitriptyline, nortriptyline (41, R_1 = CH_3, R_2 = H) and desmethylnortriptyline (41, R_1=R_2=H)

(41, * = ^{13}C)

hydrochlorides were added is shown in Figure 7[40]. The signals due to the
labelled carbons can be clearly seen and particularly the effect of nitrogen
substituents on the chemical shifts of the side-chain carbon. This spectrum was
obtained using an instrument time of only 1 hour. In a similar urine extract
from a dog administered an oral dose of the same drug a major metabolite was
detected with a γ-C resonance lower than that in the parent drug. This
metabolite was subsequently identified as an N-oxide.

Figure 7

^{13}C-n.m.r. spectrum of a 0-24 h rat urine extract
containing ^{13}C$_2$-labelled amitriptyline (AT),
nortriptyline (NT) and desmethylnortriptyline(DNT)
(about 1 mg of each)

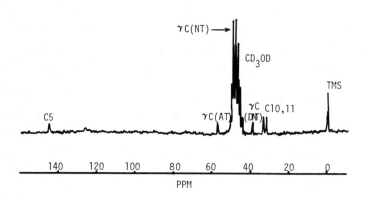

^{13}C-N.m.r. provides an effective, non-destructive method of detecting drug
metabolites in urine, involving only minimal isolation and purification, while
at the same time giving a useful insight into the structural nature of the
metabolites present. Another application would be in the comparison of patterns
of metabolites from different animal species. This could be complementary
to conventional chromatographic comparisons when radioisotopes are used, for
example, in the study of species differences in the extent of N-hydroxylation
of amines. In cases where the use of radioisotopes was precluded, ^{13}C-n.m.r.

of urine extracts after administration of the ^{13}C-labelled drug could provide a way of indicating which samples contained metabolites. Detection and identification of metabolites by m.s. would then be facilitated by the use of a mixture of ^{13}C-labelled and unlabelled drugs and the techniques previously discussed.

References

[1]D.R. Hawkins, Chem. in Britain, 1976, 12, 379.

[2]D.R. Hawkins, "Progress in Drug Metabolism", J.W. Bridges and L.F. Chasseaud (Eds.) Wiley, London, 1977, Vol.2, p.163.

[2a]"Stable Isotopes", T.A. Baillie (Ed.), Macmillan, London, 1978.

[3]Liquid Scintillation Counting", Heyden, New York, 1971-1978, Vols. 1-5.

[4]B.J. Millard, "Progress in Drug Metabolism", J.W. Bridges and L.F. Chasseaud (Eds.), Wiley, London, 1975, Vol.1, p.1.

[5]B.J. Millard, Mass Spectrom., 1975, 3, 339.

[6]T.R. Roberts, "Radiochromatography", Elsevier, Amsterdam, 1978.

[7]V.M.A. Chambers, E.A. Evans, J.A. Elvidge and J.R. Jones, Radiochemical Centre, Amersham, Review No.19.

[8]F.A.L. Anet and G.C. Levy, Science, 1973, 180, 141.

[9]C.A. Bache and D.J. Leske, Anal.Chem., 1967, 39, 786.

[10]W.R. McLean, D.L. Stanton and G.E. Penketh, Analyst, 1973, 98, 432.

[11]L.F. Elsom and D.R. Hawkins, J.Labelled Cpds., 1978, XIV, 799.

[12]I.Midgley, R.W. Pryor and D.R. Hawkins, J.Labelled Cpds, 1978, XV, 511.

[13]D.R. Hawkins and I.Midgley, J.Labelled Cpds., 1974, X, 663.

[14]W. Soudijn, I.Van Wijngaarden and F. Allewijn, Europ.J.Pharmacol., 1967, 1,47.

[15]L. Vereczkey and L. Szporny, Arzneim-Forsch., 1976, 26, 1933.

[16]L.F. Chasseaud, D.R. Hawkins, D.H. Moore and D.E. Hathway, Arzneim.-Forsch., 1972, 22, 2033.

[17]B.D. Cameron, L.F. Chasseaud, D.R. Hawkins and T.Taylor, Xenobiotica, 1976, 6, 441.

[18]"Autoradiography for Biologists", P.B. Gahan (Ed.), Academic, London, 1972.

[19]E.R. Franklin, L.F. Chasseaud and T.Taylor, Arzneim.-Forsch., 1977, 27, 1800.

[20]I. Midgley, D.R. Hawkins and L.F. Chasseaud, Arzneim.-Forsch., 1978, 28, 1911.

[21]R.R. Brodie, L.F. Chasseaud, D.R. Hawkins and I. Midgley, Arzneim.-Forsch., 1978, 28, 1907.

[22]D.R. Hawkins, D.H. Moore and L.F. Chasseaud, Xenobiotica, 1977, 7, 315.

[23]R.E. McMahon, F.J. Marshall, H.W. Culp and W.M. Miller, Biochem.Pharmacol., 1963, 12, 1207.

[24]D.E. Case, H. Illston, P.R. Reeves, B. Shuker and P. Simons, Xenobiotica, 1975, 5, 83.

[25]C. Rosenblum, "Isotopes in Experimental Pharmacology", L.J. Roth (Ed.), Univerisity of Chicago Press, Chicago, 1965, p.353.

[26]R.G. Gordon, F.M. Sinex, I.N. Rosenberg, A.K. Solomon and A.B. Hastings, J.Biol.Chem., 1949, 177, 295.

[27]D.E. Case, Xenobiotica, 1975, 5, 133.

[28]R.E.McMahon, H.R. Sullivan, S.L. Due and F.J. Marshall, Life Sci., 1973, 12, 463.

[29]A.R. Branfman, R.J. Bruni, V.N. Reinhold, D.M. Silveira, M. Chadwick and D.W. Yesair, Drug Metab. and Disp., 1978, 6, 542.

[30]M. Vore, B.J. Sweetman and M.T. Bush, J.Pharmacol.Exp.Ther., 1974, 190, 384.

[31]B.J. Millard, Biochem.Soc.Trans., 1975, 3, 462.

[32] B.J. Millard, "Quantitative Mass Spectrometry", Heyden, London, 1978.

[33] C. Lindberg, M. Berg, L.O. Boreus, P. Hartvig, K-E. Karlsson, L. Palmer and A-M. Thörnblad, Biomed.Mass Spectrom., 1978, 5, 540.

[34] T.B. Vree, J.P.M.C. Gorgels, A.Th. Muskens and J.M. Van Rossum, Clin.Chim.Acta, 1971, 33, 333.

[35] F. Marcucci, E. Mussini, P. Martelli, A. Guaitani and S. Garrattini, J.Pharm.Sci., 1973, 62, 1900.

[36] P.J. Cox, P.B. Farmer, A.B. Foster, E.D. Gilby and M. Jarman, Cancer Treatment Reports, 1976, 60, 483.

[37] P.J. Cox, P.B. Farmer, A.B. Foster, L.J. Griggs and M. Jarman, Biomed. Mass Spectrom., 1977, 4, 374.

[38] K.N. Scott, M.W. Couch, B.J. Wilder and C.M. Williams, Drug Metab.Disp., 1973, 1, 506.

[39] F.I. Carroll, D. Smith and M.E. Wall, J.Med.Chem., 1974, 17, 985.

[40] D.R. Hawkins and I. Midgley, J.Pharm.Pharmacol., 1978, 30, 547.

Applications of Isotopes in Biosynthesis

D.W. Young

School of Molecular Sciences, University of Sussex, Brighton BN1 9QJ

1   Introduction

Biosynthesis, the study of how natural products are synthesised by the
plants, micro-organisms and animals in which they are found, had its beginnings
in educated speculation about biosynthetic pathways based on the structural
relations between natural products.   This approach is elegantly summarised in
the seminal work of Sir Robert Robinson.[1]   The field of experimental biosynthesis, however, began with the advent of radioactive isotopes which were
used to label proven and putative intermediates in a biosynthetic pathway.   By
feeding such labelled compounds to the organism and isolating the resultant
natural products, features of both the biosynthetic pathway itself and the
mechanisms of reactions in the pathway have been elucidated.   More recently
stable isotopes have been used in such studies.   The area has been covered
by several general books and reviews[2-12] and a useful series of Chemical
Society Specialist Periodical Reports updates work in this area very
thoroughly.[13]

In this chapter I shall first consider a variety of the techniques which
have been used to determine the nature of biosynthetic pathways with isotopically labelled compounds.   I shall keep structural and biosynthetic
detail to a minimum in this section, concentrating on one pathway in the
second half of the chapter.   By this approach I hope to reduce the complexity
of the subject whilst illustrating as many of the techniques as possible.   It
is not irrelevant that the area chosen for detailed study is the biosynthesis
of β-lactam antibiotics where we have made some contribution in our own
research.

The sequence of steps in a typical biosynthetic experiment would be to
guess which precursors might be involved in the biosynthetic pathway and to
prepare these in a suitably labelled form.   The nature of the labelling
would be based on factors such as the ease of synthesising the labelled
precursor, the ease of determining the distribution of the label in the final
natural product and, of course, on the biosynthetic points which one wished to
solve.   The chemical and isotopic purity and isotope distribution in the

compound would be checked and it would be fed to the organism which produced
the natural product.   The organism would be allowed to grow and the natural
product and any other labelled compounds would be isolated and purified.   The
distribution of the label in these compounds would then be determined.

It is evident from this that there are many purely biological problems
associated with biosynthetic work and this is an area which has been reviewed
fairly thoroughly.[11,14,15]   In a feeding experiment the amount of compound
fed should ideally be small enough not to affect the normal physiological
state of the organism since changes in physiology would encourage aberrant
pathways.   Further, if a radioactive isotope is used, radioactivity should
be kept at levels which will not cause radiation damage to the organism.   It
is important to feed the precursor at the appropriate stage in the development
of the organism and contact of the precursor with the organism should be as
short as possible consistent with good incorporation.   This latter point
should minimise metabolism of product and precursor and thus prevent scrambling
of the label due to reincorporation of the metabolic products.   Where large
precursors are likely to be metabolised to smaller molecules which may then
be incorporated in their own right, it is important doubly to label the
molecule with one label in each of the portions likely to result from the
metabolism.   If the ratio of the labels in the product is the same as in the
precursor then it is usually assumed that the precursor is incorporated intact.
Although this is in fact usually the case, it is appropriate to note that when
the glycoside (1) was labelled separately in the glucose and cinnamic acid
portions of the molecule then the ratio of the labels remained the same in the
relevant portions of skimmin (2) obtained five days after feeding the glycoside
(1) to *Hydrangea macrophylla*.[16]   This retention of the ratio was, however,
fortuitous since, one day after feeding, the ratio of labels was such as to
indicate that hydrolysis of the glycoside into its component parts had
occurred before incorporation.[16]

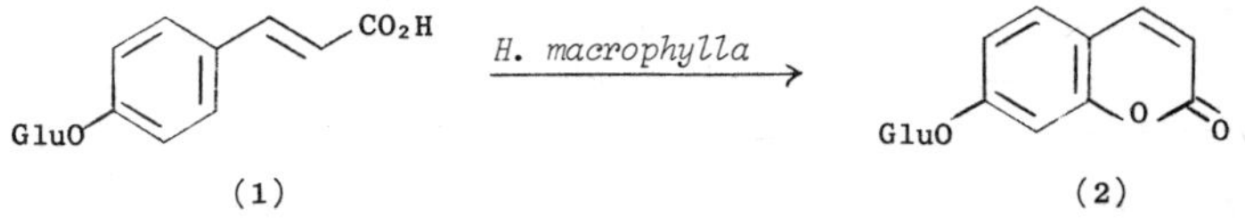

$$\text{(1)} \xrightarrow{\textit{H. macrophylla}} \text{(2)}$$

Scheme 1

Even though a compound may have been shown to be incorporated by an
organism in a biosynthetic experiment it might be argued that the incorporation
was in fact the result of an abnormal pathway.   Thus proof of the true
intermediacy of a compound in a pathway is never totally certain.   If,
however, incorporation is high, isotopic distribution is as expected and the
result is consistent with other incorporation experiments on the pathway, then

it is usually assumed that the experiment is valid.

Having discovered the overall nature of a biosynthetic pathway by incorporation experiments, details of the steps of the pathway may be investigated by use of isotopically labelled compounds.   One such detail is the
stereospecificity of the reactions involved in a biosynthetic scheme.   Biosynthetic processes are not only stereospecific with respect to chiral centres,
Cabcd, but also with respect to prochiral centres, Ca_2bc, and Ca_3b.   Stereospecific replacement of one of two identical groups 'a' at a centre Ca_2bc
with a group which contains an isotopic atom but which otherwise is identical
to 'a' converts the prochiral centre into a chiral centre, Caa'bc.   Similarly
use of more than one different isotopic atom converts Ca_3b into a chiral centre
Caa'a"b.   Isotopic methods may thus be used to follow the fate of prochiral
centres in biosynthetic processes.   Work in this area is covered by two books
on stereochemistry in biology[17,18] and several reviews which are listed in a
recent article of my own.[10]

(a) Radioactive Isotopes.   The isotopes ^{14}C, ^3H, ^{32}P and ^{35}S have been
used in biosynthetic studies.   In early work, the ease of obtaining quantitative data with ^{14}C using Geiger-Muller counting techniques made this the
most widely used isotope.   As liquid scintillation counting became more
available, however, other isotopes, especially ^3H, came to be used.   The
different radiation energies of ^{14}C and ^3H have allowed the activities of
these two isotopes to be measured separately in the same sample and double
labelling with these isotopes has proved most useful as we shall see in our
discussion of β-lactam antibiotic biosynthesis.

A major advantage in using radioactive isotopes is that the sensitivity
of counting methods allows very small amounts of the isotope to be used and
fairly low incorporations of the precursor to be observed.   Incorporations
of 0.01% have been claimed to be meaningful[19] and lower incorporations than
this are suspect and may result from an aberrant pathway.   Radiochromato-
graphic techniques and autoradiography have proved useful in biosynthetic
work both in detecting the presence of metabolites and in checking for radioactive impurities.   The importance of radiochemical purity in biosynthetic
work need hardly be stressed and it would be invidious to cite examples where
supposed incorporation was due to the presence of a radiochemical impurity.
An example of some work[20] where the fortunate incorporation of a radiochemical
impurity led to the discovery of a biosynthetic pathway is, however, worth
recording.   It had been found that while [3-^{14}C]- and [2-^{14}C]-DL-tyrosine (3)
were not incorporated into ubiquinone (4), [U-^{14}C]-L-tyrosine did appear to be
incorporated.[20]   The incompatibility of these results led to the authors
discovering that their commercial sample of [U-^{14}C]-L-tyrosine contained

labelled *para*-hydroxybenzaldehyde as an impurity and later work established
this compound as the biosynthetic intermediate.

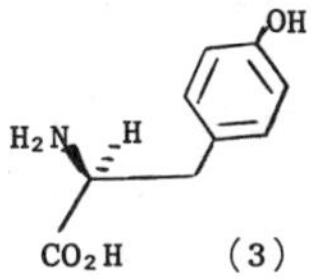

Since in reputable biosynthetic work it is essential to know the distribution of the label in both the precursor and the final product, a disadvantage of radiochemical methods is that time-consuming chemical degradations
must be carried out on these compounds.   A monograph on useful degradative
methods in this area is available[12] and, although such methods are on the whole
reliable, care must be taken to ensure that labels are not removed inadvert-
ently in the degradation and that isotope effects do not give misleading
results.   This latter point is exemplified by the degradation of $[6-^3H]$-
glucose to formaldehyde by periodate oxidation.[21]   It was found that when
excess periodate was used the molar activity in the formaldehyde was higher
than in the original glucose.   This was due to oxidation of some of the
formaldehyde produced to formic acid as in Scheme 2.   In this oxidation, an
isotope effect discriminated in favour of oxidation of C-H bonds rather than
C-T bonds, so that the molar activity in the remaining formaldehyde was increased:

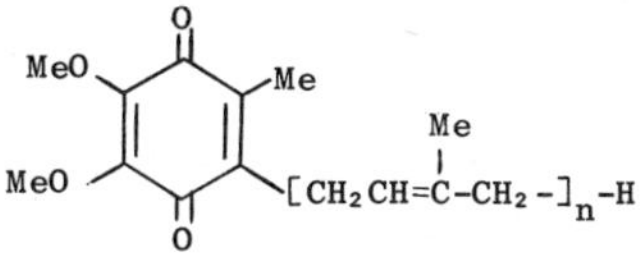

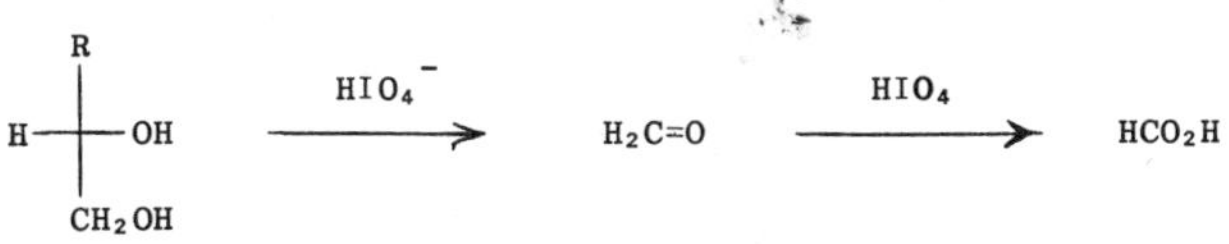

Scheme 2

Liquid scintillation counting is the major technique for detecting radioactive isotopes in biosynthetic studies, although observation of M + 2 and
M + 4 ions in the mass spectra of compounds derived from feeding very radioactive samples of $[2-^{14}C]$-mevalonic acid to cell-free systems has been used
to detect incorporation in one instance.[22]   Direct mass spectrometry of
radioactive compounds using a graphite plate as a collector may yet have some
promise.[23]   Autoradiography of the graphite plate in such an experiment showed
up radioactive fragment ions when the method was applied to $[U-^{14}C]$-benzene.[23]

Tritium n.m.r. spectroscopy,[24,25] a technique discussed by Professor
Elvidge in this symposium, looks like being a potentially useful technique for
biosynthetic work especially as improvements in instrumental design allow

smaller amounts to be detected.   The distribution and location of the label
can be found directly by this method without recourse to degradative techniques and the method has been used to elucidate some points in the biosynthesis of penicillic acid (6).[26]   The ^{1}H-n.m.r. spectrum of this compound
was elucidated with the aid of an europium shift reagent.   The presence of
various peaks was then noted in the ^{3}H-n.m.r. spectrum of penicillic acid (6)
derived from feeding experiments with [^{3}H]-acetate and with [3,5-^{3}H$_2$]-
orsenillic acid (5).   The distribution of label in the acid (6) was estimated
by integration and is shown in Table 1.   The results are consistent with
biosynthesis *via* fission of the C$_4$-C$_5$ bond of orsenillic acid (5), and any
detailed biosynthetic scheme will have to account for the stereospecific
labelling of the (Z)-hydrogen at C$_5$ of penicillic acid.

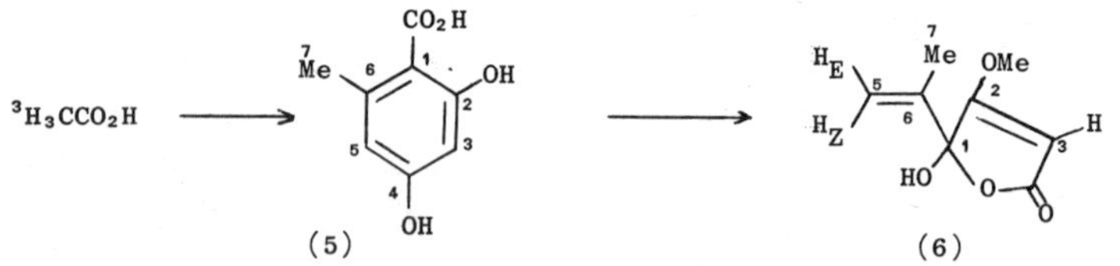

$$^3H_3CCO_2H \longrightarrow$$

(5)   (6)

Scheme 3

	Atom	H$_7$	H$_3$	H5Z	H5E
%^{3}H	from [^{3}H]-acetate feeding	43	34	19	4
	from [2,3-^{3}H$_2$]-orsenillate feeding	-	57	31	12

Table 1   Incorporation of ^{3}H into Penicillic Acid (6)

Although infrared spectroscopy can be used to detect tritium,[27] the method
has not been used in biosynthetic studies.

(b) <u>Stable Isotopes</u>.   A variety of stable isotopes has been used in biosynthetic studies, but since the physical methods required to detect these
isotopes are much less sensitive than radioactive counting techniques, considerably higher percentages of the compound containing the unnatural isotope
must be fed.   It has been suggested that the change in the metabolic pool
size resulting from the higher precursor concentrations introduces the risk
of metabolic distortion, but in fact this has seldom been demonstrated.   In
earlier studies infrared and mass spectrometric methods were normally used
but, with the advent of Fourier transform n.m.r. spectrometers, ^{13}C, ^{15}N, ^{2}H
and even ^{17}O can be observed directly.   Incorporation can be noted by peak
enhancement and the appearance of coupling in these spectra and the distri-

bution of the labels can be assigned directly without time-consuming chemical degradation.

2   Isotopes of Carbon

We have discussed the use of the radioisotope ^{14}C in Section 1(a) of this chapter.   The isotope was very widely used in early biosynthetic studies and is still much used to-day.   The isotope ^{13}C which has a relative natural abundance of 1.1% and a nuclear spin of $^1\!/_2$ can be detected by n.m.r. spectroscopy as we have seen in Dr. Rackham's contribution to this symposium.   By use of precursors enriched in this isotope, incorporations and distributions of the labels may be determined from the ^{13}C-n.m.r. spectra of the resultant natural products.   There are two excellent recent reviews on the application of ^{13}C-n.m.r. spectroscopy to biosynthesis.[28,29]

Early work using ^{13}C as an isotopic label in biosynthetic studies made use of mass spectrometric methods to detect isotopic enrichment in the resultant natural products.   Thus Cornforth[30] was able to show that a 1,2-methyl migration from C_{14} to C_{13} had occurred in the biosynthesis of cholesterol (8) by feeding a mixture of $[^{12}C]$- and $[3',4-^{13}C_2]$-mevalonolactone (7) to a cell-free preparation of rat liver and isolating the cholesterol produced.   By Kuhn-Roth oxidation, the CH_3-C groups were degraded to acetic acid which was then further degraded so that the isotopic content of the total carbon and the carboxylate carbon could be determined separately using mass spectrometry.   A statistical treatment was then used to show the nature of the methyl migration.

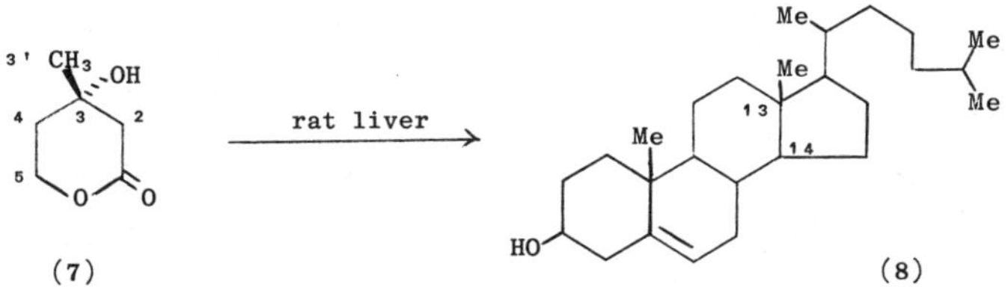

(7)   rat liver →   (8)

Scheme 4

In later work, satellites due to ^{13}C-H coupling in the ^1H-n.m.r. spectra of samples enriched with ^{13}C from biosynthetic studies were used to estimate incorporation and distribution of ^{13}C directly.[31]   Only carbon bound to hydrogen could be detected in this way.   The advent of ^{13}C-n.m.r. spectroscopy has now made this the method of choice for biosynthetic studies using ^{13}C-enriched precursors.   Since the natural abundance of ^{13}C is 1.1%, spectra may be analysed at natural abundance.   Enrichments obviously have to be noticeably higher than the background natural abundance and this is especially true because of the differences in spin lattice relaxation times of different carbons and also the variable nuclear Overhauser enhancements.   These latter factors cause a lack of uniformity in peak heights so that to be sure of peak enhancements, minimum enrichments of $ca.$ 1% above natural abundance should be aimed at.   In some instances where low incorporation has led to difficulty in applying peak enhancement data, the problem has been overcome by use of gated decoupling techniques[32] or relaxation reagents.[33]   Both of these methods have been discussed by Dr. Rackham.

The very low probability of two ^{13}C atoms being adjacent in a molecule at natural abundance means that satellites due to [13]C-^{13}C coupling will be very evident in the spectra of products biosynthesised from compounds doubly labelled with adjacent ^{13}C atoms.   In such experiments very low incorporations may be used with confidence.   The determination from the ^{13}C-n.m.r. spectrum that ^{13}C atoms which were adjacent in the precursor remain adjacent[34] or are no longer adjacent[35] in the biosynthetic product can provide useful information about a biosynthetic pathway.   Further, the appearance of [13]C-^{13}C coupling in natural products derived from precursors which were either singly labelled[36] or non-contiguously labelled[37] can be equally informative.   In a study of the biosynthesis of protoporphyrin IX (10), [2,11-[13]C_2]-porphobilinogen (9) was fed diluted with unlabelled material so that any one molecule of the resultant porphyrin only contained <u>one</u> doubly labelled unit.[37]   The ^{13}C-n.m.r. spectrum of the enriched porphyrin[37] not only showed that the γ-carbon and the carbon C_2 of ring D were now adjacent with $^1J_{CC}$ = 72 Hz, but that the other *meso* carbons (α, β and δ) were still in a 1,4-relationship with the carbon C_2 of rings B, C and A respectively, with $^3J_{CC}$ = 5 Hz.

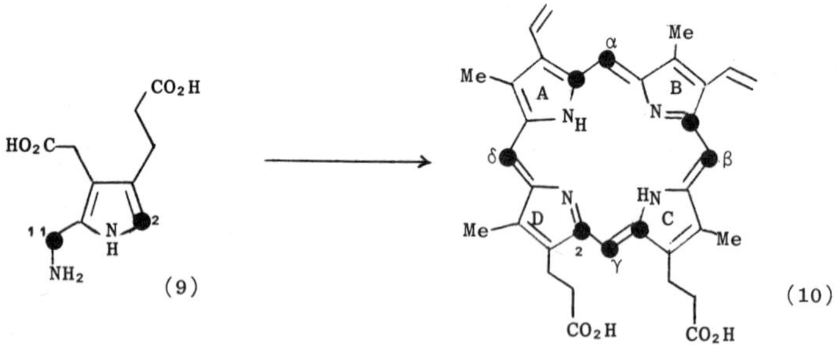

Scheme 5

In one biosynthetic experiment the long relaxation time of a nitrile group did not allow this carbon to be observed in the ^{13}C-n.m.r. spectrum and use was made[38] of an isotopic shift[27] in the infrared absorption of the nitrile to detect incorporation of the label.[38]

3   Isotopes of Hydrogen

Use of the radioactive isotope tritium has been dealt with in Section 1(a) of this chapter and will also be discussed in later sections. The stable isotope deuterium has also proved to be an excellent isotope for biosynthetic work. Early reports of infrared spectroscopy[27] being used to distinguish between isomers of deuterated succinic acids[39] and other compounds have not led to this method of detection of the isotope finding widespread use in biosynthetic studies. Early work made use of ions in the mass spectra due to deuterated species to determine the number of deuterium atoms incorporated in biosynthetic studies.[40-42]   This was taken with absences in the ^{1}H-n.m.r. spectrum[40-42] to determine the distribution of the label. In recent studies of the metabolism of glutamic acid, we have prepared (2S,3S)-[3-^{2}H$_1$]- and (2S,3R)-[2,3-^{2}H$_2$]-glutamic acids (11) by stereospecific synthesis.[43]   The stereospecificity of deuteration could be seen from the ^{1}H-n.m.r. spectra in NaO^{2}H/^{2}H$_2$O shown in Figure 1. The protons being substituted by deuterium are evidently an AB system with further additional coupling. Substitution of the pro-R hydrogen removes the 'B proton' specifically. The stereospecificity was confirmed by degradation of the glutamic acids to [^{2}H$_1$]-succinic acids (12) and the O.R.D. curves of these compounds were consistent with the assumed stereochemistry.[44]

In the method using absences in the ^{1}H-n.m.r. spectrum discussed above, it will be noted that we have been able to define the stereochemistry of an iso-topically substituted prochiral centre because the chiral centre in the same molecule causes the protons at the prochiral centre to become diastereotopic. When chirality is due solely to isotopic substitution then other methods must be used. Optical properties are useful as we have seen in the case of the [^{2}H$_1$]-succinic acids (12) above and there are many instances of the use of such methods.[10]   Other methods include the use of enzymic reactions of known specificity[10] and the use of a chiral reagent to convert the molecule to a chiral compound in which the prochiral protons will now be diasterotopic. Thus in ethanol the prochiral -CH$_2$- group normally appears in the ^{1}H-n.m.r.

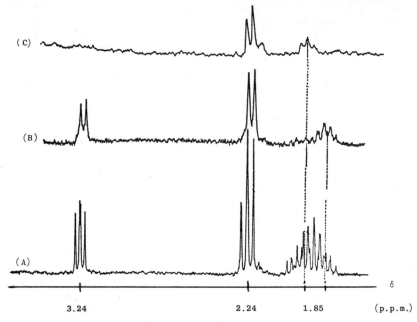

3.24	2.24	1.85	(p.p.m.)

Figure 1 ^{1}H-n.m.r. spectra of glutamates in $NaO^2H/^2H_2O$

    (A) (2S)-Glutamic acid (11)

    (B) (2S,3S)-[3-^{2}H$_1$]-Glutamic acid [(11), H_S = ^{2}H]

    (C) (2S,3R)-[2,3-^{2}H$_2$]-Glutamic acid [(11), H_R = H_X = ^{2}H]

spectrum as the A$_2$ part of an A$_2$X$_3$ system as shown in Figure 2(a). Esterification with (-)camphanoyl chloride (13) causes these protons to become diastereotopic[45] and use of an europium shift reagent enhances the resolution of
the resultant AB system,[45] [Figure 2(b) and (c)]. Stereospecifically
deuterated ethanols will now show absences in this latter spectrum.[45]   The
method has also been used with amines which can be converted to camphanoyl
amides.[46]

    As well as using absences in the ^{1}H-n.m.r. spectra to detect deuterium,
this isotope may be detected by ^{2}H-n.m.r. spectroscopy, a topic dealt with by
Professor Elvidge in this symposium.   Deuterium shifts are identical to
proton shifts so that the hydrogens of the molecule may be assigned on the
basis of ^{1}H-n.m.r. spectroscopy.   The location and, if necessary, stereochemistry of an incorporated deuterium in a biosynthetic product may then be
made on the basis of the ^{2}H-n.m.r. spectrum.   Although in its infancy, the
technique has been applied successfully to biosynthetic problems[47-50] and
the lack of nuclear Overhauser and relaxation problems allows integration to
be used to determine the amount of label at each atom.   The low natural

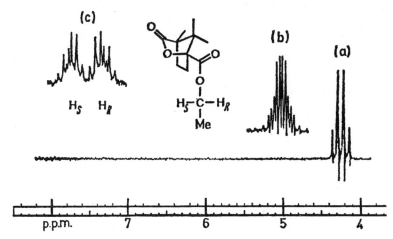

Figure 2 ¹H-n.m.r. spectra of the CH_2-O protons of (-)ethyl camphanate

(a) without shift reagent

(b) with 10 mol% Eu(dpm)₃

(c) with 35 mol% Eu(dpm)₃

abundance of deuterium makes this method useful for work with low enrichments of deuterium where the use of absences in the proton spectrum would be impracticable.

¹³C-N.m.r. spectroscopy has also been used to detect deuterium in biosynthetic studies. When ¹³C is substituted with ²H then the longer ¹³C-relaxation time and the absence of proton nuclear Overhauser effects results in suppression of the intensity of the ¹³C-absorption. An isotope effect also results in an upfield shift of $ca.0.3$ p.p.m. for each ²H atom directly bound to the ¹³C and of $ca.0.1$ p.p.m. for each ¹³C-¹²C-²H. In using methionine highly enriched (>90%) in both ¹³C and ²H in the methyl group for incorporation studies in vitamin B₁₂, a 0.7 to 0.9 p.p.m. upfield shift[51] in the methyl ¹³C-absorptions indicated intact incorporation of ¹³CD₃. Further, a deuterium-decoupled ¹³C-n.m.r. spectrum showed no evidence of the ¹³C-H coupling which would result from protium incorporation.[51,52] In another experiment highly enriched (>90%) [3-¹³C, 4-²H₂]-mevalonate (14) in which the labels were non-contiguous was incorporated into a natural product. This had a ¹³C-n.m.r. spectrum with two markedly suppressed carbon signals[53] which implied that two 1,2-hydrogen migrations had occurred during the biosynthetic sequence.

(14)

(15)

When [2-^{13}C, 2-^{2}H$_3$]-acetate was incorporated into terrein (15) not only did absorption intensity suppression and isotopic shift allow the detection of intact ^{13}C-^{2}H bonds, but ^{13}C-^{2}H coupling was present for the C$_8$ carbon.[54] This coupling resulted in a triplet (c) which was partly obscured in the proton-decoupled spectrum [Figure 3 (A) and (B)], but was much clearer in the undecoupled spectrum [Figure 3 (C) and (D)].[54]

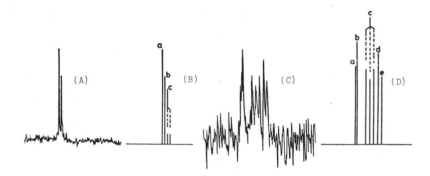

Figure 3 ^{13}C-N.m.r. spectra of terrein (15) derived from [2-^{13}C, 2-^{2}H$_3$]-
           acetate

(A) proton noise decoupled - actual spectrum

(B) proton noise decoupled - line diagram (c = triplet)

(C) proton coupled - actual spectrum

(D) proton coupled - line diagram (c = triplet)

We have seen how the isotopes deuterium and tritium may be used separately to discover structural and stereochemical aspects of biosynthetic mechanisms. A methyl group substituted in a stereospecific manner with **both** of these isotopes would be chiral, and two enantiomeric forms would be possible as in (2R)- and (2S)-[2-^{3}H$_1$, 2-^{2}H$_1$]-acetic acids [(16) and (17) in Scheme 6]. Careful stereospecific synthesis of the (R)- and (S)-enantiomers of [2-^{3}H$_1$, 2-^{2}H$_1$]-acetic acid has been achieved[55-58] and the stereochemistry was un-ambiguously defined by the method of synthesis.   Chiral acetates and compounds derived from them are therefore available for use in biosynthetic studies.   It is evident, however, that none of the methods we have so far discussed in this chapter would allow us to deduce the chirality of a stereospecifically labelled methyl group in a biosynthetic product.   An assay for the stereochemistry of

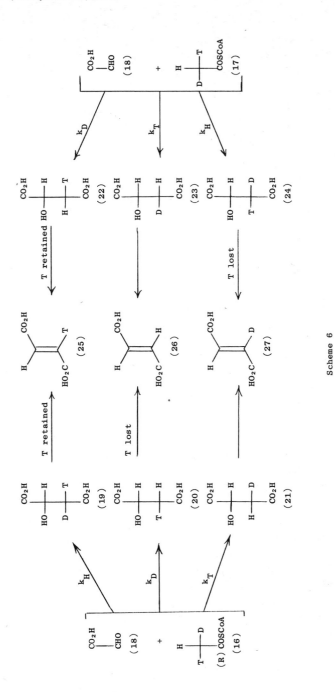

Scheme 6

the methyl group of 'chiral acetate' has been developed[55],[57] using isotope
effects in an extremely elegant manner.   Since Kuhn-Roth oxidation of CH_3-C
groups in molecules gives acetic acid, the assay is applicable to a variety of
compounds containing a chiral methyl group.

   The assay relies on two reactions catalysed respectively by the enzymes
malate synthase and fumarase.   Malate synthase catalyses the condensation of
acetate with glyoxylate (18).   If this condensation is stereospecific and
occurs with inversion of stereochemistry then $(2R)$-$[2-^3H_1, 2-^2H_1]$-acetate (16)
should yield three possible samples of malate (19), (20), and (21) and $(2S)$-
$[2-^3H_1, 2-^2H_1]$-acetate (17) should yield three other malates, (22), (23), and
(24), as shown in Scheme 6.

   It was important for the assay that complete deuteration was obtained so
that each molecule which contained tritium should also contain deuterium.   A
relatively small number of the molecules would contain tritium.   Assuming
both inversion of configuration by malate synthase and a normal isotope effect,
(R)-$[2-^3H_1, 2-^2H_1]$-acetate (16) should yield the malate (19) as the major
product since this arises from C-H bond fission while malate (20) would involve
C-D bond fission.   Fumarase is known[10] to cause *trans* dehydration of malate
to fumarate so that the major isomer (19) should yield the fumarate (25) in
which tritium has been retained.   By the same argument (S)-$[2-^3H_1, 2-^2H_1]$-
acetate (17) should yield the malate (24) as the major product of the malate
synthase reaction and this will dehydrate to fumarate (27) with fumarase,
losing tritium in the process.   Thus the expectation is that 'R-acetate' (16)
will yield malate which predominantly retains tritium on treatment with
fumarase.   It was found[55] that the malate derived from authentic (R)-
$[2-^3H_1, 2-^2H_1]$-acetate (16) yielded fumarate which retained 69% of the carbon-
bound tritium whilst the malate from authentic (S)-$[2-^3H_1, 2-^2H_1]$-acetate (17)
yielded fumarate which retained only 31% of the carbon-bound tritium.   This
is in keeping with there being inversion of stereochemistry in the malate
synthase reaction, and the assay may be used for samples of acetate of unknown
stereochemistry.

   ^{3}H-N.m.r. spectroscopy has been used to assay the stereochemistry of a
biosynthetic reaction involving a chiral methyl group.[59]   In this work oxidosqualene (28) containing a chiral methyl group was synthesised using reactions
of well-defined stereochemistry.   This compound was fed to a cell-free
fraction from *O. malhamensis* to yield cycloartenol (30).   This biosynthesis
proceeds *via* an intermediate (29).   The ^{3}H-n.m.r. spectrum of the cycloartenol produced in this process was rationalised in terms of the major isomer
in the product having an *exo*-tritium **and an** *endo*-deuterium as shown in (30) in
Scheme 7, long range coupling in the ^{1}H-n.m.r. spectrum helping with assignment

of the spectrum.   This result implied that the cyclisation of the carbonium
ion (29) which led to cycloartenol (30) had proceeded with retention of
stereochemistry.

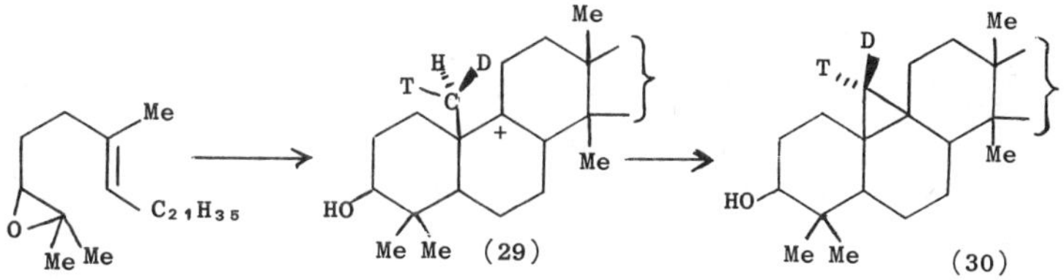

(28)   Scheme 7

4   Isotopes of Oxygen

Isotopic oxygen has been the subject of a review.[60]   In biosynthetic
studies, ^{18}O has been the isotope most commonly used and, in general,
mass spectrometric methods[3] have been used to detect the label in the biosynthetic products.[61,62]   Mass spectral fragmentation has even been used to
determine the location of ^{18}O in a biosynthetic experiment.[63]   In the metabolism of 2-fluorobenzoic acid (31) by a pseudomonad, a mixture of $^{18}O_2$
(>90% ^{18}O) and $^{18}O_2$ was used.[64]   The resultant catechol (33) was subjected
to mass spectrometry when a $(P+4)^+$ ion and a $(P)^+$ ion were observed, but no
$(P+2)^+$ ion was apparent.   This indicated incorporation of an _intact_ oxygen
molecule and so the mechanism outlined in Scheme 8 was suggested for the
process.[64]

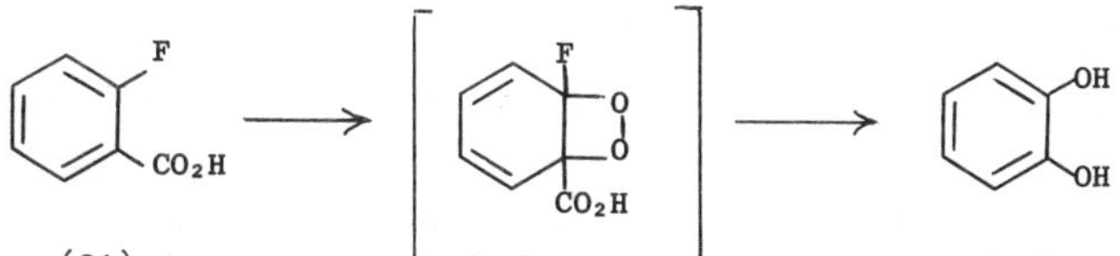

(31)   (32)   (33)

Scheme 8

When ^{18}O is substituted for ^{16}O a very distinct shift is observed in
absorptions in the infrared.[27]   It has been estimated[65] that a lower limit
of *ca.*5% of the isotope would have to be present in a carbonyl group for this
method to be used with confidence in biological studies.

The predominant isotope of phosphorus, ^{31}P, has a spin of $\frac{1}{2}$ and so ^{31}P-
n.m.r. spectra are possible.[66]   Since on substitution of each ^{18}O for ^{16}O
bound to phosphorus there is a shift of 0.02 p.p.m. to higher field of the
phosphorus resonance, it has been possible to use ^{31}P-n.m.r. isotopic shifts
by ^{18}O to study phosphate ester reactions in nature.[67]

The isotope [17]O has a natural abundance of 0.04% and a nuclear spin of $^5/_2$ so that [17]O-n.m.r. spectroscopy has been realised.[66,68,69] The technique has, however, yet to be exploited in biosynthetic studies.

5   Isotopes of Nitrogen

The less abundant of the two isotopes of nitrogen, ^{15}N, has been used as a label in biosynthetic studies. In most of the work with this isotope to date, mass spectrometry has been used to detect incorporation and, with care, fragmentation patterns have been used to detect the location and distribution of the ^{15}N label in the product of biosynthesis.[70,71] Double labelling with ^{15}N and ^{13}C,[72] and with ^{15}N and ^{18}O,[73] has been used in some studies and again mass spectrometry was used to analyse isotopic distribution.

Since ^{15}N has a natural abundance of 0.36% and a nuclear spin of $^1/_2$, ^{15}N-n.m.r. spectroscopy is possible.[66,68] The method has been used in biosynthetic studies[74] and we shall discuss these in more detail in Section 6 of this chapter. ^{13}C-N.m.r. spectroscopy has also been used to detect ^{15}N in biosynthetic studies. Observation of [15]N-^{13}C coupling in a product obtained by incorporation of a precursor containing a [15]N-^{13}C bond showed that the N-C bond remained intact during the biosynthetic sequence.[75,76] Further by using K[15]NO$_3$ as the sole nitrogen source in incubating a fungus with ^{13}C-labelled precursors, a compound was obtained which exhibited [13]C-^{15}N coupling in the ^{13}C-n.m.r. spectrum.[77] This allowed the authors to obtain structural information on their natural product.[77]

6   Biosynthesis of Penicillins and Cephalosporin C

Having discussed the techniques used in biosynthetic experiments in a general way, it is now appropriate to discuss at least one biosynthetic pathway in some detail. This should give the reader some idea of the problems involved in unravelling a pathway. The major antibiotics the penicillins (34) and cephalosporin C (35) have many common features, not least of which is the β-lactam ring. It was evident that likely precursors for the parts of the molecules labelled 'a' and 'b' below were the amino acids L-cyst(e)ine (36) and D- or L-valine (37). The unnatural D-amino acid stereochemistry was present in the 'b' part of the penicillin molecule.

This speculation was confirmed by radioactive incorporation experiments. DL-[4-^{14}C]-Valine[78] and DL-[1-^{14}C]-valine[79] were found to be incorporated into penicillin G [(34), R = PhCH$_2$CO] by the mould *Penicillium chrysogenum* and degradation of the penicillin G showed the label to be in the thiazolidine portion of the molecule[78] and in the carboxyl group[79] respectively. [3-^{14}C]-Cystine was also shown[78] to be incorporated into penicillin G with the label being localised at C$_5$. DL-[1-^{14}C]-Valine and DL-[3-^{14}C]-cystine have been

(34)   (35)

(36)   (37)

fed to *Cephalosporium* species to yield labelled cephalosporin C.[80] The
product from the valine incorporation experiment was radioactive in the di-
hydrothiazine ring whilst the product from the cystine incorporation experiment
was labelled mainly in the β-lactam ring.[80] In this early work, counting was
done using Geiger–Muller counters with samples mounted in polythene discs.

Having established that the amino acids were incorporated, it was now of
interest to determine whether they were incorporated intact. L-[3-^{14}C, ^{35}S, ^{15}N]-
Cystine has been fed to *P. chrysogenum* and, although there was some increase in
the ^{35}S:^{14}C ratio on incorporation, the ^{15}N:^{14}C ratio remained the same, and
the ^{35}S:^{14}C:^{15}N ratio was the same in the penicillin G produced as it was in
the cystine of mycelial protein.[81] In these experiments, which indicated
intact incorporation of nitrogen, carbon, and sulphur, the ^{15}N was estimated
by degradation to nitrogen gas which was then estimated by mass spectrometry.
L-[^{15}N]-Valine has also been shown to be incorporated into penicillin G,
although with some dilution of label compared to incorporation of [^{14}C]-
valine.[82,83] The partial loss of nitrogen is probably due to transamination.
More modern work using valine labelled with both deuterium and ^{15}N has given
a similar result for incorporation into both penicillin and cephalosporin C
using direct mass spectrometry to assay for the label.[84] The fate of the
valine nitrogen has been further confirmed using ^{15}N-n.m.r. spectroscopy to
detect the label.[74] In this work penicillin G was first labelled on both
nitrogen atoms using an industrial high-yielding strain of *P. chrysogenum* and
a medium containing Na^{15}NO$_3$ as the only source of nitrogen. The ^{15}N-n.m.r.
spectrum [Figure 4 (A) and (B)] showed two nitrogen absorptions one of which
was a doublet in the undecoupled spectrum [Figure 4 (A)] and a singlet in the
proton decoupled spectrum [Figure 4 (B)]. This was evidently the side-chain
nitrogen absorption and the other absorption was due to the nitrogen of the

nucleus.   When $[^{14}C, ^{15}N]$-valine was incorporated into penicillin G then the
^{15}N-n.m.r. spectrum [Figure 4 (C)] showed good incorporation into the nitrogen
of the nucleus.   Integration against an $[^{15}N]$-urea standard indicated
84 ± 10% incorporation of the valine nitrogen at this position.[74]

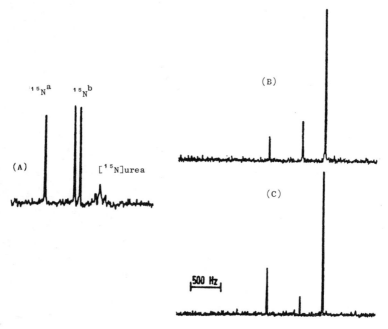

Figure 4 ^{15}N-N.m.r. spectra of potassium penicillin G and urea
(A) sample with 33.5 atoms % excess ^{15}N on both nitrogens -
undecoupled
(B) sample with 0.5 atoms % excess ^{15}N on both nitrogens - decoupled
(C) sample from feeding $[^{15}N]$-valine - noise decoupled

The realisation that L-cyst(e)ine and L-valine were incorporated intact
into penicillins and cephalosporin C catalysed speculation as to the nature
of the biosynthetic pathway from these precursors to the final antibiotics.
A tripeptide δ-[L-α-aminoadipyl)-L-cysteinyl-D-valine (L,L,D-ACV) (39) had
been found in cultures which produce penicillins and cephalosporin C[85],[86]
and it was felt that this was a very likely intermediate on the biosynthetic
pathway.   It has also been shown that L,L,D-ACV (39) arises from the combination of δ-(L-α-aminoadipyl)-L-cysteine (38) and L-valine (but not D-valine)
in cell extracts[87],[88] so that the pathway in Scheme 9 may be suggested.

(36) → (38) → L, L, L-ACV

[(39), L, L, D-ACV]

[(34), R = aminoadipyl]   Scheme 9   (35)

It will be observed from this scheme that there is a change in the
stereochemistry of the valine moiety on incorporation into penicillins.
There are also two changes in oxidation level between the tripeptide (39)
and penicillins (34) and three changes in oxidation level between the tripeptide (39) and cephalosporin C (35).   One suggestion for completion of
the biosynthesis from the tripeptide (39) which is particularly appealing to
the organic chemist is that oxidation of the tripeptide gives the thio-
aldehydes (40) and (41).[89]   An ene-reaction would convert the thioaldehyde
(40) to the β-lactam (42) which could afford penicillins by an internal
Michael reaction.   Diels-Alder addition in the thioaldehyde (41) would afford
the cephalosporin C ring system.   This suggestion is outlined in Scheme 10.
It is evident that some of these ideas might be explored by feeding experiments using precursors labelled with the isotopes of hydrogen.   The use of
such precursors might afford further details of the biosynthetic pathway.
Early work with tritium made use of windowless flow-type Geiger counters.
Parallel experiments with ^{14}C-labelled compounds gave ratios, although these
are obviously not as reliable as the ratios from the double-labelling experiments which can now be performed using liquid scintillation counting.
Thus Arnstein[90] fed both [2,2'-³H₂]- and [3,3,3',3'-³H₄]-cystine to
P. chrysogenum and isolated the penicillin G produced.   Although dilution
of the tritium label was greater than that of the ^{14}C label in the parallel
experiment, incorporation of tritium was observed, and degradation showed

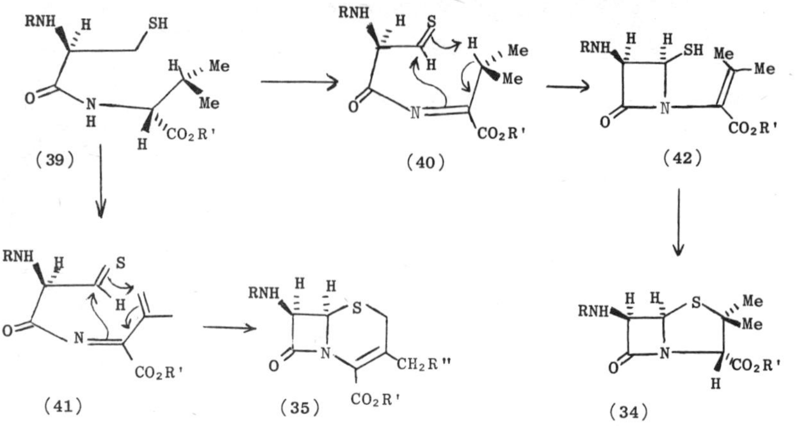

Scheme 10

the label to be in the appropriate part of the penicillin molecule.   The
retention of the [2-³H]-label of cystine in the 6-position of penicillin G
ruled out the possibility of any α,β-dehydrocysteine-containing intermediate
such as (43) in the biosynthesis and this result has been confirmed using
double ¹⁴C,³H-labelling and liquid scintillation counting[91] as well as by
feeding L-[2-²H₁]-cystine and observing incorporation using ²H-n.m.r. spectroscopy.[91]   The incorporation of the 3-³H label of cyst(e)ine into C₅ of
penicillin G has more recently[92-94] been made quantitative by use of double
labelling and liquid scintillation counting and the fact that <50% of the
tritium was incorporated may indicate a secondary isotope effect.   The
incorporation has also been observed in cephalosporin C biosynthesis.[95]

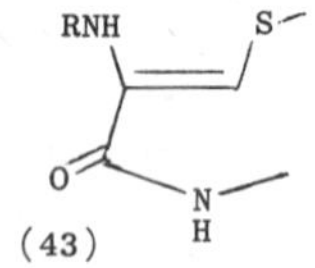

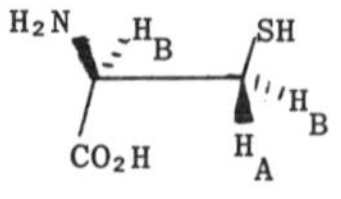

Having ascertained that the C₂-hydrogen and one of the C₃-hydrogens of
cyst(e)ine (36) were retained in the biosynthesis of penicillins and of
cephalosporin C, it was now of interest to find out which of the two prochiral hydrogens at C₃ of cyst(e)ine (36) was lost in the biosynthetic
sequence leading to the closure of the β-lactam ring.   We addressed ourselves
to this problem by synthesising cysteines (36) in which H_A and H_B were
separately and stereospecifically labelled with tritium.[92,93]   This was

achieved by first synthesising the olefin (49) in which H_A was either protium or tritium by the sequence outlined in Scheme 11.   The geometry of the olefin was defined by its being part of a thiazoline ring.   Catalytic reduction would be expected to occur with *cis*-addition of hydrogen, there being no thermodynamic preference for the *trans*-product.   Reduction of the thiazoline [(49), H_A = ^{3}H] using hydrogen gas should eventually (Scheme 11) lead to one stereochemistry at C_3 of cysteine whilst catalytic tritiation of the thiazoline [(49), H_A = H] should yield the C_3 epimer.

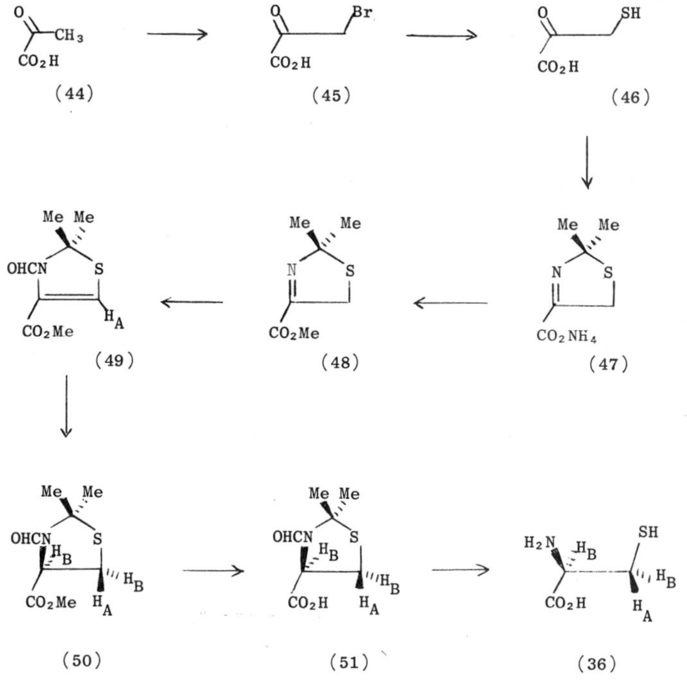

Scheme 11

Since in this scheme we have introduced the labels in well-defined relative stereochemistry, it is essential that this relative stereochemistry be maintained at least until after resolution of the true chiral centre C_2 if we are to separate the C_3 epimers.   It was therefore reassuring to find that when the optically active thiazolidine ester (50) was prepared independently from optically pure L-cysteine, it could be hydrolysed to the acid (51). Further hydrolysis yielded L-cysteine which was optically pure.   No racemisation had therefore occurred in the latter stages of the synthesis.

Having synthesised the 4-thiazoline [(49), H_A = H], it was necessary also
to have the tritiated compound [(49), H_A = ^{3}H].   We had hoped to exchange the
protons α- to the imine group in the 3-thiazoline (48), but this exchange did
not occur and so it was necessary to prepare tritiated pyruvic acid and use
the sequence of reactions in Scheme 11 to prepare the thiazoline [(49), H_A = ^{3}H].
Catalytic hydrogenation to the thiazolidine [(50), H_A = ^{3}H], hydrolysis to the
acid [(51), H_A = ^{3}H], resolution of this as the strychnine salt and further
hydrolysis afforded (2R,3S)[3-^{3}H₁]-cysteine [(36), H_A = ^{3}H].

In order to obtain the (3R)-[3-^{3}H]-cysteine we required that the 4-
thiazoline [(49), H_A = H] be catalytically reduced using tritium gas.   We
therefore sent the compound to the Radiochemical Centre, Amersham for tritiation by their service TR3.   The product was returned and from the ^{1}H-n.m.r.
spectrum was seen to consist mainly of starting material.   We therefore completed the reduction using hydrogen to obtain material which had a ^{1}H-n.m.r.
spectrum consistent with its being the desired thiazolidine (50).   Although
pure by t.l.c. developed in the normal way, a radioactive scan of the plate,
shown in Figure 5, indicated that the radioactivity of the desired product
accounted for only 20% of the total activity and that there were two very
radioactive impurities present.

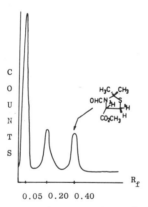

C
O
U
N
T
S

0.05 0.20 0.40

Figure 5   Radioactive scan of t.l.c. plate of the product [(50), H_B = ^{3}H]
from catalytic tritiation and further reduction

It was found convenient to purify by hydrolysis to the highly crystalline
acid (51) which could be crystallised to constant activity.   The acid was
subsequently resolved and converted to (2R,3R)-[2,3-^{3}H₂]-cysteine [(36),
H_B = ^{3}H] as before.

The two stereospecifically tritiated samples of L-cysteine were now mixed with L-[U-^{14}C]-cysteine and crystallised to constant activity and ratio. Since there were, in the literature,[10] a few instances of catalytic tritiation not leading to uniform distribution of tritium in the product, it was deemed necessary to find the distribution of tritium in the sample of (2S,3R)-[U-^{14}C, 2,3-^{3}H$_2$]-cysteine [(36), H$_B$ = ^{3}H]. This was achieved by exchange of the C$_2$ hydrogen using the racemisation sequence outlined in Scheme 12. The ratio of ^{3}H:^{14}C in the product (54) was half what it had been in the original cysteine [(36), H$_B$ = ^{3}H] and so the distribution of tritium in this compound was uniform.

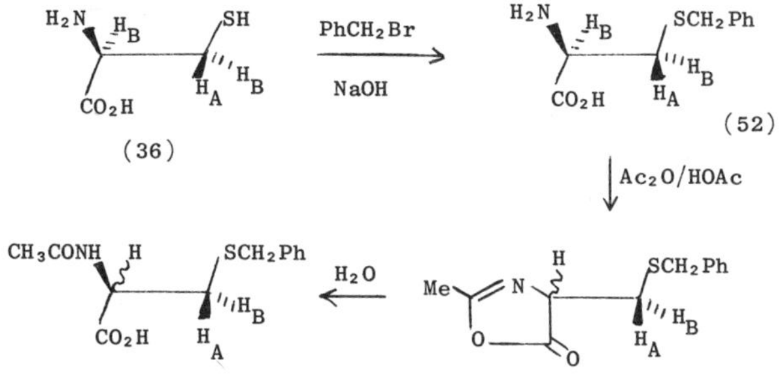

H$_2$N H$_B$ SH →(PhCH$_2$Br / NaOH)→ H$_2$N H$_B$ SCH$_2$Ph

(36) H$_A$ H$_B$ CO$_2$H   CO$_2$H H$_A$ (52)

↓ Ac$_2$O/HOAc

CH$_3$CONH H SCH$_2$Ph ←(H$_2$O)← Me N SCH$_2$Ph

CO$_2$H H$_A$ (54)   O O H$_A$ (53)

Scheme 12

Both of the stereospecifically tritiated cysteines and also [U-^{14}C, 3,3,-3',3'-^{3}H$_4$]-cystine were separately fed to cultures of *P. chrysogenum*. Penicillin G was isolated as the N-ethylpiperidinium salt and crystallised to constant activity and ^{3}H:^{14}C ratio. Comparison of the ^{3}H:^{14}C ratio found in the penicillin G with the original ratio in the cyst(e)ine fed showed that (a) for non-stereospecific labelling at C$_3$, 42% of the tritium was retained, (b) when (2R,3S)-[U-^{14}C, 3-^{3}H$_1$]-cysteine was fed, only 14% of the tritium remained in the resultant penicillin G, and (c) when (2R,3R)-[U-^{14}C, 2,3-^{3}H$_2$]-cysteine was fed, 58% of the tritium label was retained in the penicillin G. It would appear likely that these results indicate retention of the (3-pro-R)-hydrogen of cyst(e)ine in penicillin biosynthesis but it is evident that the distribution of the tritium label in the penicillin G derived from (2R,3R)-[U-^{14}C, 2,3-^{3}H$_2$]-cysteine would have to be determined. This was achieved using the degradation sequence outlined in Scheme 13.

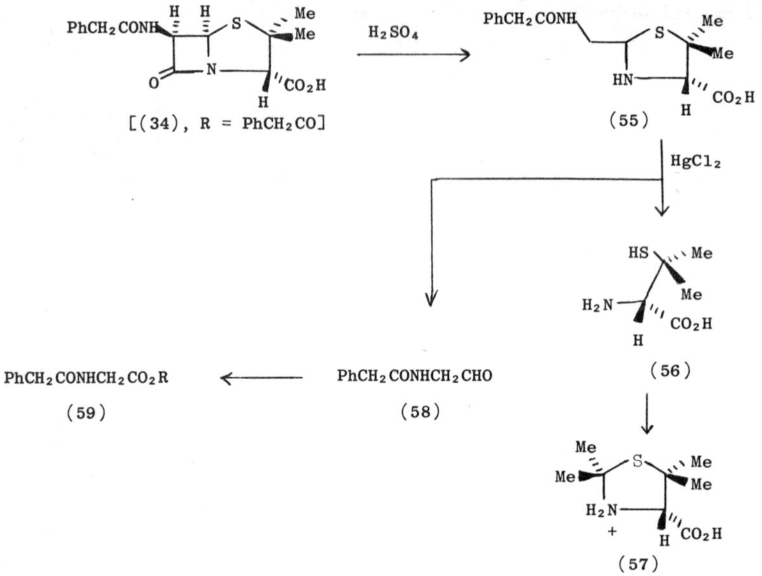

Scheme 13

Hydrolysis of the penicillin G [(34), R = $PhCH_2CO$] to the amine (55) would be expected to lose one ^{14}C atom. The product (55) was reacted *in situ* with mercuric chloride and penicillamine (56) was isolated as the mercury salt, freed and converted to the thiazolidinium chloride (57). This salt had no radioactivity. The solution, after removing the mercury salt of penicillamine, proved to contain the aldehyde (58) which could be converted to the oxime. This had a $^3H:^{14}C$ ratio which indicated no loss of tritium at this point. Oxidation of the aldehyde with silver oxide in ammonia followed by esterification gave the ester [(59), R = Me]. The $^3H:^{14}C$ ratio indicated that *ca.*75% of the tritium had been lost in the oxidation. This should represent the amount of tritium at C_5 of penicillin G provided that no exchange α to the aldehyde had occurred in the oxidation-esterification sequence. This point could be verified by conducting the oxidation in AgO/ND_3 when the n.m.r. spectrum (Figure 6) of the ester (59) showed that, although there was some exchange of the $PhC\underline{H}_2$ protons which appeared as a singlet in the spectrum of the undeuterated molecule, there was no exchange of the $NHC\underline{H}_2$ protons which appeared as a doublet which simplified to a singlet on addition of D_2O (Figure 6).

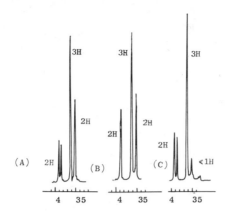

Figure 6 ^{1}H-N.m.r. spectra of $PhCH_2CONHCH_2CO_2Me$ [(59), R = Me]

(A) in $CDCl_3$

(B) in $CDCl_3$ - D_2O added

(C) from oxidation/esterification of $PhCH_2CONHCH_2CHO$ (58) using deuterated solvents

It was therefore concluded that $ca.75\%$ of the total tritium in the penicillin G was at C_5, indicating an overall retention of 87% tritium at this position from the original (3R)-tritium atom in L-cysteine.

Subsequent to our work, a second group[96] have synthesised both (2RS,2'RS,-3R,3'R)- and (2RS,2'RS,3S,3'S)-[3,3'-^{3}H$_2$]-cystines using an entirely different method from our own. In their work the isotope was introduced by di-imide reduction since catalytic deuteration had proceeded with but 95% *cis* stereo-specificity.[96] Their results confirmed that the β-lactam ring was formed with loss of the (3-pro-S)- and retention of the (3-pro-R)-hydrogen of cyst(e)ine. In work done in collaboration with Professor E.P. Abraham we have shown that the same stereospecificity applies to the biosynthesis of cephalosporin C,[97] and so β-lactam ring closure proceeds with net retention of configuration in the biosynthesis of both antibiotics.

Turning to the amino acid valine, it is known that L-valine is the precursor of the antibiotics although D-valine is incorporated more slowly, presumably *via* transamination and conversion to L-valine. It has been shown by double labelling with ^{3}H and ^{14}C that both D- and L-[2-^{3}H]-valine lose all tritium on incorporation into penicillin G.[98] When '[Me-d$_6$]-valine' was in-corporated into penicillin V [(34), R = PhOCH$_2$CO] it was noted that, for the methyl ester, the base peak in the mass spectrum corresponding to the ion (60) appeared as a doublet of m/e = 174 and 180 with no intermediate ions.[99] The valine had therefore been incorporated without loss of any of the hydrogen

atoms and so intermediates such as (61) below were not involved in the biosynthesis of penicillins.

(60)   (61)

An important question now was the fate of the two diastereotopic methyl
groups of valine.   Stereospecific syntheses of valine with the (4R)- and
(4S)- groups separately labelled with $^{13}C,^{100-102}$ $^2H,^{102-105}$ or $^3H^{102}$ were
therefore completed.   The samples of stereospecifically [Me-^{13}C]-labelled
valines were used to study the fate of the methyl groups in the biosynthesis
of penicillins and cephalosporin C, using ^{13}C-n.m.r. spectroscropy to locate
the label in the product.   Whilst the analysis of the ^{13}C-n.m.r. spectrum of
cephalosporin C had been relatively straightforward,[106] the differentiation of the
2α- and 2β-methyl groups in the ^{13}C-n.m.r. spectrum of penicillin was achieved
because nuclear Overhauser effects had allowed the ^1H chemical shifts of these
methyl groups to be assigned with certainty.[107]   Off-resonance decoupling and
application of the Ernst equation allowed these shifts to be used to assign the
^{13}C-n.m.r. spectral shifts.[108]   It was found that when the (4R)-methyl group
of valine was labelled then there was a five-fold enhancement of intensity in
the resonance for C_2 of cephalosporin C and a 1.7-fold enhancement in the
intensity of the resonance for the 2β-methyl group of penicillin V.[109]   Corresponding results were obtained in studies using (4S)-^{13}C-labelled valine[101,102]
and the observation of ^{13}C-H coupling in the appropriate methyl signal in the
proton spectrum[101] further corroborated the assignments.   The results of these
experiments indicate that the (4R)-methyl became the ring carbon in cephalosporin C and that the thiazolidine ring of the penicillins was formed with
overall retention of configuration.

Most questions about the biosynthesis which could be answered using isotopically substituted amino acids had now been answered and it became necessary
to consider feeding experiments with larger precursors.   Unfortunately cell
permeability becomes a problem with larger precursors and although isotopically
labelled peptides have been fed to β-lactam antibiotic-producing intact
organisms,[110,111] results are far from being definitive.   Use of cell-free
systems and related techniques are now regarded as the way to investigate the
incorporation of larger precursors and Abraham[84] has fed the tripeptide
L,L,D-ACV (39) labelled with tritium at C_2 of valine to protoplast lysates of

C. acremonium to obtain a penicillin in which the tritium was still present. He also showed[84] that L-[2,3-^{3}H$_2$]-valine was incorporated into L,L,D-ACV (39) with retention of the C$_3$-tritium. These results finally dispelled any notion that an α,β-dehydrovaline intermediate such as the compound (42) (Scheme 10) was involved in the biosynthesis of penicillins. Other questions such as the intermediacy of 6-aminopenicillanic acid [(34), R = H] are being investigated by feeding labelled molecules to cell-free systems.[112]

Having discussed the results of various experiments with isotopically labelled precursors which have been performed, it is relevant to consider what they tell us about the biosynthetic pathway leading to penicillins and to cephalosporin C. Some ideas relating to this are outlined in Scheme 14.

It seems certain that δ-(L-α-aminoadipyl)-L-cysteine (38) reacts with L-valine (37) to yield L,L,D-ACV (39), presumably *via* the intermediacy of L,L,L-ACV. The precursors are incorporated intact in this process except for the loss of the C$_2$-hydrogen of valine. If the next step is cyclisation to the β-lactam (65) then the hydrogen at C$_2$ and the (3-pro-R)-hydrogen of cysteine are retained. Should cyclisation involve a thioaldehyde intermediate (62) then the loss of the (3-pro-S)-hydrogen is contrary to the stereospecificity of alcohol dehydrogenases which involve loss of the (pro-R)-hydrogen.[10] If the process involves hydroxylation followed by nucleophilic attack as in (63) then double retention or double inversion of stereochemistry must be involved. Since biological hydroxylations normally occur with retention of configuration, the process might imply retention of configuration in the substitution step! Formation of an anion adjacent to the sulphur followed by substitution at an oxidised amide nitrogen as in (64) would be consistent with the observed stereospecificity.

Cyclisation of the β-lactam thiol (65) to form cephalosporin C (35) involves attack at the methyl group B in forming the reduced thiazine ring. Retention of configuration in closure of the thiazolidine ring of penicillins presents similar problems to those discussed for the intermediate (63) in the β-lactam formation step if a hydroxylation-substitution sequence is assumed.

7 Concluding Remarks

In this chapter I have attempted to show how radioactive and stable isotopes are used in studying biosynthetic pathways. In Sections 1-5, I have concentrated on the methodology used, and in Section 6 I have concentrated on the experiments which were performed to gain some understanding of aspects of one particular biosynthetic pathway. I hope that I have been able to show some of the planning which has to go into setting up an experiment

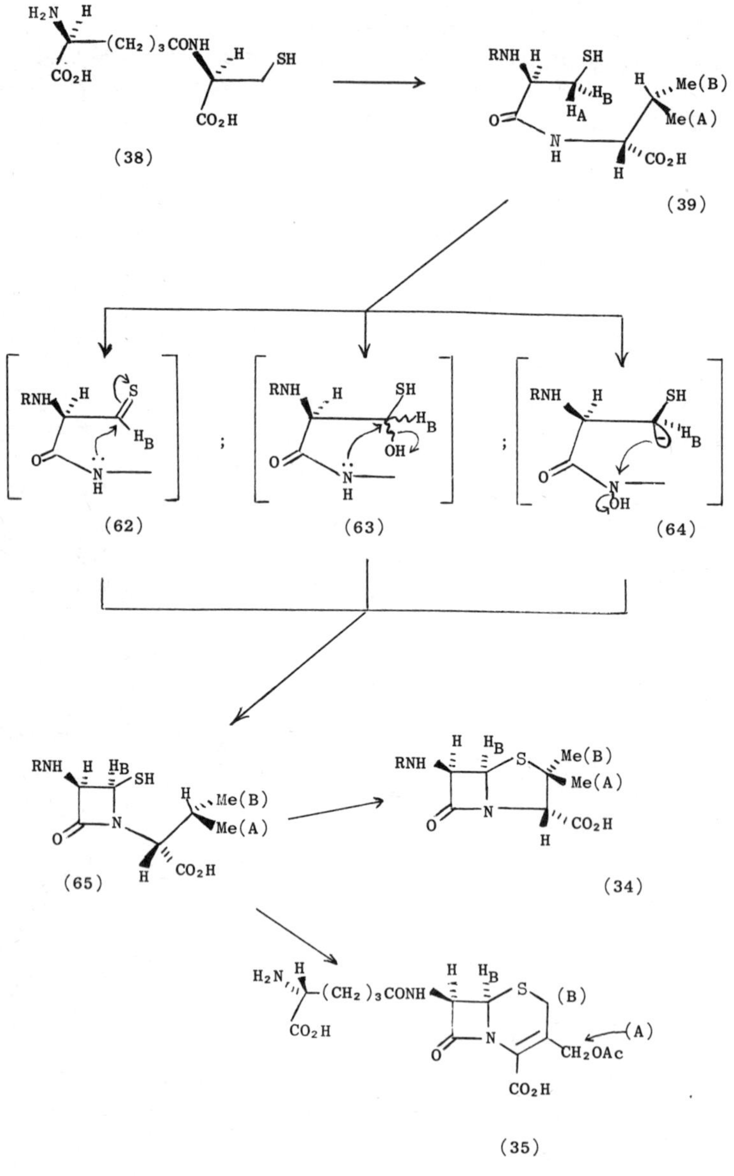

Scheme 14

to explore aspects of a biosynthesis and also some of the rigour which must
be applied if the experiment is to be meaningful.

In concluding I should like to thank Drs. D.J. Morecombe, P.K. Sen,
and S.J. Field for their efforts on the work which I have reported here which
was done in our own laboratory.

References

[1] R. Robinson, 'The Structural Relations of Natural Products',
Clarendon Press, Oxford, 1955.

[2] G. Wolf, 'Isotopes in Biology', Academic Press, New York, 1964.

[3] G. Waller (Ed.), 'Biochemical Applications of Mass Spectrometry',
Wiley, New York, 1972.

[4] E. Broda, 'Radioactive Isotopes in Biochemistry', Elsevier, Amsterdam, 1960.

[5] C.L. Comar, 'Radioisotopes in Biology and Agriculture - Principles and
Practice', McGraw Hill, New York, 1955.

[6] J.R. Catch, 'Carbon-14 Compounds', Butterworths, London, 1961.

[7] C.H. Wang and D.L. Willis, 'Radiotracer Methodology in Biological Science',
Prentice Hall, Englewood Cliffs, N.J., 1965.

[8] E.A. Evans, 'Tritium and its Compounds', Butterworths, London, 2nd. Edn,
1974.

[9] L.E. Feinendegen, 'Tritium Labelled Molecules in Biology and Medicine',
Academic Press, New York, 1967.

[10] D.W. Young, in E. Buncel and C.C. Lee (Eds), 'Isotopes in Organic Chemistry',
Elsevier, Amsterdam, 1978, Vol.4, p.177.

[11] H.G. Floss, in E.A. Evans and M. Muramatsu (Eds), 'Radiotracer Techniques
and Applications', Marcel Dekker, New York, 1977, Vol.2, p.689.

[12] H. Simon and H.G. Floss, 'Bestimmung der Isotopenverteilung in markierten
Verbindungen', Springer Verlag, Berlin, 1967.

[13] 'Biosynthesis', Specialist Periodical Reports, The Chemical Society, London;
T.A. Geissman (Snr Rptr), (a) Vol.1, 1972, (b) Vol.2, 1973, (c) Vol.3, 1975;
J.D. Bu'lock (Snr Rptr), (d) Vol.4, 1976, (e) Vol.5, 1977, (f) Vol.6, 1980.

[14] S.A. Brown in ref.13(a), p.1.

[15] T. Swain in J.B. Pridham and T. Swain (Eds), 'Biosynthetic Pathways in
Higher Plants', Academic Press, London, 1965.

[16] D.J. Austin and M.B. Meyers, *Chem. Comm.*, 1966, 125.

[17] R. Bentley, 'Molecular Asymmetry in Biology', Academic Press, New York, 1969, Vol.1; and 1970, Vol.2.

[18] W.L. Alworth, 'Stereochemistry and its Application in Biochemistry', Wiley-Interscience, New York, 1972.

[19] D.H.R. Barton, *Fourth International Symposium on the Biosynthesis and Physiology of Alkaloids*, Halle, DDR, June 1969.

[20] W.W. Parson and H. Rudney, *J. Biol. Chem.*, 1965, 240, 1855.

[21] H. Simon and G. Heubach, *Z. Naturforsch.*, 1963, 18b, 159.

[22] D.H. Bowen, J. MacMillan, and J.E. Graebe, *Phytochemistry*, 1972, 11, 2253.

[23] H. Knöppel and W. Beyrich, *Tetrahedron Letters*, 1968, 291.

[24] J.A. Elvidge, J.R. Jones, V.M.A. Chambers, and E.A. Evans, in E. Buncel and C.C. Lee (Eds), 'Isotopes in Organic Chemistry', Elsevier, Amsterdam, 1978, Vol.4, p.1.

[25] V.M.A. Chambers, E.A. Evans, J.A. Elvidge, and J.R. Jones, 'Tritium Nuclear Magnetic Resonance Spectroscopy', The Radiochemical Centre, Amersham, 1978.

[26] J.A. Elvidge, D.K. Jaiswal, J.R. Jones, and R. Thomas, *J.C.S. Perkin I*, 1977, 1080.

[27] S. Pinchas and I. Laulicht, 'Infrared Spectra of Labelled Compounds', Academic Press, London, 1971.

[28] T.J. Simpson, *Chem. Soc. Rev.*, 1975, 4, 497.

[29] A.G. McInnes and J.L.C. Wright, *Accounts Chem. Res.*, 1975, 8, 313.

[30] J.W. Cornforth, R.H. Cornforth, A. Pelter, M.G. Horning, and G. Popják, *Tetrahedron*, 1959, 5, 311.

[31] M. Tanabe and G. Detre, *J. Amer. Chem. Soc.*, 1966, 88, 4515.

[32] L. Cattel, J.F. Grove, and D. Shaw, *J.C.S. Perkin I*, 1973, 2626.

[33] M. Tanabe, K.T. Suzuki, and W.C. Janowski, *Tetrahedron Letters*, 1973, 4723.

[34] H. Seto, L.W. Carey, and M. Tanabe, *J.C.S. Chem. Comm.*, 1973, 867.

[35] M. Tanabe and K.T. Suzuki, *J.C.S. Chem. Comm.*, 1974, 445.

[36] D.M. Wilson, A.L. Burlingame, T. Cronholm, and D. Sjövall, *Biochem. Biophys. Res. Comm.*, 1974, 56, 828.

[37] A.R. Battersby, E. Hunt, and E. McDonald, *J.C.S. Chem. Comm.*, 1973, 442.

[38] K. Ajisaka, H. Takeshima, and S. Omura, *J.C.S. Chem. Comm.*, 1976, 571.

[39] C.R. Childs and K. Bloch, *J. Org. Chem.*, 1961, 26, 1630.

[40] L.M. Jackman, I.G. O'Brien, G.B. Cox, and F. Gibson, *Biochim. Biophys. Acta*, 1967, 141, 1.

[41] A.I. Scott and M. Yalpani, *Chem. Comm.*, 1967, 945.

[42] A.R. Battersby, J. Baldas, J. Collins, D.H. Grayson, K.J. James, and E. McDonald, *J.C.S. Chem. Comm.*, 1972, 1265.

⁴³S.J. Field and D.W. Young, unpublished results.

⁴⁴Thanks are due to Dr. George Rybach, Shell Biosciences Laboratories, Sittingbourne, Kent, for performing these O.R.D. experiments.

⁴⁵H. Gerlach and B. Zagalak, *J.C.S. Chem. Comm.*, 1973, 274.

⁴⁶W.L.F. Armarego, B.A. Milloy, and W. Pendergast, *J.C.S. Perkin I*, 1976, 2229.

⁴⁷P.M. Dewick and D. Ward, *J.C.S. Chem. Comm.*, 1977, 338.

⁴⁸D.E. Cane and S.L. Buchwald, *J. Amer. Chem. Soc.*, 1977, 99, 6132.

⁴⁹D.E. Cane and P.P.N. Murthy, *J. Amer. Chem. Soc.*, 1977, 99, 8327.

⁵⁰Y. Sato, T. Oda, and H. Saito, *J.C.S. Chem. Comm.*, 1978, 135.

⁵¹M. Imfield, C.A. Townsend, and D. Arigoni, *J.C.S. Chem. Comm.*, 1976, 541.

⁵²A.R. Battersby, R. Hollenstein, E. McDonald, and D.C. Williams, *J.C.S. Chem. Comm.*, 1976, 543.

⁵³A. Banerji, R. Hunter, G. Mellows, K-Y. Sim, and D.H.R. Barton, *J.C.S. Chem. Comm.*, 1978, 843.

⁵⁴M.J. Garson, R.A. Hill, and J. Staunton, *J.C.S. Chem. Comm.*, 1977, 624.

⁵⁵J.W. Cornforth, J.W. Redmond, H. Eggerer, W. Buckel, and C. Gutschow, *Eur. J. Biochem.*, 1970, 14, 1.

⁵⁶H. Lenz, W. Buckel, P. Wunderwald, G. Biedermann, V. Buschmeier, H. Eggerer, J.W. Cornforth, J.W. Redmond, and R. Mallaby, *Eur. J. Biochem.*, 1971, 24, 207.

⁵⁷J. Luthy, J. Retey, and D. Arigoni, *Nature*, 1969, 221, 1213.

⁵⁸C.A. Townsend, T. Scholl, and D. Arigoni, *J.C.S. Chem. Comm.*, 1975, 921.

⁵⁹L.J. Altman, C.Y. Han, A. Bertolino, G. Handy, D. Laungani, W. Muller, S. Schwartz, D. Shanker, W.H. de Wolf, and F. Yang, *J. Amer. Chem. Soc.*, 1978, 100, 3235.

⁶⁰M. Dole, *Chem. Rev.*, 1952, 51, 263.

⁶¹M. Akhtar, D. Corina, J. Pratt, and T. Smith, *J.C.S. Chem. Comm.*, 1976, 854.

⁶²J.R. Bearder, J. MacMillan, and B.O. Phinney, *J.C.S. Chem. Comm.*, 1976, 834.

⁶³R. Caprioli and D. Rittenberg, *Biochemistry*, 1969, 8, 3375.

⁶⁴G.W.A. Milne, P. Goldman, and J.L. Holtzman, *J. Biol. Chem.*, 1968, 243, 5374.

⁶⁵M. Byrn and M. Calvin, *J. Amer. Chem. Soc.*, 1966, 88, 1916.

⁶⁶T. Axenrod and G.A. Webb, 'Nuclear Mangetic Resonance Spectroscopy of Nuclei other than Protons', Wiley, New York, 1974.

⁶⁷G. Lowe and B.S. Sproat, *J.C.S. Chem. Comm.*, 1978, 565 and 783.

⁶⁸H. Suhr, 'Anwendung der Kernmagnetischen Rezonanz in der Organischen Chemie', Springer Verlag, Berlin, 1965.

⁶⁹W.G. Klemperer, *Angew. Chem. Int. Ed. Engl.*, 1978, 17, 246.

⁷⁰K.S. Yang and G.K. Waller, *Phytochemistry*, 1965, 4, 881.

[71] G.R. Waller, R. Ryhage, and S. Meyerson, *Anal. Biochem.*, 1966, 16, 277.

[72] A.K. Bose, K.G. Das, P.T. Funke, I. Kugajevsky, O.P. Shukla,
K.S. Khanchandani, and R.J. Suhadolnik, *J. Amer. Chem. Soc.*, 1968, 90, 1038.

[73] P.D. Shaw and J.A. McCloskey, *Biochemistry*, 1967, 6, 2247.

[74] H. Booth, B.W. Bycroft, C.M. Wels, K. Corbett, and A.P. Maloney,
J.C.S. Chem. Comm., 1976, 110.

[75] A. Haber, R.D. Johnson, and K.L. Rinehart, *J. Amer. Chem. Soc.*, 1977,
99, 3541.

[76] S.J. Gould and C.C. Chang, *J. Amer. Chem. Soc.*, 1978, 100, 1624.

[77] A.G. McInnes, D.G. Smith, C.-K. Wat, L.C. Vining, and J.L.C. Wright,
J.C.S. Chem. Comm., 1974, 281.

[78] H.R.V. Arnstein and P.T. Grant, *Biochem. J.*, 1954, 57, 353.

[79] C.M. Stevens, P. Vohra, and C.W. De Long, *J. Biol. Chem.*, 1954, 211, 297.

[80] P.W. Trown, B. Smith, and E.P. Abraham, *Biochem. J.*, 1963, 86, 284.

[81] H.R.V. Arnstein and P.T. Grant, *Biochem. J.*, 1954, 57, 360.

[82] C.M. Stevens and C.W. De Long, *J. Biol. Chem.*, 1958, 230, 991.

[83] H.R.V. Arnstein and M.E. Clubb, *Biochem. J.*, 1957, 65, 618.

[84] F.-C. Huang, J.A. Chan, C.J. Sih, P. Fawcett, and E.P.Abraham,
J. Amer. Chem. Soc., 1975, 97, 3858.

[85] H.R.V. Arnstein and D. Morris, *Biochem. J.*, 1960, 76, 357.

[86] P.B. Loder and E.P. Abraham, *Biochem. J.*, 1971, 123, 471.

[87] P.B. Loder and E.P. Abraham, *Biochem. J.*, 1971, 123, 477.

[88] E.P. Abraham, 'Biosynthesis and Enzymatic Hydrolysis of Penicillins and
Cephalosporins', University of Tokyo Press, Tokyo, 1974, p.9.

[89] J.E. Baldwin, S.B. Haber, and J. Kitchin, *J.C.S. Chem. Comm.*,
1973, 790.

[90] H.R.V. Arnstein and J.C. Crawhall, *Biochem. J.*, 1957, 67, 180.

[91] B.W. Bycroft, C.M. Wels, K. Corbett, and D.A. Lowe, *J.C.S. Chem. Comm.*,
1975, 123.

[92] D.J. Morecombe and D.W. Young, *J.C.S. Chem. Comm.*, 1975, 198.

[93] D.W. Young, D.J. Morecombe, and P.K. Sen, *Eur. J. Biochem.*,
1977, 75, 133.

[94] P. Adriaens, H. Vanderhaege, B. Meesschaert, and H. Eyssen,
Antimicrob. Agents and Chemotherapy, 1975, 8, 15.

[95] J.G. Whitney, D.R. Brannon, J.A. Mabe, and J.K. Wicker,
Antimicrob. Agents and Chemotherapy, 1972, 1, 247.

[96] D.J. Aberhart, L.J. Lin, and J.Y.-R. Chu, *J.C.S. Perkin I*,
1975, 2517.

[97] J.A. Huddleston, E.P. Abraham, D.W. Young, D.J. Morecombe, and P.K. Sen,
Biochem. J., 1978, 169, 705.

[98]B.W. Bycroft, C.M. Wels, K. Corbett, A.P. Maloney, and D.A. Lowe, *J.C.S. Chem. Comm.*, 1975, 923.

[99]D.J. Aberhart, J.Y.-R. Chu, N. Neuss, C.H. Nash, J. Occolowitz, L.L. Huckstep, and N. De La Higuera, *J.C.S. Chem. Comm.*, 1974, 564.

[100]J.E.-Baldwin, J. Loliger, W. Rastetter, N. Neuss, L.L. Huckstep, and N. De La Higuera, *J. Amer. Chem. Soc.*, 1973, 95, 3796 and (corr.) 6511.

[101]H. Kluender, C.H. Bradley, C.J. Sih, P. Fawcett, and E.P. Abraham, *J. Amer. Chem. Soc.*, 1973, 95, 6149.

[102]D.J. Aberhart and L.J. Lin, *J.C.S. Perkin I*, 1974, 2320.

[103]R.K. Hill, S. Yan, and S.M. Arfin, *J. Amer. Chem. Soc.*, 1973, 95, 7857.

[104]H. Kluender, F.C. Huang, A. Fritzberg, H. Schnoes, C.J. Sih, P. Fawcett, and E.P. Abraham, *J. Amer. Chem. Soc.*, 1974, 96, 4054.

[105]D.J. Aberhart and L.J. Lin, *J. Amer. Chem. Soc.*, 1973, 95, 7859.

[106]N. Neuss, C.H. Nash, P.A. Lemke, and J.B. Grutzner, *Proc. Roy. Soc.*, 1971, 179B, 335; *J. Amer. Chem. Soc.*, 1971, 93, 2337 and (corr.) 5314.

[107]R.D.G. Cooper, P.V. De Marco, J.C. Cheng, and N.D. Jones, *J. Amer. Chem. Soc.*, 1969, 91, 1408.

[108]R.A. Archer, R.D.G. Cooper, P.V. De Marco, and L.F. Johnson, *J.C.S. Chem. Comm.*, 1970, 1291.

[109]N. Neuss, C.H. Nash, J.E. Baldwin, P.A. Lemke, and J.B. Grutzner, *J. Amer. Chem. Soc.*, 1973, 95, 3797 and (corr) 6511.

[110]P. Adriaens and H. Vanderhaeghe, *FEMS Microbiol. Letters*, 1978, 4, 19.

[111]P. Adriaens, B. Meesschaert, H. Eyssen, and H. Vanderhaeghe, *FEMS Microbiol. Letters*, 1978, 4, 15.

[112]P.A. Fawcett, J.J. Usher, and E.P. Abraham, *Biochem. J.*, 1975, 151, 741.

Isotopes and Organic Reaction Mechanisms

by Frank Hibbert

Department of Chemistry, Birkbeck College, Malet Street,

London, WC1E   7HX

A.   Introduction

In the study of reaction mechanisms in organic chemistry, techniques involving isotopes are particularly important and many research papers describing studies using these techniques are published every year.   In the Journal of the American Chemical Society for 1978 each of the twenty-six issues contains at least one and usually more than one example of the use of isotopes in studies of organic reaction mechanisms;   in the Journal of the Chemical Society (Chemical Communications) there is at least one mechanism study using isotopes in each of the twenty-four issues for 1978.   In the space of a short review it would be impossible to give complete coverage of such a large research area.   However it is possible to illustrate the most important ways in which isotopes are used with a few examples.   Our choice of examples is restricted to reactions in solution.

Isotopes are used in two main ways.   For a reactant containing several atoms of the same element, one atom is made distinct from the others by using a different isotope of that element and the way in which this isotopic label becomes distributed in the products of reaction or redistributed in the reactant may provide information about the sequence of steps between the reactant and product.   It is sometimes also valuable to measure the rate at which the isotope label is redistributed.   An isotope is a relatively minor structural modification and the assumption is therefore made that the mechanism of reaction is unchanged by the isotopic substitution.   Examples of this sort of experiment where the isotopic label may be present in the reactant or solvent and experiments involving the use of more than one label are given in Section B.   An example is also given in this Section of a mechanism study with an optically active reactant which owes its chirality to the presence of three isotopes of oxygen attached to one phosphorus atom.

The other main way in which isotopes are used to provide mechanistic information involves measurement of the effect on the rate of reaction of replacing an atom in the reactant or in the solvent by an isotopically different atom (kinetic isotope

effect) and examples of this approach are described in Section C.
Again it is assumed that the mechanism of reaction is unchanged
by the isotopic substitution even though the rate of reaction may
be changed by an order of magnitude in some cases. Since isotopic
substitution involves an extremely minor perturbation of the
reactant structure this assumption holds in almost all examples.

B. Studies of Mechanism Using Isotope Labels

(i) Benzyne Intermediate. A typical example of the use of
an isotopic label is in the determination of one of the mechanisms
of nucleophilic aromatic substitution, reaction (1). The different

$$\text{(benzene ring with X)} \quad + \quad Y^- \quad \longrightarrow \quad \text{(benzene ring with Y)} \quad + \quad X^- \qquad (1)$$

mechanisms observed for this class of reaction have been
described[1]. An addition-elimination mechanism is most commonly
observed. However for some nucleophilic aromatic substitutions
the effect of groups in the aromatic nucleus in directing the
position of attack of the nucleophile could not be explained by
this mechanism. For these reactions formation of a benzyne intermediate was proposed.[2] The addition-elimination and the mechanism
involving the benzyne intermediate are shown in equations (2) and
(3) respectively for the aminolysis of chlorobenzene with potassium
amide in liquid ammonia. The correct mechanism was identified[2]
by using a reactant in which one carbon atom (*) was labelled with
^{14}C. If reaction occurred by the addition-elimination mechanism
(2), the product aniline would be formed with the ^{14}C label at the
same position in the aromatic nucleus as the amino group. For
mechanism (3) the benzyne intermediate will be attacked almost
equally at each side of the triple bond and therefore 50% of the
^{14}C label will be present in the product at the carbon atom next
to the site of attachment of the amino group.[2] The position of
the label was determined by degradation of the product aniline
and since almost equal amounts of label were found at two carbon
atoms the results were compatible with mechanism (3). Subsequent
work has confirmed the formation of benzyne intermediates in a
number of other reactions.[3] The structure of these species is
best written as shown in equation (3) although the precise nature
of the bonding is still not certain.[4]

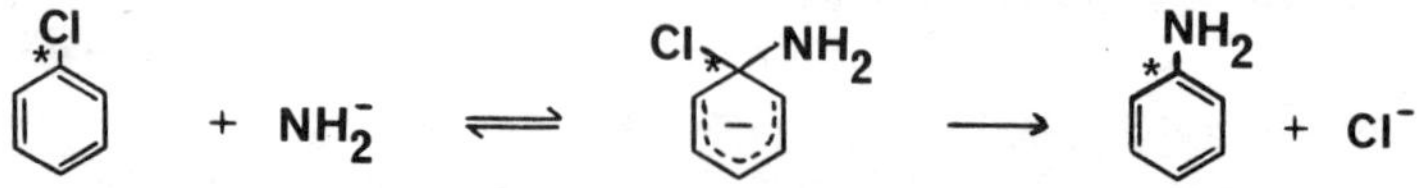

$$\overset{*}{C_6H_4}Cl + NH_2^- \;\rightleftharpoons\; [\text{intermediate}]^- \;\longrightarrow\; \overset{*}{C_6H_4}NH_2 + Cl^- \qquad (2)$$

$$\overset{*}{C_6H_4}Cl \xrightarrow{\;KNH_2\;} Cl^- + \underset{\text{benzyne}}{\overset{*}{C_6H_4}} \;\xrightarrow[\;NH_3\;]{\;NH_3\;}\; \overset{*}{C_6H_4}NH_2 \;/\; \overset{*}{C_6H_4}\!-\!NH_2 \qquad (3)$$

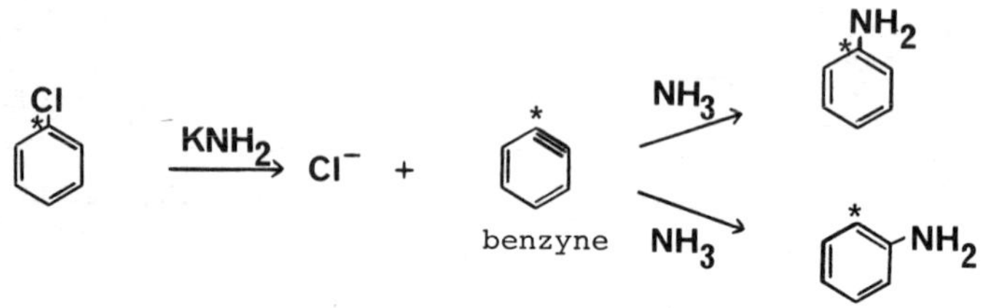

(ii) <u>Hydrolysis of Esters</u>. In the example described in (i), mechanistic information was obtained by determining the location of the isotope label in the products of reaction. In some cases it is useful not only to measure the distribution of the label in the reactants and products but also to measure the rate at which the label becomes distributed and to compare this rate with the overall rate of reaction. This technique is applied widely in studies of the mechanism of hydrolysis of esters and related compounds and was first used by Bender in investigations of the alkaline hydrolysis of ethyl benzoate in aqueous solution.[5,6] The reaction is first-order in hydroxide ion concentration and in ester concentration as shown in equations (4) and (5). The

$$C_6H_5\overset{O}{\overset{\|}{C}}.OC_2H_5 + OH^- \;\longrightarrow\; C_6H_5CO_2^- + C_2H_5OH \qquad (4)$$

$$-d[C_6H_5CO.OC_2H_5]/dt \;=\; k[C_6H_5CO.OC_2H_5][OH^-] \qquad (5)$$

hydrolysis was studied for ethyl benzoate in which the carbonyl oxygen atom was enriched with ^{18}O and it was found[5] that when unreacted ester was isolated from the reaction solution during the hydrolysis some loss of the ^{18}O label had occurred; exchange of the carbonyl oxygen of the ester with oxygen in the solvent (water) was occurring. This result has subsequently been observed in the hydrolysis of many other esters and related compounds[6,7,8]

and is of major significance in identifying the mechanism of
hydrolysis. The experiment can also be carried out by allowing
unlabelled ester to hydrolyse in water enriched with ^{18}O. In this
case when unreacted ester is isolated from the reaction solution
partial enrichment of the carbonyl oxygen with ^{18}O is observed.
This procedure has recently been used in studying the hydrolysis
of O-acetylsalicylaldehyde[9] and in the oxygen exchange of carboxylic acids.[10] Oxygen exchange is not observed in the hydrolysis
of all esters and there are very few examples in which the rate of
exchange is faster than the rate of hydrolysis.[11] An explanation
of the results must account for all these observations.

If the hydrolysis occurs by a single-step nucleophilic
attack on the carbonyl carbon atom as in equation (6), oxygen
exchange cannot be explained by the same mechanism. However a

$$
\begin{array}{c}
\underset{HO^{-}}{\overset{\overset{\displaystyle O}{\parallel}}{R\!-\!C\!-\!OR'}}
\;\longrightarrow\;
\left[\overset{\overset{\displaystyle O}{\parallel}}{\underset{OH}{R\!-\!C\cdots OR'}}\right]^{\ddagger}
\;\longrightarrow\;
\overset{\overset{\displaystyle O}{\parallel}}{R\!-\!C\!-\!OH}\;+\;R'O^{-}
\end{array}
\qquad (6)
$$

transition
state

$$
\overset{\overset{\displaystyle O}{\parallel}}{R\!-\!C\!-\!O^{-}}\;+\;R'OH
$$

mechanism for hydrolysis which will also account for the
occurrence of oxygen exchange and for the observation that the
hydrolysis is first-order in hydroxide ion and ester concentrations
is shown in equation (7) and involves reversible formation of a
tetrahedral addition intermediate in low concentration. The

$$
\overset{\overset{\displaystyle O}{\parallel}}{R\!-\!C\!-\!OR'}\;+\;OH^{-}\;\underset{k_{-1}}{\overset{k_{1}}{\rightleftarrows}}\;\underset{OH}{\overset{\overset{\displaystyle O^{-}}{|}}{R\!-\!C\!-\!OR'}}\;\overset{k_{2}}{\longrightarrow}\;\overset{\overset{\displaystyle O}{\parallel}}{R\!-\!C\!-\!OH}\;+\;R'O^{-}
\qquad (7)
$$

intermediate

$$
\overset{\overset{\displaystyle O}{\parallel}}{R\!-\!C\!-\!O^{-}}\;+\;R'OH
$$

mechanism of oxygen exchange from ^{18}O labelled ester (8) involves
the same intermediate. In mechanism (6) the tetrahedral species
formed by nucleophilic attack at the carbonyl carbon atom is
written as a transition state whereas in (7) and (8) the species
is written as an intermediate. In mechanism (8) oxygen exchange
occurs because the intermediate can collapse back to reactants in

$$
\overset{\overset{\textstyle 18}{\displaystyle O}}{R-\overset{\parallel}{C}-OR'} + OH^- \underset{k_{-1}}{\overset{k_1}{\rightleftharpoons}} \quad R-\overset{\overset{\textstyle 18}{\displaystyle O^-}}{\underset{\underset{\textstyle OH}{|}}{\overset{|}{C}}}-OR' \overset{k_2}{\longrightarrow} R-\overset{\overset{\textstyle O}{\parallel}}{C}-O^- + R'OH
$$

$$\Updownarrow$$ (8)

$$
R-\overset{\overset{\textstyle O}{\parallel}}{C}-OR' + \,^{18}OH^- \underset{k_{-1}}{\overset{k_1}{\rightleftharpoons}} \quad R-\overset{\overset{\textstyle 18OH}{|}}{\underset{\underset{\textstyle O^-}{|}}{\overset{|}{C}}}-OR' \overset{k_2}{\longrightarrow} R-\overset{\overset{\textstyle O}{\parallel}}{C}-O^- + R'OH
$$

two ways, one of which leads to ester with loss of the ^{18}O label.
By applying the steady-state approximation to mechanisms (7) and
(8), expressions (9) and (10) are obtained for the rate coefficients
for hydrolysis (k_h) and exchange (k_e). In deriving these
expressions it is necessary to assume that breakage of the $C-^{18}O$

$$k_h = k_1 k_2/(k_{-1} + k_2) \tag{9}$$

$$k_e = k_1 k_{-1}/2(k_{-1} + k_2) \tag{10}$$

and $C-^{16}O$ bonds in the intermediate occurs at similar rates. The
difference in rates will be a few percent which is small in this
context (see Section C). The factor which determines whether
isotope exchange will be observed in the hydrolysis of a particular ester is the value of the ratio of rate coefficients for
breakdown of the intermediate to reactants and to products
(k_{-1}/k_2), which is determined by the quality of the leaving group
in the ester. If the value of the ratio k_{-1}/k_2 is small, no
exchange will be observed.

Evidence confirming these conclusions was provided by
investigating oxygen exchange[12] in the hydrolysis of ethyl
trifluorothiolacetate.[13] For this ester in aqueous solution
the rate-pH profile shown in Figure 1 was obtained and the

mechanism in equations (11) and (12) is therefore identified.[13]

$$CF_3\text{-}\overset{\overset{\displaystyle O}{\|}}{C}\text{-}SC_2H_5 \;+\; 2H_2O \;\underset{k_{-1}}{\overset{k_1}{\rightleftharpoons}}\; CF_3\text{-}\underset{\underset{\displaystyle OH}{|}}{\overset{\overset{\displaystyle O^-}{|}}{C}}\text{-}SC_2H_5 \;+\; H_3O^+ \qquad (11)$$

$$CF_3\text{-}\underset{\underset{\displaystyle OH}{|}}{\overset{\overset{\displaystyle O^-}{|}}{C}}\text{-}SC_2H_5 \;\xrightarrow{\;k_2\;}\; CF_3\text{-}\overset{\overset{\displaystyle O}{\|}}{C}O^- \;+\; C_2H_5SH \qquad (12)$$

$$-d[CF_3COSC_2H_5]/dt = k_1k_2[CF_3COSC_2H_5][H_2O]^2/(k_2 + k_{-1}[H_3O^+]) \quad (13)$$

$$k_{obs} = k_1k_2[H_2O]^2/(k_2 + k_{-1}[H_3O^+])$$

If $k_{-1}[H_3O^+]> k_2$, $\;k_{obs} = (k_1k_2/k_{-1})[H_2O]^2/[H_3O^+]$ \qquad (14)

If $k_{-1}[H_3O^+]< k_2$, $\;k_{obs} = k_1[H_2O]^2$ \qquad (15)

Figure 1.   Rate-pH Profile for Hydrolysis of Ethyl Trifluorothiolacetate

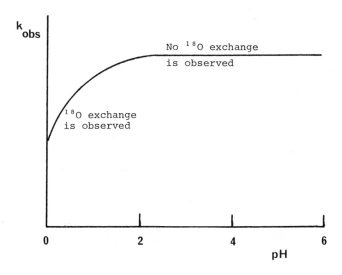

Below pH ca. 2 expression (14) applies and this corresponds to rate-determining breakdown of the tetrahedral addition inter-mediate to products. Above pH ca. 2 in the plateau region the mechanism involves rate-determining nucleophilic attack by water at the carbonyl carbon atom, equation (15). By comparison of (14) and (15) a value for the ratio k_2/k_{-1} was obtained for the hydrolysis reaction.

The rate of oxygen exchange from ethyl trifluorothiol-acetate with ^{18}O label in the carbonyl group was also measured.[12] The rate coefficient (k_e) for exchange by mechanism (16) is given by equation (17). Mechanism (16) involves the same intermediate as occurs in the mechanism of hydrolysis.

$$
\left.
\begin{aligned}
&\overset{\overset{\displaystyle ^{18}O}{\|}}{CF_3C}SC_2H_5 + 2H_2O \underset{k_{-1}}{\overset{k_1}{\rightleftharpoons}} \overset{\overset{\displaystyle ^{18}O^-}{|}}{\underset{\underset{\displaystyle OH}{|}}{CF_3C}}SC_2H_5 + H_3O^+ \\
&\qquad\qquad\qquad \Updownarrow \\
&\overset{\overset{\displaystyle O}{\|}}{CF_3C}SC_2H_5 + 2H_2^{18}O \underset{k_{-1}}{\overset{k_1}{\rightleftharpoons}} \overset{\overset{\displaystyle ^{18}OH}{|}}{\underset{\underset{\displaystyle O^-}{|}}{CF_3C}}SC_2H_5 + H_3O^+
\end{aligned}
\right\} \xrightarrow{k_2} \text{Products} \qquad (16)
$$

$$k_e = k_1 k_{-1}[H_3O^+][H_2O]/2(k_{-1}[H_3O^+] + k_2) \qquad (17)$$

Using equation (17) it is possible to derive values for k_2/k_{-1} from the oxygen exchange measurements. The agreement between these results and those obtained from the hydrolysis reaction over the range pH ca. 1-5 was excellent.[12] At low pH considerable oxygen exchange was observed during hydrolysis but at pH ca. 5 oxygen exchange is not observed. At low pH breakdown of the intermediate to products is slow compared with the reverse breakdown to reactants and oxygen exchange competes effectively with hydrolysis. At high pH breakdown of the intermediate to products occurs more rapidly than breakdown to reactants and oxygen exchange does not compete with the hydrolysis. Hence the observations are explained by assuming a common intermediate for hydrolysis and oxygen exchange.

(iii) Rearrangement of Carbocations. A large number of reactions in organic chemistry occur through the intermediate formation of carbocations.[14,15] In some cases in order to explain the nature of the products of reaction it is necessary to postulate that the intermediate carbocations can undergo rearrangement. For example in the dehydration of pinacolyl alcohol, reaction (18), the product of reaction suggests that migration of a methyl group occurs in the intermediate carbocation. This 1,2-methyl shift is known as a Wagner-Meerwein rearrangement and is a common reaction of carbocations.[16] Since the development by Olah[17] of super acids

$$\underset{\substack{Me \\ Me}}{Me} C - \underset{\substack{Me}}{C} \underset{H}{\overset{OH}{|}} \xrightarrow{H_2SO_4} \underset{Me}{\overset{Me}{\diagdown}} C = C \underset{Me}{\overset{Me}{\diagup}} + H_2O \qquad (18)$$

such as SbF_5-HSO_3F and SbF_5-SO_2 it has been possible to prepare and study the reactions of simple carbocations[18] of the type that have been postulated as intermediates in other reactions. For example isopropyl cation has been prepared[19,20] in SbF_5-HSO_3F. Isotopic labelling techniques have been particularly useful in studying the reactions of carbocations and this will be illustrated by reference to the isopropyl cation.

The nmr spectrum of the isopropyl cation in SbF_5-SO_2ClF shows that the two types of proton which are present are rapidly exchanged by an intramolecular process.[20] This could be explained by a series of 1,2-hydride shifts through the n-propyl cation, equation (19). However, when isopropyl cation labelled at the

$$\underset{CH_3}{\overset{H}{\underset{|}{C}}} \overset{+}{\underset{CH_2}{\diagdown}} H \rightleftharpoons \underset{CH_3}{\overset{H}{\underset{+}{C}}} \overset{H}{\underset{CH_2}{}} \rightleftharpoons \underset{CH_3}{\overset{H}{\underset{+}{C}}} \underset{CH_2}{\diagdown} H \qquad (19)$$

2-carbon position with ^{13}C was generated from labelled 2-chloro-propane in SbF_5-SO_2ClF, it was found that after several hours the ^{13}C label was distributed equally between the three carbon atoms.[21] This process was accounted for by scheme (20) involving protonated cyclopropane as an intermediate or transition state. In further experiments[22] the relative rates of deuterium and carbon scrambling were measured by simultaneously observing

$$\underset{\text{CH}_3\quad\text{CH}_3}{{}^{13}\overset{+}{\text{C}}\text{H}} \longrightarrow \underset{\text{CH}_3\quad\text{CH}_2}{{}^{13}\text{CH}_2{}^{+}} \longrightarrow \underset{+\text{CH}_3-\text{CH}_2}{{}^{13}\text{CH}_2} \quad\text{(20)}$$

protonated
cyclopropane

$$\underset{\text{CH}_3-\overset{+}{\text{C}}\text{H}}{{}^{13}\text{CH}_3} \longleftarrow \underset{\text{CH}_3-\text{CH}_2}{{}^{13}\overset{+}{\text{C}}\text{H}_2}$$

$$\text{CD}_3\overset{+}{\text{C}}\text{HCH}_3 \xrightarrow{\text{SbF}_5-\text{SO}_2\text{ClF}} \text{deuterium scrambling} \quad\text{(21)}$$

$$\text{CH}_3{}^{13}\overset{+}{\text{C}}\text{HCH}_3 \xrightarrow{\text{SbF}_5-\text{SO}_2\text{ClF}} \text{carbon scrambling} \quad\text{(22)}$$

reactions (21) and (22) in the same solution and it was found
that the results could be explained by a combination of schemes
(19) and (20). The possibility that the protonated cyclopropane
shown in mechanism (20) could undergo proton exchange by a corner-
to-corner proton migration through an edge-protonated species (24)
was also considered, but the evidence was thought to be more in

$$\underset{+\text{CH}_3-\text{CH}_2}{\text{CH}_2} \qquad\qquad \underset{\text{CH}_2-\text{CH}_2}{\text{H}\cdots\text{CH}_2}$$

(23)   (24)

favour of mechanisms (19) and (20). The existence of protonated
cyclopropanes in these reactions is well established. However it
is not yet firmly established whether the protonated cyclopropane
takes part as a transition state or as an intermediate and whether
the species is edge-protonated or corner-protonated.[22] The
protonated cyclopropanes as written in (23) and (24) involve
pentacovalent carbon atoms and are therefore of considerable
theoretical interest.

An edge-protonated cyclopropane was proposed to account
for the products of reaction of cyclopropane in deuteriosulphuric
acid.[23] The reaction of cyclopropane with sulphuric acid gives
1-propyl hydrogen sulphate as the initial product (25) and one
mechanistic possibility for this reaction consists of a single-

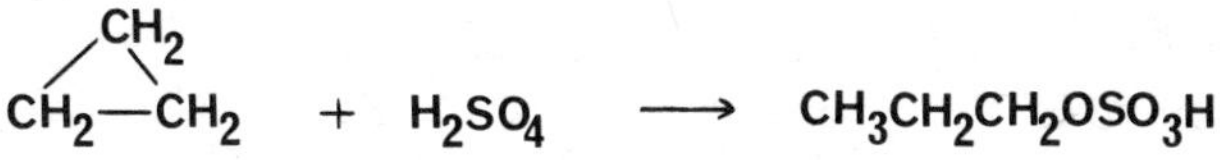

CH2
CH2—CH2 + H2SO4 ⟶ CH3CH2CH2OSO3H

(25)

step concerted attack by acid with simultaneous ring opening and 1,3-addition. In this case the product of reaction in deuterio-sulphuric acid would be as shown in equation (26). In practice

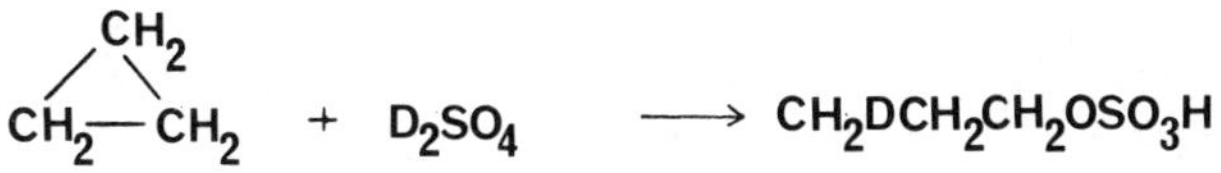

CH2
CH2—CH2 + D2SO4 ⟶ CH2DCH2CH2OSO3H

(26)

it was found[23] that the product from reaction in deuteriosulphuric acid contained deuterium on each of the three carbon atoms. Similar results were obtained by allowing cyclopropane to react in sulphuric acid containing tracer levels of tritium.[24] The results of the deuterium labelling experiments were explained by scheme (27) in which it is the edge-protonated cyclopropanes which lead to the observed products.[23]

CH2DCH2CH2OSO3H
CH2DCH2CH2OSO3H
CH3CH2CHDOSO3H
CH3CHDCH2OSO3H

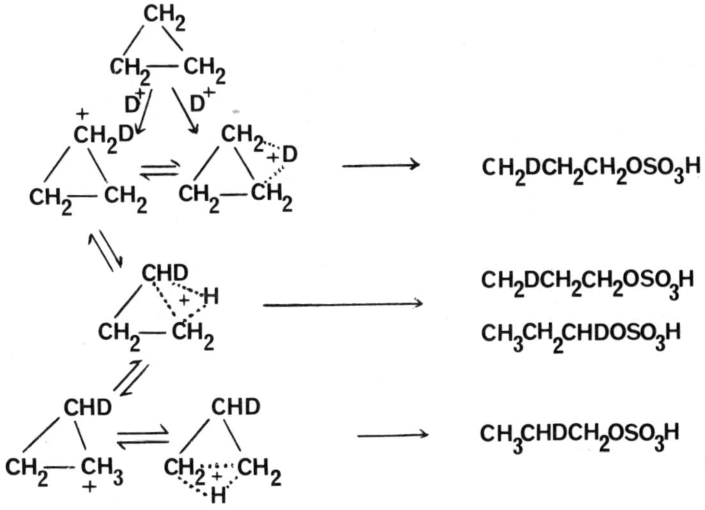

(27)

Although further experiments are necessary to identify the precise
nature of the intermediates in these reactions it is significant
that isotopic labelling experiments have provided information about
these unstable species.[22] The processes in (21) and (22) involve
no overall chemical reaction since only exchanges of protons or
carbon atoms are involved. Isotopic labelling is of great value
in probing this sort of process and further examples are described
in Section B(v).

(iv)   Enzyme Catalysed Phosphoryl Transfer. An interesting
example of the application of isotopic labelling to the mechanism
of an enzyme catalysed reaction is provided by work carried out
by the research groups of Lowe at Oxford and Knowles at Harvard.
The reactions which were studied are shown in equations (28)-(30).
The enzymes belong to the group known as kinases which catalyse
phosphoryl transfer from adenosine triphosphate (ATP). It has
been argued[26] that the mechanism of the reaction catalysed by

$$\text{ATP + pyruvate} \xrightleftharpoons[\text{kinase}]{\text{pyruvate}} \text{ADP + phosphoenolpyruvate} \qquad (28)$$

$$\text{ATP + D-glucose} \xrightleftharpoons{\text{hexokinase}} \text{ADP + glucose-6-phosphate} \qquad (29)$$

$$\text{ATP + glycerol} \xrightarrow[\text{kinase}]{\text{glycerol}} \text{ADP + sn-glycerol-3-phosphate}^{25} \qquad (30)$$

pyruvate kinase is likely to be one of the four processes shown in
equations (32)-(35) in which $AOPO_3{}^{2-}$ refers to adenosine
triphosphate and XO^- represents pyruvate. The mechanisms show
phosphoryl transfer between AO^- and XO^- within the active site of
the enzyme after formation of the enzyme-substrate complex.
Mechanism (32) consists of a single-step transfer of the phosphoryl
group through a pentacovalent phosphorus species. In mechanism
(33) the transfer occurs by at least two steps with intermediate
formation of metaphosphate ion ($PO_3{}^-$). It is assumed that this
highly reactive species is tightly bound by the enzyme although
the binding is not specified. Mechanism (34) illustrates adjacent
attack on phosphorus. Studies of the hydrolysis of esters in
aqueous solution[27] have shown that the pentacovalent species so
formed will undergo a pseudorotation to a second intermediate from
which the phosphoryl group is transferred to XO^-. The fourth
possibility (35) allows for phosphoryl transfer from ATP to a
group on the enzyme before transfer to the substrate occurs.
Evidence in favour of one of the possible mechanisms has been

$$AOPO_3{}^{2-} + XO^- \underset{kinase}{\overset{pyruvate}{\rightleftharpoons}} AO^- + XOPO_3{}^{2-} \tag{31}$$

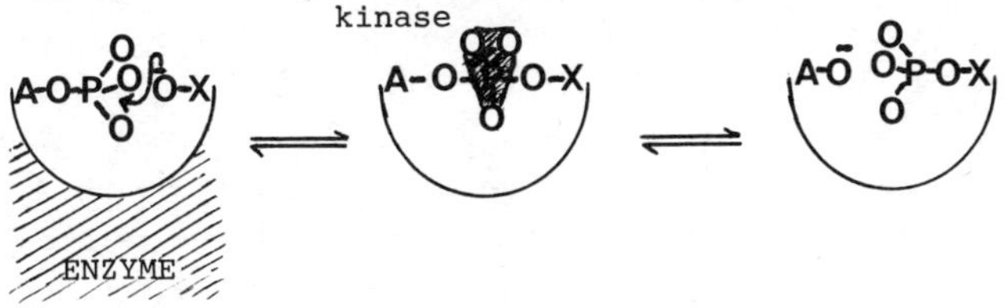

(32)

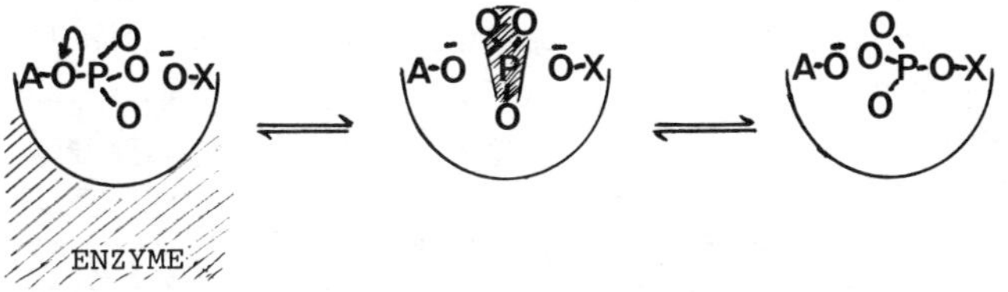

(33)

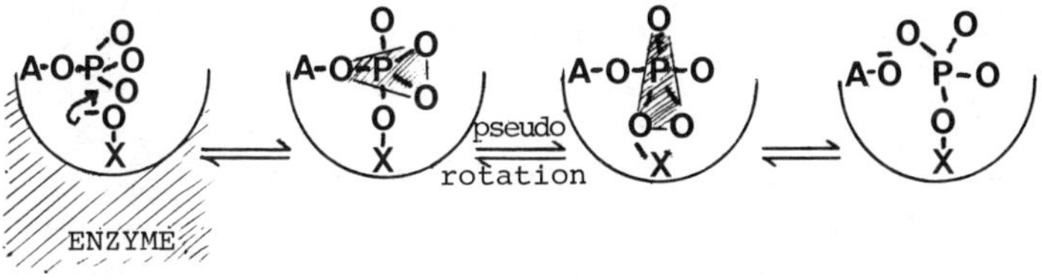

pseudo
rotation

(34)

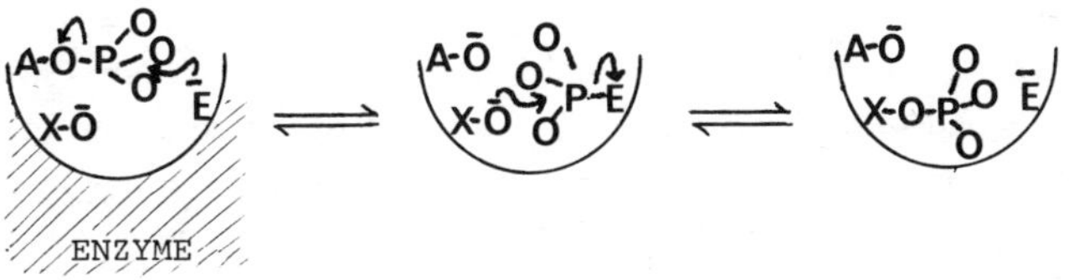

(35)

obtained from two different isotopic labelling experiments. In the first experiment[26] adenosine triphosphate was prepared

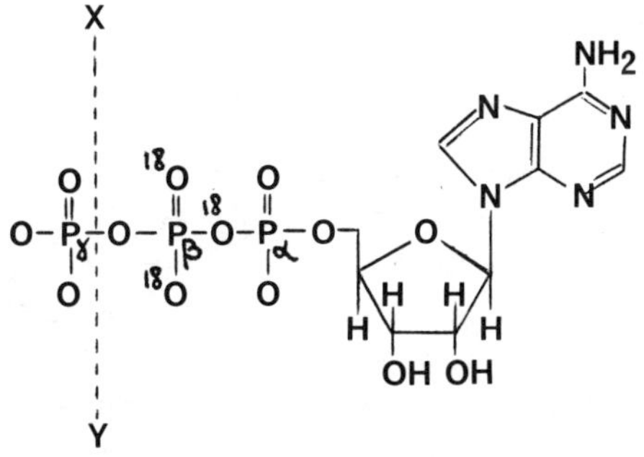

(36)

containing ^{18}O labels as shown in (36). Labelled ATP was then
incubated with pyruvate kinase in the absence of pyruvate. Even
in the absence of pyruvate, mechanisms (33) and (35) predict that
ATP will undergo bond cleavage at X-Y, see (36). For mechanism
(33) the metaphosphate ion will be generated and for (35) the
enzyme will be phosphorylated. In both cases the three terminal O
atoms bound to P_β in the product ADP will become equivalent by
rotation about the P_β-O-P_α bond and therefore recombination to
form ATP will introduce ^{18}O label into the P_γ-O-P_β bridge in the
reformed ATP. Mechanisms (32) and (34) would not lead to redistribution of the label in ATP in the absence of pyruvate. It was
observed that when labelled ATP was incubated with pyruvate kinase
in the presence and in the absence of pyruvate the ATP which was
isolated from solution possessed ^{18}O label in the P_γ-O-P_β bridge.[26]
This was established by nmr measurements on the isolated ATP[26,28]
and the result provides evidence in favour of mechanisms (33) and
(35). It was also observed that ^{18}O scrambling occurred when
pyruvate was replaced by the inhibitor oxalate. In identifying
mechanisms (33) and (35) in this way it is necessary to assume
that the interaction between ATP and the enzyme is not changed
by leaving out pyruvate.

          The final choice of a single mechanism (either (33)
or (35)) can be made on the basis of a further experiment using
isotopes. It is predicted that inversion of configuration at
phosphorus occurs in going from reactant to product by mechanisms
(32) and (33) whereas for (34) and (35) the configuration at
phosphorus in ATP is retained in the product phosphoenolpyruvate.
With this experiment in mind Lowe has prepared methyl (R)-phosphate
(37) which owes its chirality to the presence of the three

$$\left[MeO - P \overset{\overset{16}{O}}{\underset{\underset{18}{O}}{\overset{17}{O}}} \right]^{2-} \qquad (37)$$

different isotopes of oxygen.[29] The configuration was established
by circular dichroism. There are previous examples in the literature of the use of three isotopes to generate a chiral molecule
(for example hydrogen, deuterium, tritium and one other group
attached to a single carbon atom). However the phosphate ester
(37) is the first example in which three oxygen isotopes have been
used and represents a subtle use of isotopes in labelling

the configuration of a molecule. Following the preparation of (37)
it is necessary to use the same route to chirally labelled ATP.
These experiments have not yet been completed by Lowe's group but
chiral $[\gamma(S)-^{16}O,^{17}O,^{18}O]$ATP has been prepared by Knowles[30] using
the same synthesis as was originally used for the preparation of
$[1(R)-^{16}O,^{17}O,^{18}O]$phospho-(S)-propane-1,2-diol.[31] The configuration
at phosphorus was determined by chemical conversion to a cyclic
ester which could be analysed by mass spectrometry. Labelled ATP
was used by Knowles to investigate the stereochemical course of
the reaction catalysed by glycerol kinase (30). However, the
results apply to the reactions catalysed by pyruvate kinase and
hexokinase as these have been shown to occur with similar stereochemistry.[32] When the reaction of chiral $[\gamma(S)-^{16}O,^{17}O,^{18}O]$ATP
was studied it was found that the product sn-glycerol-3-phosphate[25]
labelled with ^{16}O, ^{17}O, and ^{18}O possessed the opposite configuration at phosphorus to that of the original ATP. This is explained
by mechanisms (32) and (33) and therefore taken together with the
isotope distribution experiment identifies mechanism (33) as the
course of phosphoryl transfer catalysed by pyruvate kinase. When
considered in the light of X-ray crystallography studies[33] these
conclusions will provide a very detailed picture of the mechanism
of action of this enzyme.

    (v)   <u>Acid-Base Catalysed Proton Exchange in Carbon-Hydrogen</u>
<u>Bonds</u>. One of the most frequent uses of isotopes in organic
chemistry is in the study of the acid and base catalysed exchange
of protons attached to carbon. The information which these studies
provide is outlined in this section. The general topic of hydrogen
isotope exchange is dealt with more fully in a separate chapter of
this volume.

        For weak carbon acids which dissociate to a negligible
extent in aqueous solution rates of dissociation of the C-H bonds
cannot be measured directly. In these cases rates of dissociation
can be measured by following base catalysed isotope exchange. For
example the base catalysed loss of tritium from chloroform labelled
with tracer levels of tritium occurs by the mechanism[34] shown in
equations (38) and (39). Since the reaction is carried out in

$$C^3HCl_3 + OH^- \underset{k_{-1}}{\overset{k_1}{\rightleftharpoons}} {}^-CCl_3 + {}^3HOH \qquad slow \qquad (38)$$

$$ {}^-CCl_3 + H_2O \xrightarrow{k_2} CHCl_3 + OH^- \qquad fast \qquad (39)$$

aqueous solution[35] or in dimethyl sulphoxide-water mixtures with
low concentrations of hydroxide ion[34] the chloroform is negligibly

dissociated and the mechanism can be treated by the steady state approximation. The tritium is present at tracer levels and it is therefore assumed that the reverse step of reaction (38) is much slower than step (39). Under these conditions the rate expression is given by equation (40). The loss of tritium from the labelled

$$-d[C^3HCl_3]/dt \quad = \quad k_1[C^3HCl_3][OH^-] \quad\quad\quad (40)$$

chloroform was followed by extracting the chloroform and measuring the level of activity with a liquid scintillation counter.[34] The rate coefficient (k_1) obtained in this way is the rate coefficient for ionisation of the carbon acid. This conclusion applies to the base catalysed exchange of most carbon acids in aqueous solution and therefore isotope exchange is widely used as a method of measuring the rates of ionisation of weak carbon acids for which ionisation cannot be studied directly. The procedure has been applied for example to the ionisation of acetylenes,[36] olefins,[37] heterocyclic compounds,[38] and to hydrocarbons activated with cyano,[39] sulphone,[40] alkoxy,[41] thioalkoxy,[41] halogen,[34,35,42] and phenyl groups.[43] Comparisons of the activating effect of the various groups have been made.[44] For very weak acids more severe conditions are needed to bring about exchange; for example the exchange of toluene (pK ca. 41) has been studied in cyclohexyl-amine containing lithium cyclohexylamide.[45]

In some cases, for example in solvents of low dielectric constant where ion pair formation is important, the assumption in mechanism (38) and (39) that $k_2 > k_{-1}$ cannot be made. If collapse of the intermediate back to reactants cannot be neglected rate expression (41) applies and the rate of exchange does not correspond to the rate of ionisation of the carbon acid. The reaction is then said to involve internal return and this is most

$$RC^3H + B \underset{k_{-1}}{\overset{k_1}{\rightleftarrows}} RC^- + B^3H \xrightarrow{k_2} RCH$$

$$-d[RC^3H]/dt \quad = \quad k_1k_2/(k_{-1} + k_2) \ [RC^3H][B] \quad\quad (41)$$

likely to be found for weak carbon acids where non-aqueous solvents containing bases stronger than hydroxide ion are used to bring about isotope exchange. Methods are available for estimating the importance of internal return.[46] Information about internal return has been obtained by Cram and his group[44a,47] by comparing the rate of racemisation of an optically active carbon acid with the rate of exchange of deuterium attached to the asymmetric carbon atom, equations (42) and (43). Both processes involve

$$\underset{R_3}{\overset{R_1}{\underset{\diagup}{R_2 \!-\!\! C \!-\! D}}} \quad + \quad B \quad \xrightarrow{\ k_\alpha\ } \quad \text{racemisation} \tag{42}$$

$$\underset{R_3}{\overset{R_1}{\underset{\diagup}{R_2 \!-\!\! C \!-\! D}}} \quad + \quad B \quad \xrightarrow[\text{solvent}]{\ k_e\ } \quad \underset{R_3}{\overset{R_1}{\underset{\diagup}{R_2 \!-\!\! C \!-\! H}}} \tag{43}$$

ionisation of the carbon acid to give a carbanion or carbanion ion pair and details of the mechanism of racemisation and isotope exchange and of the stereochemistry and properties of the intermediate carbanions can be obtained from these studies. The values of k_e/k_α vary widely depending upon the carbon acid, base catalyst, and solvent.

Measurement of the rates of base catalysed isotope exchange can be a very useful procedure for obtaining estimates of the acidity of protons attached to carbon and this procedure is used widely for very weak carbon acids which dissociate to a negligible extent even under the most severely basic conditions. For isotope exchange to occur it is only necessary for the carbanion to be formed in steady-state concentrations and this occurs under mildly basic conditions. The procedure can also be used for estimating the pK values of acids which are unstable under the conditions necessary to bring about measurable ionisation but which are stable under the relatively mild isotope exchange conditions. In one approach it is assumed that the rate of exchange is related to the acidity of the carbon acid so that relative acidities for two acids are estimated by comparing the rates of exchange under the same conditions. The stronger acid is assumed to undergo isotope exchange more rapidly. In a more quantitative approach, the Brönsted relation,[44c] equation (44), is used to correlate the rates of exchange (k_e) with acidity. In

$$k_e = G(K_{RCH})^\alpha \tag{44}$$

equation (44) K_{RCH} is the acid dissociation constant of the carbon acid and it is usually found that G and α are constants for a series of closely related acids. Equation (44) predicts that the

rate of exchange increases as the acidity of the carbon acid
increases. For example the rates of detritiation (k^T) of the
hydrocarbons shown in Table 1 were measured in cyclohexylamine
containing lithium cyclohexylamide and correlated with pK values
determined under similar conditions.[45] A Brönsted plot of log k^T
against pK was linear with slope α = 0.31. This plot was extrapolated to toluene for which the rate of exchange could be
measured but for which a pK value could not be obtained. The
estimated pK value for toluene is shown in the Table. The data in

Table 1.   Rates of Isotope Exchange and Acidity of Hydrocarbons

	Relative rate of tritium exchange	pK
p-Ph.C_6H_4.CH(Ph)$_2$	2.3	30.2
(p-Ph.C_6H_4)$_2$CH$_2$	1.39	30.83
Ph$_3$CH	1.0	31.45
Ph$_2$CH$_2$	0.28	33.38
p-CH$_3$.C_6H_4.Ph	0.0048	38.73
m-CH$_3$.C_6H_4.Ph	0.0021	(39.95)[a]
PhCH$_3$	0.00104	(40.91)[a]

[a]Extrapolated from Brönsted plot.

the Table illustrate how best to estimate acidity from rates of
exchange. Unless very similar acids are compared the assumptions
involved in this procedure are not valid. In some cases the rates
of exchange and pK values are not well correlated by the Brönsted
relation and then the estimated pK values for an acid of unknown
acidity are very uncertain.[48]

Acid catalysed isotope exchange is observed for aromatic
compounds and the mechanism is shown in equations (45) and (46)
for trimethoxybenzene. In this case[49,50] the loss of tritium from
initially labelled trimethoxybenzene was followed by observing the
loss of activity in the substrate, or increase in activity of the
water. However, uptake of tritium into the aromatic compound from
initially labelled solvent could have been measured. The activity
was measured by liquid scintillation counting and the technique
is sufficiently sensitive that very slow reactions were followed
accurately by measuring only the first few per cent of reaction.
For example the slowest exchange reactions observed for trimethoxybenzene occurred with t½ ca. 30 years.[50] Since the trimethoxybenzenonium ion is present in low concentrations the steady-state
approximation can be applied to mechanism (45) and (46) and
equation (47) is obtained. The ratio k_{-1}/k_2 comprises a primary

$$\text{MeO} \underset{\text{OMe}}{\overset{\text{T}}{\bigcirc}} \text{OMe} \ + \ H_3O^+ \ \underset{k_{-1}}{\overset{k_1}{\rightleftharpoons}} \ \text{MeO} \underset{\text{OMe}}{\overset{\text{T} \quad \text{H}}{\bigoplus}} \text{OMe} \ + \ H_2O \qquad (45)$$

$$\text{MeO} \underset{\text{OMe}}{\overset{\text{T} \quad \text{H}}{\bigoplus}} \text{OMe} \ + \ H_2O \ \xrightarrow{k_2} \ \text{MeO} \underset{\text{OMe}}{\overset{\text{H}}{\bigcirc}} \text{OMe} \ + \ TOH_2^+ \qquad (46)$$

and secondary isotope effect and a value can be estimated

$$-d[\text{TMB-T}]/dt = (k_1/(1+k_{-1}/k_2))[\text{TMB-T}][H_3O^+] \qquad (47)$$

TMB = tritiated trimethoxybenzene

(see Section C). Therefore the rate coefficient for protonation of the aromatic carbon can be calculated from the rate coefficient for exchange. From the pK value of the trimethoxybenzenonium ion and the value of k_1, a value for the rate coefficient for dissociation of the trimethoxybenzenonium ion can be determined. Rates of proton transfer to and from benzene have been measured using this technique.[49] This illustrates the usefulness of isotope exchange methods in measuring rates of proton transfer to and from carbon over a very wide range of reactivity. Rate coefficients for proton transfer from toluene[45] (pK ca. 41) have been calculated from base catalysed exchange and rates of proton transfer from the benzenonium ion (pK ca. -23) have been calculated from acid catalysed exchange.[49] In many cases these reactions could not be followed in any other way. For schemes (38)-(39) and (45)-(46) in the absence of the isotopic label no overall chemical reaction occurs.

C. Kinetic Isotope Effects

(i) Introduction. The magnitude of the ratio of rates of reaction of isotopically different reactants (kinetic isotope effect) is determined by, among other things, the mechanism of the reaction. This observation forms the basis for using kinetic isotope effects to reach conclusions about reaction mechanisms. Possible mechanisms which can be envisaged for the reaction may

lead to different predictions of the size of the kinetic isotope
effect and comparison with the experimental value will then lead to
the elimination of some of the possible mechanisms. It is useful
to consider three sorts of isotope effect.

(1) A primary isotope effect occurs when bonds to the isotopically
substituted atom are made or broken in the rate-determining step
of the reaction. Primary isotope effects are often quite large.
For example in the decomposition of diphenyldiazomethane catalysed
by benzoic acid in dibutyl ether a rate difference k^H/k^D = 4.5 was
observed[51] for catalysis by $C_6H_5CO_2H$ and $C_6H_5CO_2D$. The results
were explained by the mechanism in schemes (48) and (49).

$$C_6H_5CO_2H + Ph_2CN_2 \xrightarrow{\ \ k^H\ \ } Ph_2\overset{+}{C}H.N_2 + C_6H_5CO_2^-\quad slow$$
$$Ph_2\overset{+}{C}HN_2 + C_6H_5CO_2^- \longrightarrow Ph_2CH.O.CO.C_6H_5 + N_2\quad fast$$

(48)

$$C_6H_5CO_2D + Ph_2CN_2 \xrightarrow{\ \ k^D\ \ } Ph_2\overset{+}{C}D.N_2 + C_6H_5CO_2^-\quad slow$$
$$Ph_2\overset{+}{C}HN_2 + C_6H_5CO_2^- \longrightarrow Ph_2CD.O.CO.C_6H_5 + N_2\quad fast$$

(49)

The reaction between p-methyl-N,N-dimethylaniline and benzyl
benzenesulphonate in acetone gave a kinetic isotope effect
k^{12}/k^{14} = 1.162 ± 0.008 and as we shall see later this represents
quite a large primary effect for carbon isotopes. The single-step
reactions in equation (50) were used to explain this result.[52]

(50)

(2)   A secondary isotope effect refers to the effect on the rate
of reaction of isotopic substitution of an atom which does not
take part directly in bond formation or bond breakage in the rate-
determining step of the reaction. In equations (48) and (49) if
the second step shown had been rate-determining a secondary
isotope effect but no primary effect would have been observed on
the overall rate. Since secondary effects are always much

smaller than the observed isotope effect $k^H/k^D = 4.5$, the mechanism
with a rate-determining second step can be rejected. A typical
secondary hydrogen isotope effect was observed[53] in the neutral
(pH-independent) hydrolysis of O-ethyl-S-phenylbenzaldehyde acetal
and its 1-D derivative, equation (51). It was argued that the

$$Ph-\overset{H(D)}{\underset{SPh}{\overset{|}{C}}}-OEt \quad \xrightarrow{k^H(k^D)} \quad Ph-\overset{H(D)}{\underset{OEt}{\overset{+}{C}}} \quad + \quad PhS^- \quad \text{slow}$$

$$\tag{51}$$

$$Ph-\overset{+}{\underset{OEt}{C}}\overset{H(D)}{} \quad + \quad H_2O \quad \longrightarrow \quad Ph-C\overset{H(D)}{\underset{O}{}} \quad + \quad EtOH + H^+ \quad \text{fast steps}$$

observed difference in rates $k^H/k^D = 1.13 \pm 0.02$ is consistent
with considerable C–S bond breakage in the transition state. Since
a much lower secondary isotope effect ($k^H/k^D = 1.038 \pm 0.008$) was
observed for the acid catalysed reaction it was concluded that in
this case there is relatively little C–S bond breakage when the
transition state is reached.

(3)   A solvent isotope effect is the effect on the rate of reaction
of isotopic substitution of an atom in the solvent and may include
primary and secondary isotope effects as well as contributions
from the isotope effect on the solvation of the reactants and
transition state. Measurements of solvent isotope effects are
restricted to the substitution of deuterium for hydrogen as for
example in water (k_{H_2O}/k_{D_2O}) and methanol (k_{CH_3OH}/k_{CH_3OD}). The
solvent isotope effect ($k_{H_2O}/k_{D_2O} = 3.2$) on the acid catalysed
hydrolysis of ethyl vinyl ether, equation (52), includes contributions from primary, secondary, and solvation effects.[54] The

$$CH_2\!\!=\!\!CH.OC_2H_5 + H_3O^+ \longrightarrow CH_3\overset{+}{C}H.OC_2H_5 + H_2O \quad \text{slow}$$

$$CH_3\overset{+}{C}H.OC_2H_5 + H_2O \longrightarrow CH_3\underset{\overset{|}{OH}}{CH}.OC_2H_5 + H^+ \quad \text{fast} \quad \tag{52}$$

$$CH_3\underset{\overset{|}{OH}}{CH}.OC_2H_5 \longrightarrow CH_3.CHO + C_2H_5OH \quad \text{fast}$$

primary effect arises because in the rate-determining step of the reaction in H_2O H^+ is transferred from H_3O^+, and in D_2O D^+ is transferred from D_3O^+. In the same step the two non-reacting protons in H_3O^+ give rise to a secondary isotope effect when the rate in H_2O is compared with the rate in D_2O. To obtain mechanistic information the overall solvent isotope effect is separated into these individual components.

In using kinetic isotope effects to reach conclusions about reaction mechanisms   it is necessary to be able to predict the isotope effects for a particular mechanism. This can be achieved from a knowledge of measured isotope effects for single-step reactions or for reactions of known mechanism. Isotope effects for single-step reactions can be calculated theoretically[55,56] and this approach can be used to estimate isotope effects for a particular reaction mechanism. The calculations rely heavily on the original theory developed by Bigeleisen[57] and involve the application of transition state theory. The application of transition state theory to this problem is more successful than in the calculation of absolute rates of reaction since in calculating the ratio of rates for reactants which only differ isotopically considerable cancellation of unknown terms occurs. The theory of isotope effects is based on the principle that the potential energy surface for a reaction is unchanged by isotopic substitution and in this respect isotopic substitution differs fundamentally from the effect of changing substituent groups in the reactant. Isotopic substitution is a much more subtle change. The magnitude of the isotope effect for a reaction step is determined by the changes in vibrational force constants in going from the reactants to the transition state, and in order to calculate the isotope effect the changes in force constants must be estimated. Thus measured isotope effects provide a probe for vibrational force constants in transition states and therefore give detailed information on transition state structure.

The maximum isotope effects to be expected for isotopic substitution of different atoms can be estimated[58] from the theory and some results are given in Table 2. The maximum primary effects were estimated by assuming that there is no bonding to the isotopically substituted atom in the transition state. The secondary isotope effects were calculated by assuming that the force constant to the isotopically substituted atom changes by a factor of two in going from the reactants to the transition state; a decrease gives rise to the regular isotope effect and an increase gives rise to the inverse effect. The results in Table 2 give a rough idea of

Table 2. Estimated Maximum Values for Isotope Effects[58]

Isotopic Substitution		Primary Isotope Effect (k^{light}/k^{heavy})	
^1H	^2H	18	
^1H	^3H	60	
^{12}C	^{13}C	1.25	
^{12}C	^{14}C	1.5	
^{14}N	^{15}N	1.14	
^{16}O	^{18}O	1.19	

		Secondary Isotope Effect (k^{light}/k^{heavy})	
		Regular	Inverse
C-H	C-D	1.74	0.46
O-H	O-D	2.02	0.37
$C-^{12}C$	$C-^{13}C$	1.012	0.983
$C-^{12}C$	$C-^{14}C$	1.023	0.968

the relative magnitude of isotope effects to be expected for various isotopic substitutions although in real cases the assumptions will not be valid and the actual isotope effects will be lower than the values given in Table 2.

Because of the large difference in relative masses of the hydrogen isotopes, the isotope effects observed with these atoms are very much larger than those observed with other atoms as shown in Table 2. Hence deuterium and tritium isotope effects can be obtained by measuring rates for the isotopically different reactants in separate experiments. For other atoms, however, the isotope effects are obtained by competitive methods. For example if X_1 and X_2 in equation (53) are isotopically different reactants the ratio of rate coefficients (k_1/k_2) can be determined by

$$X_1 + Y \xrightarrow{k_1} Products_1$$
$$X_2 + Y \xrightarrow{k_2} Products_2 \quad (53)$$

observing the change in the isotope ratio in the reactant $([X_1]/[X_2])$ or in the products $([product_1]/[product_2])$ as reaction proceeds.

It is apparent from Table 2 that the maximum values expected for primary and secondary isotope effects are very different. Hence if a kinetic isotope effect is observed which is larger than the maximum value expected for a secondary effect, the observed isotope effect can be identified as primary. This means that a bond to the isotopically substituted atom is being

formed or broken in the rate-determining step of the reaction and
the postulated mechanism must include this. However the converse
of this argument cannot be used in making deductions about
mechanisms.[59] If the observed isotope effect is small it cannot
be concluded that the rate-determining step of the reaction does
not involve bond-breaking or bond-making to the isotopically
substituted atom. For example the breaking of a bond to the
isotopically substituted atom may not be appreciably advanced in
the transition state of the rate-determining step even though the
bond may be broken in that step. In this case the primary isotope
effect would be small. However this brings us to a second way in
which isotope effects may be used to provide detailed information
about the reaction. If the mechanism of reaction is known it may
be possible to deduce the transition state structure from the size
of the measured isotope effect. The sizes of primary and secondary
isotope effects are determined by the extent of bonding changes in
going from the reactant to the transition state and isotope effects
will therefore vary in magnitude according to whether the transition state is reactant-like or product-like. These various
approaches to obtaining mechanistic information and detailed
information about transition state structure are best described
with examples. Two main topics will be discussed : hydrogen
isotope effects on proton transfers and isotope effects on
elimination reactions.

     (ii)   Hydrogen Isotope Effects on Proton Transfer Reactions.
     (a)   Primary Isotope Effects. Isotope effects on proton
transfer reactions have been widely studied. The interest in
proton transfers has arisen because these processes are important
elementary steps in many reactions in organic chemistry. Proton
transfers are convenient reactions for testing theories of
isotope effects because the observed isotope effects are large
and easy to measure.
     A theoretical treatment[60] of the relationship between
transition state structure and the size of the primary isotope
effect for a proton transfer was proposed many years ago and is
still of considerable value.[61] It was predicted[60] that for a
reaction like (54) the primary isotope effect will be largest

$$RCH \ + \ B \ \longrightarrow \ [RC \ .... \ H \ .... \ B]^{\ddagger} \longrightarrow RC^- \ + \ BH^+ \qquad (54)$$

when the transition state is symmetrical, corresponding to the
proton being about half-transferred. In order to explain this

conclusion it is necessary to consider the various contributions to the isotope effect (k^H/k^D) for reactions (55) and (56). The difference in zero-point energies between the C-H and C-D bonds in the reactant will contribute a factor of 6-7 to the primary

$$RCH + B \xrightarrow{\ k^H\ } RC^- + BH^+ \qquad (55)$$

$$RCD + B \xrightarrow{\ k^D\ } RC^- + BD^+ \qquad (56)$$

effect.[62] In the transition state the antisymmetric stretching frequency (57) corresponds to motion along the reaction coordinate and need not be considered in calculating the isotope effect. The symmetric vibration (58) may give rise to zero-point energy differences between the transition states containing H or D. If

$$\overleftarrow{RC} \dots \overrightarrow{H} \dots \overleftarrow{B} \qquad\qquad \overleftarrow{RC} \dots H \dots \overrightarrow{B}$$

$$\text{(57)} \qquad\qquad\qquad\qquad \text{(58)}$$

the force constants of the C .... H and B .... H bonds in the transition state are equal (symmetrical transition state) this vibration will not lead to motion of H or D and there will be no difference in the zero-point energies of the two transition states. In this case the isotope effect will be determined mostly by the zero-point energy difference between the isotopically different reactant molecules and will have a value close to 6-7. However when the transition state is highly reactant-like or product-like the symmetric vibration (58) will lead to a difference in zero-point energies in the transition states for reaction of H and D and the isotope effect will be lower than 6-7. Hence the theory predicts that the isotope effect will depend upon the degree of proton transfer in the transition state and a maximum isotope effect of 6-7 will be observed if the proton is about half-transferred. The maximum value may be as high as 10 if bending vibrations are taken into account.[62a] In this treatment[60] tunnelling corrections[62a] have been ignored.

There have been a large number of studies in which it has been found that the isotope effect goes through a maximum value with changing transition state structure, thus confirming the theoretical predictions. Most of the studies have involved proton transfer to or from carbon and the transition state structure has been varied by changing the base strength of B or the acidity of RCH by means of substituent groups. For a strong base B it is usually argued that the transition state will be reactant-like whereas for a weak base B the proton will be almost

fully transferred to B in the transition state.   These predictions
are shown in Figure 2.   The range of basicity needed to achieve
these situations may be very large indeed, depending upon the

<u>Figure 2</u>.   <u>Transition State Structure and Primary Isotope Effects</u>
<u>for a Proton Transfer Reaction</u>

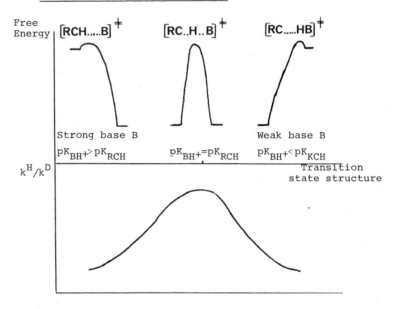

reaction studied.   When BH^+ and RCH have similar pK values the
transition state is assumed to be symmetrical and it is for this
reaction that the maximum isotope effect should be found.   This
prediction has been verified for proton transfer from nitro-
paraffins,[63] from ketones,[64] and for proton transfer to aromatic
substrates.[65]   Maxima in the primary isotope effects have also
been observed for proton transfer from oxygen and nitrogen acids[66]
and for the hydride ion transfer (59).   In this case substituents
in the aromatic rings were varied.[67]   In a few cases the isotope
effect is invariant[68] over a large range of reactivity of the
substrate and this means either that the transition state structure
does not change very much or that the isotope effect in these
cases is insensitive to a change in the structure of the transition state.   Examples have also been found for which the
variation in the isotope effect parallels the variation of other

$$\text{Ar}_3\text{C}^+ \underset{\underset{\text{DCO}_2^-}{k^D}}{\overset{\overset{\text{HCO}_2^-}{k^H}}{\diagup}} \begin{array}{c} \nearrow \text{Ar}_3\text{CH} + \text{CO}_2 \\ \\ \searrow \text{Ar}_3\text{CD} + \text{CO}_2 \end{array} \qquad (59)$$

parameters which relate to transition state structure. For example the exponent β in the Brönsted relation is often used as an index of the degree of proton transfer, in the transition state for a base catalysed reaction.[44c] An exponent $\beta = 0.5$ corresponds to a transition state in which the proton is symmetrically placed and $\beta = 0$ and $\beta = 1.0$ correspond to reactant- and product-like transition states respectively. It was found[69] for a series of ketones that as the Brönsted exponent varied along the series from $\beta = 0.8$ to $\beta = 0.5$, the primary isotope effect (k^H/k^D) changed from 2.0 to 4.5.

On the whole the experimental evidence for a variation in isotope effect with transition state structure is convincing even though, theoretically, this variation may not be fully understood.[70] It is therefore possible to reach conclusions about the degree of proton transfer in the transition state from the size of the primary isotope effect. For a complex mechanism the magnitude of the isotope effect can be used to decide whether a slow proton transfer is involved. If an isotope effect $k^H/k^D >$ ca. 2 is measured it can be concluded that the mechanism involves a rate-determining proton (or possible hydride ion) transfer. If a low isotope effect is observed $(k^H/k^D <$ ca. 2) a slow proton transfer step cannot necessarily be ruled out. A strongly thermodynamically favourable or unfavourable proton transfer will lead to a low isotope effect. Reactions (60) and (61) involve rate-determining proton transfer but, because the reactions are very strongly thermodynamically unfavourable in the forward direction, the transition states are highly unsymmetrical and the primary isotope effects are low. The isotope effects for reactions (60) and (61) were obtained from measurements of isotope exchange[71] and racemisation[72] respectively as described in Section B(v). If a low isotope effect is observed for a complex mechanism and it is thought that a rate-determining proton transfer step is involved an estimate of the pK values of the acid and base species which

$$\underset{\substack{\displaystyle CN \\ |}}{\text{t-Bu-}\overset{\displaystyle CN}{\underset{|}{C}}\text{-H(or D)}} \quad + \quad AcO^- \xrightarrow[k^H/k^D=1.47]{H_2O} \quad \underset{\substack{\displaystyle CN \\ |}}{\text{t-Bu-}\overset{\displaystyle CN}{\underset{}{C^-}}} + AcOH(\text{or D}) \qquad (60)$$

$$\underset{\substack{\displaystyle CN \\ |}}{\langle \bigcirc \rangle CH_2\overset{\displaystyle CN}{\underset{\displaystyle CH_3}{C}}\text{-H(or D)}} \quad + \quad MeO^- \xrightarrow[k^H/k^D=1.15]{MeOH} \quad \underset{\substack{\displaystyle CN \\ |}}{\langle \bigcirc \rangle CH_2\overset{\displaystyle CN}{\underset{\displaystyle CH_3}{C^-}}} + MeOH(\text{or D}) \qquad (61)$$

are thought to take part in the proton transfer will make it possible to decide whether the isotope effect is low because of an unsymmetrical transition state or whether a different mechanism not involving rate-determining proton transfer must be chosen.

For some reactions isotope effects have been found which are greater than the theoretical maximum values given in Table 2. For example[73] for the proton transfer shown in equations (62) and (63) in which B is $(Me_2N)_2$ C=NH a value k^H/k^D = 45±2 was observed for reaction in toluene. The experimental evidence for anomalously high isotope effects has been surveyed.[62a,74] The results are

$$NO_2.C_6H_4.CH_2NO_2 \quad + \quad B \xrightarrow{k^H} NO_2.C_6H_4.\bar{C}HNO_2 \quad + \quad BH^+ \qquad (62)$$

$$NO_2.C_6H_4.CD_2NO_2 \quad + \quad B \xrightarrow{k^D} NO_2.C_6H_4.\bar{C}DNO_2 \quad + \quad BD^+ \qquad (63)$$

accounted for by postulating that a special mechanism is available in these cases. There is considerable theoretical foundation for the idea that the proton can undergo quantum mechanical tunnelling through the energy barrier and this mechanism strongly favours the proton over the deuteron and leads to an exceptionally high isotope effect. Some disagreement exists in the literature as to the importance of tunnelling in proton transfers generally. Reactions which give anomalously high isotope effects and involve tunnelling are usually considered as exceptions to the general behaviour. However it has been suggested[70] that tunnelling could account for the maximum in the variation of k^H/k^D with transition state structure. The explanation which we have described involving the zero-point energy contribution of a symmetric vibration in the transition state is generally accepted.

(b)   <u>Solvent Isotope Effects</u>. The application of
solvent isotope effects to studies of reaction mechanisms has been
reviewed.[75] Solvent isotope effects are most useful when it is
thought that a proton transfer step may be involved although
important information may also be obtained in other cases, for
example in studies of solvolysis reactions.[76] Three major uses
will be considered here.

(1) For many reactions measurement of the solvent isotope effect
may be the only way to obtain values for primary and secondary
isotope effects. This is the case when the proton for which the
isotope effect is to be measured by substitution with deuterium
undergoes rapid exchange with the solvent. For example the primary
and secondary isotope effects for reaction (52) involving hydronium
ion were measured in this way.

(2) We have described the way in which the magnitude of a primary
isotope effect may be diagnostic of the occurrence of a rate-
determining proton transfer step and experimentally this may be
observed by measuring rates in H_2O and D_2O. When the reaction is
carried out in H_2O, the rate-determining step involves H^+ transfer
and in D_2O D^+ transfer is involved. In some cases a solvent
isotope effect may also be used to identify a rapid pre-equilibrium
proton transfer occurring before the rate-determining step. If
the second step in scheme (64) does not involve proton transfer,

$$S + H_3O^+ \xrightleftharpoons{K} SH^+ + H_2O \qquad \text{fast}$$

$$SH^+ \xrightarrow{k} \text{Products} \qquad \text{slow} \tag{64}$$

$$d[\text{Prods}]/dt = kK[S][H_3O^+]$$

the reaction will not show a primary isotope effect. However a
difference in rates of reaction in H_2O and D_2O may be observed
because the value of the equilibrium constant K may be different
in the two solvents. For most acids (SH^+) it is found[77] that the
ratio of acid dissociation constants in H_2O and D_2O ($K^{SH^+}_{H_2O}/K^{SD^+}_{D_2O}$) is
greater than unity. The reaction (64) will then occur more
rapidly in D_2O than in H_2O because SD^+ is present in higher concentration than SH^+. The difference in rates may be as high as a
factor of 5. The kinetic solvent isotope effect is in the
opposite direction ($k_{H_2O}/k_{D_2O} < 1$) to the result which would be
observed for a primary isotope effect ($k^H/k^D > 1$).

(3) Finally, the magnitude of the solvent isotope effect may
give information about solvation changes between the reactant and

transition state for a reaction.[78]

The use of solvent isotope effects in the determination
of reaction mechanisms can be illustrated by reference to reactions
catalysed by hydronium ion in aqueous solution. The reactions
chosen will show how measurement of the solvent isotope effect can
give broad mechanistic information, for example whether the
mechanism includes a rate-determining proton transfer. In addition
the examples will show how fine-detailed information of the
structure of a transition state (reactant- or product-like) can be
obtained for a reaction for which the mechanism is already
established.

Mechanisms (65), (66) and (67) are observed for
hydronium ion catalysed reactions. For the A-1 mechanism the
solvent isotope effect is given by equation (68) in which

A-1 $S + H_3O^+ \rightleftharpoons SH^+ + H_2O$

$SH^+ \xrightarrow{\text{slow, } k_1}$ Products   (65)

A-2 $S + H_3O^+ \rightleftharpoons SH^+ + H_2O$
                                                                    (66)
$SH^+ + H_2O \xrightarrow{\text{slow}}$ Products

A-S_E2 $S + H_3O^+ \xrightarrow{\text{slow}}$ Products   (67)

$K_{SH+}^{H_2O}$ and $K_{SD+}^{D_2O}$ are the acid dissociation constants of SH^+ in H_2O
and SD^+ in D_2O. For most acids[77] $K_{SH+}^{H_2O}/K_{SD+}^{D_2O}$ has a value in the

$$k_{H_2O}/k_{D_2O} = (K_{SD+}^{D_2O}/K_{SH+}^{H_2O})(k_1^{H_2O}/k_1^{D_2O}) \tag{68}$$

range 2-5 and if the second step does not involve proton transfer
the value of $k_1^{H_2O}/k_1^{D_2O}$ will be close to unity. Hence the
solvent isotope effect k_{H_2O}/k_{D_2O} will be in the range 0.2-0.5.
Experimental results[77] confirm this expectation. The A-S_E2
mechanism (67) involves one primary and two secondary effects on
hydronium ion. The primary isotope effect is usually dominant and
values of k_{H_2O}/k_{D_2O} ca. 3 have been measured.[79] The A-1 and
A-S_E2 mechanisms can therefore be distinguished on the basis of
the measured solvent isotope effects. This procedure is illustrated by the results for reactions (69) and (70) for which solvent
isotope effects $k_{H_2O}/k_{D_2O} = 0.34$ and 2.3 respectively have been
measured.[80] The mechanism of hydrolysis of acetals, ketals, and
orthoesters has also been identified in this way. The A-1 and A-2
mechanisms (65) and (66) are expected to give similar solvent
isotope effects $k_{H_2O}/k_{D_2O} < 1$. However in the hydrolysis of ethyl

$$\text{(69)}$$

$$\text{(70)}$$

diazoacetate the two mechanisms have been distinguished by the
detailed dependence of rate on solvent composition in H_2O-D_2O
mixtures.[81] The A-1 mechanism is shown in equation (71) and the
A-2 mechanism which is thought to operate in this case is given in
equation (72).

$$N_2CHCO_2Et + H_3O^+ \rightleftharpoons \overset{+}{N}_2CH_2CO_2Et + H_2O \quad \text{fast}$$

$$\overset{+}{N}_2CH_2CO_2Et \longrightarrow \overset{+}{C}H_2CO_2Et + N_2 \quad \text{slow}$$

$$\overset{+}{C}H_2CO_2Et + 2H_2O \longrightarrow HOCH_2CO_2Et + H_3O^+$$

$$\text{(71)}$$

$$\overset{+}{N}_2CH_2CO_2Et + H_2O \longrightarrow HOCH_2CO_2Et + H_3O^+ + N_2 \text{ slow} \quad \text{(72)}$$

For reactions occurring by the rate-determining proton
transfer mechanism (A-S_E2) detailed information about transition
state structure has been obtained by separating[79,82] the overall
solvent isotope effect into contributions from primary, secondary,
and solvation effects. This is best carried out using fractionation factor theory.[82,83] According to this theory the solvent
isotope effect for reaction (73) is given by equation (74) in which
k_n is the rate coefficient for reaction in an H_2O-D_2O mixture
containing an atom fraction of deuterium n and k_o is the rate
coefficient in H_2O. The terms ϕ_1 and ϕ_2 are the fractionation
factors of protons at positions 1 and 2 in the transition state
and represent the fractional abundance of deuterium at these
positions relative to the atom fraction of deuterium in the
solvent. The solvent is represented as L_2O in which L=H or D.
The fractional abundance of deuterium in L_3O^+ relative to that in
the solvent also appears in equation (74) and is given the symbol
ℓ. The value $\ell=0.69$ has been measured.[82] For some reactions it
is necessary to include further product terms in the numerator of

$$S + H_3O \longrightarrow \left[\begin{array}{c} \text{position 2} \\ H \\ \diagdown \\ O \cdots H \cdots S \\ \diagup \quad \text{position 1} \\ H \\ \text{position 2} \end{array} \right]^{\ddagger} \longrightarrow SH^+ \longrightarrow \text{Products} \qquad (73)$$

$$k_n/k_o = (1-n+n\phi_1)(1-n+n\phi_2)^2 / (1-n+n\ell) \qquad (74)$$

$$n = 1, \quad k_{D_2O}/k_{H_2O} = \phi_1\phi_2^2/\ell^3 \qquad (75)$$

$$\phi_1 = (D/H)_1 / (D/H)_{L_2O} = (D/H)_1(1-n)/n \qquad (76)$$

$$\phi_2 = (D/H)_2 / (D/H)_{L_2O} = (D/H)_2(1-n)/n \qquad (77)$$

$$\ell = (D/H)_{L_3O+} / (D/H)_{L_2O} = (D/H)_{L_3O+}(1-n)/n \qquad (78)$$

equation (74) to account for isotope effects on solvating water molecules but usually these terms are close to unity.[82] The primary and secondary components of the overall isotope effect are given by equations (79) and (80) respectively. It is the values of ϕ_1 and ϕ_2 which are used to draw conclusions about the structure of the transition state. Two methods are commonly used[82,84] to

$$(k^H/k^D)_{primary} = \ell/\phi_1 \qquad (79)$$

$$(k^H/k^D)_{secondary} = \ell/\phi_2 \text{ per proton} \qquad (80)$$

determine the values of ϕ_1 and ϕ_2 . Equation (74) can be fitted to the experimental results for the variation of the rate ratio k_n/k_o with the fraction of deuterium in the solvent and best-fit values for ϕ_1 and ϕ_2 are thereby obtained. In some cases a different procedure can be used. The atom fraction of deuterium in the reaction products at a position corresponding to position 1 in the transition state will be the same as the atom fraction in the transition state providing this position does not undergo exchange in any intermediate which occurs between the transition state and the products. Hence a value for ϕ_1 can be determined from the isotopic composition of the product and this can be combined with the overall solvent isotope effect to calculate a value for ϕ_2 by using equation (75). We now turn to the interpretation of the values of ϕ_1 and ϕ_2 in terms of transition state structure. Since ϕ_1 is related to the primary isotope effect on proton transfer from hydronium ion by equation (79) deductions

about transition state structure using ϕ_1 are made in the same way
as has been described for primary isotope effects. The value of
ϕ_2 relates to a secondary isotope effect on hydronium ion and this
can be used as follows. For a reactant-like transition state (81)
the protons at position 2 will resemble the protons in H_3O^+ and a
fractionation factor $\phi_2 = 0.69$ should be observed in this case.
For a product-like transition state (82) the protons at position 2

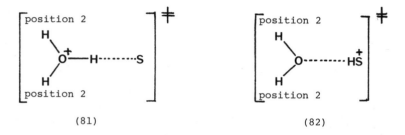

(81)   (82)

will resemble the protons in H_2O, with $\phi_2 = 1.0$. Hence the
measured value of ϕ_2 can be used to deduce whether the protons at
position 2 resemble protons in H_3O^+ or H_2O and therefore whether
the transition state is reactant- or product-like or some way in-
between. This procedure has been used quite widely and can be
checked[79,85] by comparing the conclusions with the conclusions of
other methods of predicting transition state structure. We
previously mentioned in Section C(ii)(a) that for proton transfers
the value of the exponent in the Brönsted relation determined from
studies of general acid or base catalysis has often been used to
reach conclusions about transition state structure.[44c] The degree
of proton transfer in the transition state for several A-S_E2
reactions has been deduced in this way. For several of these
reactions where a comparison is possible, the conclusions based
on the value of the Brönsted exponent and those reached from the
secondary isotope effect measurements in H_2O-D_2O mixtures are in
good agreement.[79,85]

(iii) Isotope Effects on Elimination Reactions. Elimination
reactions (83) are considered to occur by four possible mechanisms
and isotope effects are extremely useful in placing a particular
reaction into a mechanistic class.[86] When the broad mechanism has
been established isotope effects can then be used to provide more
detailed information about transition state structure. The four
mechanistic classes are shown in equations (84)-(87).

$$H-\overset{R}{\underset{R}{\overset{\beta}{C}}}-\overset{R}{\underset{R}{\overset{\alpha}{C}}}-X \ + \ B \ \longrightarrow \ \overset{R}{\underset{R}{\overset{}{C}}}{\overset{\beta}{=}}\overset{R}{\underset{R}{\overset{\alpha}{C}}} \ + \ BH^+ + \ X^- \qquad (83)$$

E1, carbonium ion mechanism

$$H-\overset{\beta}{C}-\overset{\alpha}{C}-X \ \xrightarrow[\substack{\text{rate-deter-}\\ \text{mining step}}]{k} \ H-\overset{}{C}-\overset{}{C^+} \ + \ X^-\qquad(84)$$

$$\overset{}{C}=\overset{}{C} \ + \ BH^+$$

$$\text{Rate} = k[HCR_2CR_2X]$$

E2, concerted single-step mechanism

$$H-\overset{\beta}{C}-\overset{\alpha}{C}-X \ + \ B \longrightarrow \left[\overset{\delta+}{B}\cdots H\cdots \overset{}{C}\cdots\cdots \overset{}{C}\cdots \overset{\delta-}{X}\right]^{\ddagger} \rightarrow \overset{}{C}=\overset{}{C} + \ BH^+ + \ X^- \quad(85)$$

$$\text{transition state}$$

$$\text{Rate} = k[HCR_2CR_2X][B]$$

(E1cB)$_I$, carbanion mechanism with rate-determining ionisation

$$H-\overset{\beta}{C}-\overset{\alpha}{C}-X \ + \ B\xrightarrow[\substack{\text{rate-determining}\\ \text{step}}]{k_1} \ ^-\overset{}{C}-\overset{}{C}-X \ + \ BH^+$$

$$\xrightarrow{k_2} \ \overset{}{C}=\overset{}{C} + \ X^- \qquad(86)$$

$$\text{Rate} = k_1[HCR_2CR_2X][B]$$

(E1cB)$_R$, carbanion mechanism with rate-determining decomposition of the carbanion

$$H-\overset{\beta}{C}-\overset{\alpha}{C}-X \ + \ B\underset{k_{-1}}{\overset{k_1}{\rightleftharpoons}} \ ^-\overset{}{C}-\overset{}{C}-X \ + \ BH^+$$

$$\xrightarrow[\substack{\text{rate-determining}\\ \text{step}}]{k_2} \ \overset{}{C}=\overset{}{C} \ + \ X^- \qquad(87)$$

$$\text{Rate} = (k_1k_2/k_{-1}[BH^+])[B][HCR_2CR_2X]$$

The carbanion mechanisms[87] are limiting cases of mechanism (88) for which expression (89) is derived by assuming that the carbanion is present in low concentration and can be treated by the steady-state approximation. In equation (88), B

$$H-\overset{|}{\underset{|}{\beta C}}-\overset{|}{\underset{|}{\alpha C}}-X \; + \; B \; \underset{k_{-1}}{\overset{k_1}{\rightleftharpoons}} \; -\overset{|}{\underset{|}{\bar C}}-\overset{|}{\underset{|}{C}}-X \; + \; BH^+ \qquad (88)$$

$$\overset{k_2}{\searrow}$$

$$\diagup C = C \diagdown \; + \; X^-$$

$$\text{Rate} = k_1 k_2 [HCR_2CR_2X][B]/(k_2 + k_{-1}[BH^+]) \qquad (89)$$

and BH^+ are buffer species. Under conditions where $k_2 > k_{-1}[BH^+]$, the $(ElcB)_I$ mechanism applies and when $k_2 < k_{-1}[BH^+]$, the $(ElcB)_R$ mechanism operates. A change in mechanism from $(ElcB)_I$ to $(ElcB)_R$ with buffer concentration has been observed[88] for reaction (90). Rates were measured at constant buffer ratio ($[B]/[BH^+]$) but at

$$PhSO_2CH_2CH_2\overset{+}{N}Me_2Ph + Et_3N \xrightarrow{\text{EtOH}} PhSO_2CH=CH_2 + Et_3NH^+ + PhMe_2N$$

$$(90)$$

different buffer concentrations. At low buffer concentrations the reaction is first-order in B and the mechanism is $(ElcB)_I$. At high buffer concentrations the rate of reaction is independent of buffer concentration but depends upon the buffer ratio and this identifies the mechanism as $(ElcB)_R$. These results are in contrast to the results obtained for El and E2 reactions. For the El mechanism the rate of reaction is independent of the concentration of B but for the E2 mechanism the reaction is first-order in B.

For each of the mechanisms, predictions as to whether a primary isotope effect will be observed for various isotopic substitutions are given in Table 3. Not all of these isotope effects have been observed experimentally.[86] Some examples for which measurements have been made will be described to illustrate the technique.

Hydrogen isotope effects have been used most frequently. The $(ElcB)_R$ mechanism can be readily distinguished from the E2 and $(ElcB)_I$ mechanisms on the basis of the hydrogen isotope effect. Since the E2 and $(ElcB)_I$ mechanisms involve breakage of C-H bonds in the rate-determining step primary hydrogen isotope effects should be observed for these mechanisms. For the $(ElcB)_R$ mechanism

<u>Table 3.</u> <u>Primary</u> <u>Kinetic Isotope Effects in Elimination Reactions</u>

Mechanism	α-carbon	β-carbon	H	X
E1	Yes	No	No	Yes
E2	Yes	Yes	Yes	Yes
$(\text{ElcB})_I$	No	Yes	Yes	No
$(\text{ElcB})_R$	Yes	Yes	No	Yes

no isotope effect will be observed because the deuterated
substrate will rapidly exchange deuterium with protium in the
solvent. A solvent isotope effect $(k^{H_2O}/k^{D_2O} = $ ca. 3) would be
observable for this mechanism, however, because of the difference
in acid dissociation constants of the C-H and C-D acids in H_2O and
D_2O.[77] Isotope effects resulting from substitution of deuterium
at the β-carbon atom have been measured[89] for reaction (91) in

$$PhSO_2CH_2CH_2Z + EtO^- \longrightarrow PhSO_2CH=CH_2 + EtOH + Z^- \qquad (91)$$

ethanol. With Z = SO_2Ph, SPh, SePh, and $\overset{+}{N}Me_2Ph$ the result
$k^H/k^D = 1.00 \pm 0.03$ was obtained and it was therefore concluded
that the $(\text{ElcB})_R$ mechanism was operating. With Z=halogen much
higher isotope effects in the range 3-5 were observed, indicative
of an E2 or $(\text{ElcB})_I$ mechanism.

The measured rate coefficient for $(\text{ElcB})_I$ elimination refers
to a single proton transfer step. Hence in view of our discussion
in Section C(ii)(a) a maximum kinetic hydrogen isotope effect for
the $(\text{ElcB})_I$ mechanism should be observed when the proton is
roughly half-transferred in the transition state. For reaction
(92) the hydrogen isotope effect was measured for a series of
bases (B) of varying base strength.[90] The isotope effect varied

$$Ar_2CHCCl_3 + B \longrightarrow Ar_2C=CCl_2 + Cl^- + BH^+ \qquad (92)$$

with base strength and a maximum of $k^H/k^D = 6.2$ was observed with
B = PhO^-. Lower values of the isotope effect were obtained with
weaker or stronger bases. For this $(\text{ElcB})_I$ reaction it is
possible to reach conclusions about the degree of proton transfer
in the transition state. The situation is often more complicated
for E2 elimination because proton transfer is coupled with
departure of the leaving group. It was found[91] that for reaction
(93) the value of $k^H/k^D = $ ca. 9 was largely independent of changes
in the basicity of B. This observation is apparently consistent
with theoretical calculations of the primary hydrogen isotope
effect for E2 elimination.[92] However in some examples of E2

$$CH_3COC_6H_4.Cl_2CH_2\overset{+}{S}Me_2Br^- + B$$

$$\searrow Me_2SO\text{-}H_2O \qquad\qquad (93)$$

$$CH_3COC_6H_4CL{=}CH_2 + Me_2S + Br^- + BL^+$$

elimination a maximum in k^H/k^D is to be expected.[92]

One of the aims of measuring isotope effects is to deduce structural information about the transition state for a reaction of known mechanism. Even if an elimination is known to occur by the E2 mechanism a wide variety of transition state structures is possible. At one extreme is the E2 reaction with a transition state in which considerable C-H bond breakage but negligible C-X bond breakage has occurred and in this case many of the characteristics of the (E1cB)$_I$ reaction will be observed. At the other extreme the E2 transition state may be similar to that for the E1 mechanism. In drawing conclusions about transition state structure it is helpful to make several isotopic substitutions. For example three isotope effects have been measured for the E2 reaction (94). A leaving group isotope effect $k^{32}S/k^{34}S=1.007$

$$PhCH_2CH_2\overset{+}{S}Me_2 + OH^- \longrightarrow PhCH{=}CH_2 + Me_2S + H_2O \qquad (94)$$

was observed[93] as well as isotope effects on the β-carbon atom[94] ($k^{12}C/k^{13}C = 1.02$) and on protons[94] attached to the β-carbon atom ($k^H/k^D = 4$). The way in which the β-carbon and β-hydrogen isotope effects vary with transition state structure has been treated theoretically.[95] The β-hydrogen ($k^H/k^D = 6.1$) and leaving group ($k^{35}Cl/k^{37}Cl$) isotope effects in reaction (95) have been measured and used in reaching conclusions[95] about transition state structure

$$Ph_2CH.CHCl_2 + t\text{-}BuO^- \longrightarrow Ph_2CH{=}CHCl + t\text{-}BuOH + Cl^- \qquad (95)$$

for this E2 reaction. It is realistic to expect that kinetic isotope effect measurements such as these will ultimately lead to detailed information on the structure of transition states in elimination reactions, particularly since many isotopic substitutions can be studied for the same reaction.

References

1. F. Pietra, *Quart. Rev. Chem. Soc.*, 1969, *23*, 504

2. J.D. Roberts, D.A. Semenow, H.E. Simmons Jr., and
    L.A. Carlsmith, *J. Amer. Chem. Soc.*, 1956, *78*, 601

3. H. Heaney, *Chem. Rev.*, 1962, *62*, 81; T.L. Gilchrist and
    C.W. Rees, "Carbenes, Nitrenes, and Arynes", Nelson, London
    (1969), chapter 4, p. 42

4. J.W. Laing and R.S. Berry, *J. Amer. Chem. Soc.*, 1976, *98*, 660;
    J.O. Noell and M.D. Newton, *J. Amer. Chem. Soc.*, 1979, *101*, 51

5. M.L. Bender, *J. Amer. Chem. Soc.*, 1951, *73*, 1626

6. M.L. Bender, *Chem. Rev.*, 1960, *60*, 53

7. D. Samuel and B.L. Silver, *Adv. Phys. Org. Chem.*, 1965, *3*, 123

8. W.P. Jencks, 'Catalysis in Chemistry and Enzymology',
    McGraw-Hill (1969) chapter 10, p. 463

9. J.A. Walder, R.S. Johnson and I.M. Klotz, *J. Amer. Chem. Soc.*,
    1978, *100*, 5156

10. T.D. Lomax and C.J. O'Connor, *J. Amer. Chem. Soc.*, 1978,
    100, 5910

11. Reference 8, p. 509

12. M.L. Bender and H. d'A. Heck, *J. Amer. Chem. Soc.*, 1967,
    89, 1211

13. L.R. Fedor and T.C. Bruice, *J. Amer. Chem. Soc.*, 1965, *87*,
    4138

14. D. Bethel and V. Gold, 'Carbonium Ions An Introduction'
    Academic Press, London (1967)

15. N.S. Isaacs, 'Reactive Intermediates in Organic Chemistry',
    John Wiley and Sons Ltd., London (1974) chapter 2, p. 92

16. Reference 15, p. 168

17. G.A. Olah and C.U. Pittman Jr., *Adv. Phys. Org. Chem.*, 1966,
    4, 305; G.A. Olah and D.J. Donovan, *J. Amer. Chem. Soc.*,
    1978, *100*, 5163

18. G.A. Olah and D.J. Donovan, *J. Amer. Chem. Soc.*, 1977, *99*,
    5026; J.S. Staral, I. Yavari, J.D. Roberts, G.K.S. Prakash,
    D.J. Donovan, and G.A. Olah, *J. Amer. Chem. Soc.*, 1978, *100*,
    1018, and earlier references given there

19. G.A. Olah, E. Baker, J. Evans, W. Tolgyesi, J. McIntyre, and
    I. Bastein, *J. Amer. Chem. Soc.*, 1964, *86*, 1360

20. M. Saunders and E.L. Hagen, *J. Amer. Chem. Soc.*, 1968, *90*,
    6881

21. G.A. Olah and A.M. White, *J. Amer. Chem. Soc.*, 1969 *91*, 5801

22. M. Saunders, P. Vogel, E.L. Hagen, and J. Rosenfeld,
     Accts. Chem. Res., 1973, *6*, 53

23. R.L. Baird and A. Aboderin, *J. Amer. Chem. Soc.*, 1964, *86*,
     252

24. C.C. Lee and L. Gruber, *J. Amer. Chem. Soc.*, 1968, *90*, 3775

25. This compound is named according to stereospecific numbering,
     see *Europ. J. Biochem.*, 1970, *12*, 1, R.H. Gigg; 'Rodd's
     Chemistry of Carbon Compounds', Elsevier, Amsterdam (1976),
     vol. 1E, p. 349

26. G. Lowe and B.S. Sproat, *J.C.S. Chem. Comm.*, 1978, 783;
     G. Lowe and B.S. Sproat, *J.C.S. Perkin I*, 1978, 1622

27. F.H. Westheimer, *Accts. Chem. Res.*, 1968, *1*, 70

28. G. Lowe and B.S. Sproat, *J.C.S. Chem. Comm.*, 1978, 565

29. P.M. Cullis and G. Lowe, *J.C.S. Chem. Comm.*, 1978, 512

30. W.A. Blättler and J.R. Knowles, *J. Amer. Chem. Soc.*, 1979,
     101, 510

31. S.J. Abbott, S.R. Jones, S.A. Weinman, and J.R. Knowles,
     J. Amer. Chem. Soc., 1978, *100*, 2558

32. G.A. Orr, J. Simon, S.R. Jones, G.J. Chin and J.R. Knowles,
     Proc. Natl. Acad. Sci. U.S.A., 1978, *75*, 2230

33. D.K. Stammers and H. Muirhead, *J. Mol. Biol.*, 1977, *112*, 309

34. Z. Margolin and F.A. Long, *J. Amer. Chem. Soc.*, 1973, *95*, 2757

35. D.B. Dahlberg, A.J. Kresge, and A.C. Lin, *J.C.S. Chem. Comm.*,
     1976, 35; A.J. Kresge and A.C. Lin, *J. Amer. Chem. Soc.*,
     1975, *97*, 6257

36. A.J. Kresge and A.C. Lin, *J.C.S. Chem. Comm.*, 1973, 761

37. A.I. Shatenstein, *Adv. Phys. Org. Chem.*, 1962, *1*, 155

38. J.A. Elvidge, J.R. Jones, C. O'Brien, E.A. Evans, and
     H.C. Sheppard, *J.C.S. Perkin II*, 1974, 174

39. F. Hibbert, F.A. Long, and E.A. Walters, *J. Amer. Chem. Soc.*,
     1971, *93*, 2829

40. F. Hibbert, *J.C.S. Perkin II*, 1973, 1289

41. S. Oae, W. Tagaki, and A. Ohno, *Tetrahedron*, 1964, *20*, 417

42. A. Streitwieser Jr., D. Holtz, G.R. Ziegler, J.O. Stoffer,
     M.L. Brokaw, and F. Guibe, *J. Amer. Chem. Soc.*, 1976, *98*, 5229

43. D.J. Cram and W.D. Kollmeyer, *J. Amer. Chem. Soc.*, 1968, *90*,
     1791

44. (a) D.J. Cram, 'Fundamentals of Carbanion Chemistry',
    Academic Press, New York (1965); (b) A. Streitwieser and
    J.H. Hammons, *Prog. Phys. Org. Chem.*, 1965, *3*, 41;
    (c) F. Hibbert in 'Comprehensive Chemical Kinetics',
    Elsevier, Amsterdam (1977), vol. 8, p. 97

45. A. Streitwieser Jr., M.P. Granger, F. Mares, and R.A. Wolf,
    J. Amer. Chem. Soc., 1973, *95*, 4257

46. A. Streitwieser Jr., P.H. Owens, G. Sonnichsen, W.K. Smith,
    G.R. Ziegler, H.M. Niemeyer, and T.L. Kruger, *J. Amer. Chem.
    Soc.*, 1973, *95*, 4254

47. See for example J.N. Roitman and D.J. Cram, *J. Amer. Chem.
    Soc.*, 1971, *93*, 2225

48. F.G. Bordwell, W.S. Matthews, and N.R. Vanier, *J. Amer. Chem.
    Soc.*, 1975, *97*, 442

49. A.J. Kresge, S.G. Mylonakis, Y. Sato, and V.P. Vitullo,
    J. Amer. Chem. Soc., 1971, *93*, 6181

50. A.J. Kresge and Y. Chiang, *J. Amer. Chem. Soc.*, 1961, *83*,
    2877

51. N.S. Isaacs, K. Javaid, and E. Rannala, *J.C.S. Perkin II*,
    1978, 709

52. H. Yamataka and T. Ando, *J. Amer. Chem. Soc.*, 1979, *101*, 266

53. J.P. Ferraz and E.H. Cordes, *J. Amer. Chem. Soc.*, 1979, *101*,
    1488

54. R. Eliason and M.M. Kreevoy, *J. Amer. Chem. Soc.*, 1978, *100*,
    7037

55. E.K. Thornton and E.R. Thornton, 'Isotope Effects in
    Chemical Reactions', A.C.S. monograph 167, editors
    C.J. Collins and N.S. Bowman, van Nostrand Reinhold Co.,
    New York (1971), chpt. 4, p. 213

56. M. Wolfsberg, *Accts. Chem. Res.*, 1972, *5*, 225

57. J. Bigeleisen, *J. Chem. Phys.*, 1949, *17*, 675

58. J. Bigeleisen and M. Wolfsberg, *Adv. Chem. Phys.*, 1958, *1*, 15

59. L. Melander, 'Isotope Effects on Reaction Rates', Ronald
    Press, New York (1960), p. 5

60. F.H. Westheimer, *Chem. Rev.*, 1961, *61*, 265

61. R.A. More O'Ferrall, 'Proton Transfer Reactions', editors
    E.F. Caldin and V. Gold, Chapman and Hall, London (1975),
    chpt. 8, p. 201

62. (a) R.P. Bell, *Quart. Rev. Chem. Soc.*, 1974, *3*, 513;
    (b) R.P. Bell, 'The Proton in Chemistry', 2nd edition,
        Chapman and Hall, London (1973), chpt.12, p.250

63. R.P. Bell and D.M. Goodall, *Proc. Royal Soc.*, 1966, *A294*, 273; D.J. Barnes and R.P. Bell, *Proc. Royal Soc.*, 1970, *A318*, 421; J.E. Dixon and T.C. Bruice, *J. Amer. Chem. Soc.*, 1970, *92*, 905; R.P. Bell and B.G. Cox, *J. Chem. Soc. B*, 1971, 783; J.R. Keefe and N.H. Munderloh, *J.C.S. Chem. Comm.*, 1974, 17; F.G. Bordwell and W.J. Boyle, *J. Amer. Chem. Soc.*, 1975, *97*, 3447

64. D.W. Earls, J.R. Jones, and T.G. Rumney, *J.C.S. Far. I*, 1972, 925

65. J.L. Longridge and F.A. Long, *J. Amer. Chem. Soc.*, 1967, *89*, 1292; A.J. Kresge, D.S. Sagatys, and H.L. Chen, *J. Amer. Chem. Soc.*, 1968, *90*, 4174; S.B. Hanna, C. Jermini, and H. Zollinger, *Tet. Lett.*, 1969, 4415

66. J.R. Jones and T.G. Rumney, *J.C.S. Chem. Comm.*, 1975, 995; N-Å. Bergman, Y. Chiang, and A.J. Kresge, *J. Amer. Chem. Soc.*, 1978, *100*, 5954; M.M. Cox and W.P. Jencks, *J. Amer. Chem. Soc.*, 1978, *100*, 5956

67. R. Stewart and T.W. Toone, *J.C.S. Perkin II*, 1978, 1243

68. B.C. Challis and E.M. Millar, *J.C.S. Perkin II*, 1972, 1618

69. R.P. Bell and J.E. Crooks, *Proc. Roy. Soc.*, 1965, *A286*, 285

70. Reference 62(a), p. 537; for evidence against this suggestion, see J. Banger, A. Jaffe, An-Chung Lin, and W.H. Saunders, *J. Amer. Chem. Soc.*, 1975, *97*, 7178

71. F. Hibbert and F.A. Long, *J. Amer. Chem. Soc.*, 1971, *93*, 2836

72. L. Melander and N-Å. Bergman, *Acta Chem. Scand.*, 1971, *25*, 2264

73. E.F. Caldin and S. Mateo, *J.C.S. Chem. Comm.*, 1973, 854

74. E.S. Lewis, 'Proton TransferReactions', editors E.F. Caldin and V. Gold, Chapman and Hall, London (1975), chpt. 10, p. 317

75. R.L. Schowen, *Prog. Phys. Org. Chem.*, 1972, *9*, 275

76. R.E. Robertson, *Prog. Phys. Org. Chem.*, 1967, *4*, 213

77. P.M. Laughton and R.E. Robertson, 'Solute-Solvent Interactions', editors J.F. Coetzee and C.D. Ritchie, Marcel Dekker, New York (1969), chpt. 7, p. 399

78. W.J. Albery, *Prog. Reaction Kinetics*, 1967, *4*, 353

79. J.M. Williams Jr. and M.M. Kreevoy, *Adv. Phys. Org. Chem.*, 1968, *6*, 63

80. A. Kankaanpera, *Acta Chem. Scand.*, 1969, *23*, 1465

81. W.J. Albery and M.H. Davies, *Trans. Far. Soc.*, 1969, *65*, 1066

82. V. Gold, *Adv. Phys. Org. Chem.*, 1969, *7*, 259

83. W.J. Albery, 'Proton Transfer Reactions', editors E.F. Caldin and V. Gold, Chapman and Hall, London (1975), chpt.9, p. 263

84. R.A. More O'Ferrall, G.W. Koeppl, and A.J. Kresge, *J. Amer. Chem. Soc.*, 1971, *93*, 9; W.J. Albery and A.N. Campbell-Crawford, *J.C.S. Perkin II*, 1972, 2190

85. W.J. Albery, A.N. Campbell-Crawford, and J.S. Curran, *J.C.S. Perkin II*, 1972, 2206

86. A. Fry, *Quart. Rev. Chem. Soc.*, 1972, *1*, 163

87. For a review see D.J. McLennan, *Quart. Rev.*, 1967, *21*, 490

88. K.N. Barlow, D.R. Marshall and C.J.M. Stirling, *J.C.S. Chem. Comm.*, 1973, 175. For other examples of a change in mechanism see T.I. Crowell, R.T. Kemp, R.E. Lutz, and A.A.Wall, *J. Amer. Chem. Soc.*, 1968, *90*, 4638; L.R. Fedor and W.R. Glare, *J. Amer. Chem. Soc.*, 1971, *93*, 985

89. D.R. Marshall, P.J. Thomas, and C.J.M. Stirling, *J.C.S. Perkin II*, 1977, 1898

90. D.J. McLennan, *J.C.S. Perkin II*, 1976, 932

91. L.F. Blackwell, *J.C.S. Perkin II*, 1976, 488

92. G.W. Burton, L.B. Sims, and D.J. McLennan, *J.C.S. Perkin II*, 1977, 1763

93. A.F. Cockerill and W.H. Saunders Jr., *J. Amer. Chem. Soc.*, 1967, *89*, 4985; A.M. Katz and W.H. Saunders Jr., *J. Amer. Chem. Soc.*, 1969, *91*, 4469

94. J. Banger, A. Jaffe, A-C. Lin, and W.H. Saunders Jr., *J. Amer. Chem. Soc.*, 1975, *97*, 7177

95. G.W. Burton, L.B. Sims, and D.J. McLennan, *J.C.S. Perkin II*, 1977, 1847

HYDROGEN ISOTOPE EXCHANGE REACTIONS

J.R. Jones,
Chemistry Department,
University of Surrey,
Guildford, GU2 5XH.

1. Introduction.

The main difference between isotopically labelled
compounds of carbon and hydrogen is that the carbon
label is much more stable than its hydrogen counterpart.
This is the reason why the carbon isotope is usually
introduced as part of a synthetic procedure in contrast
to the large and increasing number of methods that are
available for introducing both deuterium[1] and tritium.[2]

To the biochemist the fact that the synthesis of
a ^{14}C (or ^{13}C) labelled compound may have been
demanding in both time and skill is more than compensated
for by the knowledge that the label can not be lost
except under the most extreme conditions. The chemist
on the other hand tends to use more of the deuterium
(or tritium) labelled compounds and the probable reason
for this lies in his interest in reaction mechanisms.[3]
Many organic compounds can be easily labelled with the
isotopes of hydrogen and conditions can be readily found
where isotope exchange occurs at a convenient rate.
It follows therefore that a better understanding of the
factors responsible for inducing exchange can be of
benefit in several directions. Firstly labelling of a
compound can be carried out under optimum conditions
i.e. where exchange is both rapid and efficient and
where possible side-reactions, leading to decomposition,

are kept to a minimum. Secondly, a better appreciation
of likely pitfalls during the specific application is
possible. Finally the conditions under which it is best
to store the labelled compound can be determined. These
may, because of possible radiation decomposition in the case
of tritiated compounds, be different from those appropriate
for corresponding deuterium labelled compounds.

2. Procedures[4]

The two most widely used methods for following
hydrogen isotope exchange reactions, namely dedeuteriation
and detritiation, involve in the first place the synthesis
of an appropriately labelled compound. Rates of dedeuteriation are usually followed by measuring changes in the
^{1}H n.m.r. spectrum of the substrate (Table 1 contains
several examples); the method not only gives the rate
but also the site(s) of exchange. It is limited to
rather slow reactions and is not as accurate as some
of the other methods. The development of deuterium n.m.r.
spectroscopy[5] means that changes in the ^{2}H n.m.r. spectrum
can also be used to measure rates of dedeuteriation.
It is likely therefore that some of the older established
methods such as the measurement of infra-red absorption
will become less widely used.

The development of liquid scintillation counting
in the late 1950's greatly eased the problem of how to
detect weak β emitters; the attractions of tritium as
a tracer were thereby much enhanced. Nowadays the study
of rates of detritiation constitutes one of the most
versatile and accurate methods of following hydrogen
isotope exchange. The method is again limited to rather
slow reactions ($t\frac{1}{2}$ > 1-2 minutes) but unlike the deuter-

iation method it can, because of the extremely small
amounts of radioactivity that can be detected, be used to
measure extremely slow reactions ($t\frac{1}{2}$ in years) over a
relatively short period of time. If one is measuring
the rate of loss of tritium from the substrate
the customary first-order kinetic equation is

$$k = \frac{1}{t} \ln \left(\frac{N_o - N_\infty}{N_t - N_\infty}\right) \tag{1}$$

where N_o, N_t and N_∞ are the radioactivities (usually
expressed as counts/min) of the separated substrate at the
beginning, at time t and on completion of the reaction,
respectively. If on the other hand the reaction is followed
by measuring the increase in the radioactivity of the solvent
(usually water) the relevant equation is

$$k = \frac{1}{t} \ln \left(\frac{C_\infty - C_o}{C_\infty - C_t}\right) = \ln \left(\frac{C_\infty}{C_\infty - C_t}\right) = \frac{1}{t} \ln \left(\frac{1}{1-x}\right) \tag{2}$$

C_o, C_t and C_∞ represent the radioactivity of the water at
time $t = o$, t and the end of the reaction, respectively;
$x = C_t/C_\infty$. If we are concerned only with measuring the
increase in the radioactivity of the water over the initial
stages (< 4%) of the reaction and remember that, for small
values of x, $-\ln(1 - x) = x$ we obtain equation (3):

$$k = x/t \tag{3}$$

In other words the tritium content of the water increases
linearly with time and the zero-order rate constant obtained
from the plot of C_t against time can be converted into
a first-order rate constant (k) by dividing by C_∞. Such
an example is given in Fig. 1. Other examples refer to
exchange into mesitylene and 1,3,5- trimethoxybenzene[16] and from
dimethyl sulphoxide[18] and many heterocyclic carbon acids.[19]

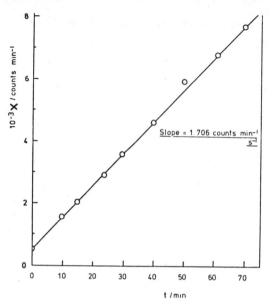

Fig. 1.   Example of the initial rate method - detritiation of
1-methyl-[8-^{3}H]guanosine at 50°C and pH = 6.0

When the tritiated compound contains label at two
exchangeable sites the separate detritiation rate constants
can be obtained by various procedures. In the simplest
case where the two sites differ greatly (> than tenfold)
in reactivity, e.g. the C-2 (H) in adenine^{22} is some 2000
times less reactive than the C-8(H), it is a simple matter
to analyse the data graphically. When sites A and B are
of similar reactivity let N_A be the radioactivity at the
more acidic site and N_B the radioactivity at site B, both
at time t. The amount of tritium remaining in the compound
at this time is therefore given by

$$N_o - N_t = (N_A)_o - N_A + (N_B)_o - N_B \qquad (4)$$

If k_A and k_B are the first-order rate constants for detritiation from sites A and B respectively equation (5) follows:

$$N_o - N_t = (N_A)_o - (N_A)_o e^{-k_A t} + (N_B)_o - (N_B)_o e^{-k_B t}$$

$$= (N_A)_o [1 - e^{-k_A t}] + (N_B)_o [1 - e^{-k_B t}] \qquad (5)$$

Dividing through by $N_o - N_\infty$ and writing n for the mole fraction of tritium at the more acidic site leads to

$$\frac{N_o - N_t}{N_o - N_\infty} = n (1 - e^{-k_A t}) + (1-n)(1 - e^{-k_B t}) \qquad (6)$$

The experimental data can therefore be fitted to this equation, using different values of n. A large number of experimental points (in the region of 30) together with an appropriate computer programme are necessary before satisfactory values of k_A and k_B can be obtained. This can be seen in Fig. 2 where a number of theoretical plots are presented using n as fixed (0.5) and varying k_A/k_B.

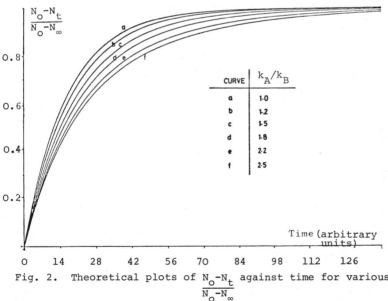

CURVE	k_A/k_B
a	1·0
b	1·2
c	1·5
d	1·8
e	2·2
f	2·5

Fig. 2. Theoretical plots of $\frac{N_o - N_t}{N_o - N_\infty}$ against time for various values of k_A/k_B, with fixed n (0.5).

Kankaanperä et. al.[20] have successfully used this method
in detritiation studies involving unsymmetrical ketones of
the type $RCOCH_3$ (e.g. R = $PhCH_2$, $(CH_3)_3C$, CH_3CH_2); a
similar study on the reactivity of the <u>exo</u> and <u>endo</u>-hydrogens
in camphors has been reported.[23] A slightly different approach
has been adopted in the author's laboratory.[24] The tritium
distribution in the two sites has been directly determined
by ^3H n.m.r. spectroscopy.[25] In this way the number of unknowns
has been reduced from three to two and in favourable circumstances it would seem possible to extend the studies to
compounds having three exchangeable sites. In both approaches
it is difficult to obtain separate rate constants when
the ratio k_A/k_B is close to unity; caution must also be
exercised in circumstances when the mole fraction of tritium
at one site is very much higher than at the other.

In some cases detritiation may be accompanied by
side-reactions such as hydrolysis. In such a situation
the first-order detritiation plot will be curved and this
curvature may sometimes be capable of analysis as can be
seen from the following example.[21] Detritiation at 85°C from
the C-8 position of [8-^{3}H]xanthosine is accompanied by
hydrolysis to xanthine which also undergoes exchange under
the experimental conditions. The situation can be depicted
as follows:

$$[8\text{-}^3H]\text{Xanthosine} \xrightarrow[\text{hydrolysis}]{k_\psi} [8\text{-}^3H]\text{Xanthine} + \beta\text{-D-ribose}$$

$k_1 \downarrow$ detritiation $k_2 \downarrow$ detritiation

After time t the concentration of tritiated xanthosine remaining and xanthine formed are $[B]e^{-k_\psi t}$ and $[B](1-e^{-k_\psi t})$ respectively; $[B]$ is the total concentration of tritiated material at time t and k_ψ the first-order hydrolysis rate-constant. The overall rate of detritiation is thus given by equation (7) which, on integration, leads to (8).

$$\text{Rate} = -d[B]/dt = k_1[B]e^{-k_\psi t} + k_2[B](1 - e^{-k_\psi t}) \qquad (7)$$

$$\ln [B]_0/[B]_t = \left(\frac{k_1-k_2}{k_\psi}\right)\left(1 - e^{-k_\psi t}\right) + k_2 t \qquad (8)$$

If the reaction is followed by measuring the increase in the radioactivity of the water, $[B]_0$ can be equated to $(C_\infty-C_0)$ and $[B]_T$ to $(C_\infty-C_t)$. Equation (8) then leads to (9) which in turn reduces to (10) when $e^{-k_\psi t} \ll 1$.

$$\ln(C_\infty-C_0) - \ln(C_\infty-C_t) = \left(\frac{k_1-k_2}{k_\psi}\right)(1 - e^{-k_\psi t}) + k_2 t \qquad (9)$$

$$\ln(C_\infty-C_0) - \ln(C_\infty-C_t)_{\text{at } e^{-k_\psi t} \ll 1} = \left(\frac{k_1-k_2}{k_\psi}\right) + k_2 t \qquad (10)$$

Under these circumstances the plot of $\ln(C_\infty-C_t)$ against time is linear (Fig.3), the slope yielding the detritiation rate constant for the hydrolysis product, xanthine. Subtraction of equation (10) from (9) yields (11):

$$\ln(C_\infty-C_t) - \ln(C_\infty-C_t)_{\text{at } e^{-k_\psi t} \ll 1} = \left(\frac{k_1-k_2}{k_\psi}\right) e^{-k_\psi t} \qquad (11)$$

Extrapolation of the linear portion of the plot to $t = 0$ enables values of $\ln(C_\infty-C_t)_{\text{at } e^{-k_\psi t} \ll 1}$ to be calculated. Subtraction of these from the measured data allows the function $\ln[\Delta(C_\infty-C_t)]$ which is equal to the term $\ln[\ln(C_\infty-C_t) - \ln(C_\infty-C_t)_{\text{at } e^{-k_\psi t} \ll 1}]$ to be calculated and plotted against

time according to the logarithmic form of equation (11),
i.e. equation (12).

$$\ln[\Delta(C_\infty - C_t)] = \ln\left(\frac{k_1 - k_2}{k_\psi}\right) - k_\psi t \qquad (12)$$

The slope of such a plot, shown in the insert to Fig.3,
gives k_ψ and k_1 can then also be obtained (from the intercept).
This procedure increases the usefulness of isotopic hydrogen
exchange studies.

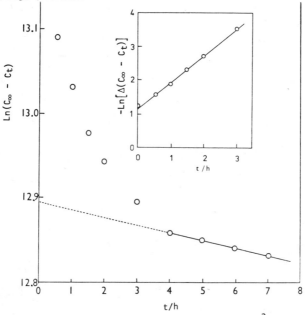

Fig. 3.   Plot of $\ln(C_\infty - C_t)$ against time for [8-^{3}H]-xanthosine
which undergoes both detritiation and hydrolysis.

Of the other methods (Table 1) of measuring rates of
hydrogen isotope exchange mention may be made of the use
of laser-Raman spectroscopy. This method has been used
by Thomas and Livramento[13] to follow the kinetics of hydrogen-
deuterium exchange in e.g. adenosine-5´-monophosphate.
A single Raman spectrum permits the 8-CH and 8-CD forms

to be distinguished from one another while the relative
concentrations are quantitatively determined by the intensities of the Raman scattering. Although the method has
been claimed to be superior to the tritium-labelling techniques it has not as yet been widely adopted.

Table 1. Procedures for following hydrogen isotope exchange
reactions

Procedure	Method	Compound	Comments	Refs.
	^{1}H n.m.r.	dihydroxyacetone phosphate		6
		L-alanine		7
Deuteriation (Dedeuteriation)		1,1-bisethyl-sulphonyl methane		8
	Infra-red	aromatic hydrocarbons		9
		L-alanine		10
	^{2}H n.m.r.	trans-decalone		11
	Mass spectrometry	1,3,5-trimethoxy benzene		12
	Laser Raman	adenine		13
Detritiation	^{13}C n.m.r.	alkenes		14
	Sample withdrawal/	phenylacetylene		15
(Tritiation)		trimethoxybenzene		16
	liquid scintillation	acetophenone		17
	counting	dimethylsulphoxide		18
		benzimidazole		19
		unsymmetrical ketones	exchange from 2 sites	20
		xanthosine	exchange accompanied by hydrolysis	21

3.   Exchange Reactions

(a) Acid-catalysed

Work in this area has centred on determining the reaction mechanism and the effects of substituents on reaction rates and inevitably some of the studies overlap. Early work by Ingold[26] established aromatic hydrogen exchange

$$ArH + {}^*HA \rightleftharpoons Ar\overset{*}{H} + HA$$

as an example of electrophilic substitution but it was only after the second world war when deuterium became more readily available that more detailed investigations were undertaken. Isotopic hydrogen exchange studies showed that the rate coefficients increased with increasing concentration of acid (H_2SO_4, HCl, and H_3PO_4 were most frequently used). In many cases a linear relationship was found to exist between the logarithm of the rate coefficient and the acidity of the medium as expressed by the acidity function H_o; moreover the slope was close to unity and on the basis of such findings it was argued that the mechanism was of the A-1 kind in which the rapid pre-equilibrium addition of a proton to the aromatic substrate was followed by a slow rate-determining rearrangement of the resulting intermediate:-

$$ArH + {}^*HA \rightleftharpoons Ar\overset{+}{H}\overset{*}{H} + A^- \quad \text{(fast pre-equilibrium)}$$
$$Ar\overset{+}{H}H \longrightarrow Ar\overset{*}{H} + \overset{+}{H} \quad \text{(slow)}$$

Aromatic hydrogen exchange should therefore exhibit specific hydrogen ion catalysis. Results of primary

and solvent isotope effect studies were also interpreted
as indicating support for the A-1 mechanism. However in
1959 Kresge and Chiang[27] measured the rates of detritiation
of $[1-^3H]$-2,4,6-trimethoxybenzene in acetate buffers and
found the first-order rate coefficients increased by a
factor of 3.3 in going from 0.01 to 0.1M acetic acid whereas
no increase would have been expected if the reaction was
specifically catalysed by hydrogen ions. The hydronium-
ion catalysed rate constant was obtained from runs in dilute
hydrochloric acid solutions and allowance for this in the
acetate buffers showed that aromatic hydrogen exchange
was indeed general-acid catalysed. The ability of the tritium
tracer technique to measure very slow rates of reaction
is shown to good advantage in this study. Subsequent studies[28,29]
covering a wide range of reactants have served to confirm
that aromatic hydrogen exchange is subject to general acid
catalysis, the generally accepted mechanism being of the
A-S_E2 type:

$$ArH + \overset{*}{H}A \underset{k_{-1}}{\overset{k_1}{\rightleftharpoons}} Ar\overset{+}{H}\overset{*}{H} + A^-$$

$$Ar\overset{+}{H}\overset{*}{H} + A^- \underset{k_{-2}}{\overset{k_2}{\rightleftharpoons}} Ar\overset{*}{H} + HA$$

If rates of detritiation (rather than tritiation) of the
aromatic are being investigated the second equilibirum can
be considered effectively irreversible, because the tritium
is present at tracer level. Also, under the experimental
conditions the benzenonium ion is present in very low
concentration compared to Ar$\overset{*}{H}$ and a steady state treatment
gives the rate of loss of tritium as

$$-d[Ar\overset{*}{H}]/dt = \frac{k_1}{1 + \dfrac{k_{-1}}{k_2}} [Ar\overset{*}{H}][HA] \qquad (13)$$

The ratio k_{-1}/k_2 (the ratio of the rate of H versus T removal
from the intermediate) can be estimated so that the rate of
detritiation from the aromatic can be calculated from the
observed rate coefficient. Table 2 contains such data for
the hydronium ion catalysed detritiation of a number of
aromatic compounds. The rates vary by a factor of 10^{16}
and parallel the ease with which the aromatics can be protonated, as reflected in the pK values. If on the other hand
we keep to a single substrate such as 1,3,5-trimethoxybenzene,
we find a similar kind of relationship (Table 3); the pK
values of the acids cover 17 units and the corresponding rate
constants vary by 9 powers of ten, the strongest acid (H_3O^+)
being the most effective catalyst.

Table 2. Hydronium ion detritiation rate constants (k_H^T+)
and protonation pK values for various aromatic compounds[28]

Substrate	$\log_{10} k_H^T + (25^{\circ})$	pK
Guaiazulene	+0.79	+1.5
4,6,8-Trimethylazulene	+0.60	+0.5
Azulene	-0.74	-1.7
1,3,5-Triethoxybenzene	-1.90	-4.8
1,3,5-Trimethoxybenzene	-2.21	-5.7
1,3-Dimethoxybenzene[a]	-5.10	-9.0
Anisole	-8.8	-15.3
Benzene	-15.2[b]	-23.0

[a] Exchange at the 4 position

[b] Rate of dedeuteriation

Table 3. Acid-catalysed detritiation rate constants (k_{HA}^T) for 1,3,5 -(ring-[^{3}H])-trimethoxybenzene[16] at 25°C

Acid	K_{HA}	$k_{HA}^T (M^{-1} min^{-1})$
H_3O^+	55	0.40
CH_2FCO_2H	2.9×10^{-3}	5.2×10^{-3}
HCO_2H	2.85×10^{-4}	6.9×10^{-4}
CH_3CO_2H	2.82×10^{-5}	3.84×10^{-4}
$H_2PO_4^-$	6.23×10^{-8}	2.32×10^{-5}
NH_4^+	5.5×10^{-10}	3.0×10^{-7}
H_2O	1.79×10^{-16}	5.7×10^{-10}

In comparing the relative reactivities of different aromatics towards isotopic hydrogen exchange the best approach is to employ a one component solvent system rather than an aqueous acid mixture of fixed composition. For this purpose anhydrous trifluoroacetic acid has proved to be a very good choice and a large number of results pertaining to this solvent have been brought together by Taylor;[30] these include benzene, alkyl benzenes, substituted anisoles, polycyclic aromatic hydrocarbons such as fluorenes, naphthalenes and phenanthrenes, as well as heterocyclic compounds. For compounds which have very high reactivity anhydrous trifluoroacetic acid will give inconveniently fast rates, even at low temperatures. In these circumstances mixtures of acetic and trifluoroacetic acids have been used.

All organic compounds in the presence of strong enough bases are acids; likewise in highly acidic media they are potential bases. The best measure of acidity under such conditions is the acidity function H_o (= $pK_{BH}^+ - \ell g[BH^+]/[B]$). Despite the fact that because of the different behaviour of indicator bases no great significance can be attached to

the absolute value of H_o, it can provide a semi-quantitative
measure of the relative acidities of various highly acidic
media. The available results show that these are very
much the same in 100% acid as in dilute aqueous solution
although HF which is a rather weak acid in these circumstances
approaches the acidity of 100% H_2SO_4 in very concentrated
solutions. Still further increases in acidity can be brought
about by the addition of a substance that is capable of
protonating the acid e.g. SbF_5 to fluorosulphonic acid

$$SbF_5 + 2HSO_3F \rightleftharpoons H_2SO_3F^+ + SbF_5(SO_3F)^-$$

The $-H_o$ values for such solutions (> 18) are more than 6
units higher than for 100% H_2SO_4. It is no surprise therefore
that the SbF_5 - DSO_3F system has been used to deuteriate such
weak bases as methane[31] and hydrogen.[32] For the latter the most
probable mechanism is:-

$$H_2 + D^+ \rightleftharpoons [H_2D^+] \rightleftharpoons HD + H^+$$

and this kind of addition - elimination mechanism is greatly
favoured in acid catalysed exchange of aliphatic systems.
Cycloalkanes[33] as well as t-butyl chloride[34] have been satisfactorily
deuteriated using this approach. The propensity of carbonium
ions to rearrange rapidly is the main reason why the method
has not been as extensively used as the corresponding base-
catalysed deprotonation reaction in the labelling of organic
compounds. An alternative to using highly acidic media
is to use conditions of high temperature ($>200^\circ C$) and dilute
acid and this approach has only recently been explored.[35]
Because of the low cost, simplicity of the procedure, high
degree of isotope incorporation and the ability to make large
scale preparations the method offers distinct advantages.

(b) Base-catalysed

Nearly all organic compounds come under the heading of carbon acids. Treatment with a sufficiently strong base leads to ionisation

$$\equiv\!C\!-\!H + B^- \rightleftharpoons \equiv\!C^- + BH$$

and in the presence of deuterium or tritium this will lead to isotopic hydrogen exchange. The extent to which exchange takes place depends on two factors (a) the acidity of the carbon acid and (b) the basicity of the medium. Conversely labelled compounds produced in this way can be safely used as tracers only when the acidity of the compound and the basicity of the medium in which they are to be employed are known.

The acidities of carbon acids vary greatly - some are more acidic than mineral acids whilst for others the measurement of the very feeble acidic properties constitutes a serious problem in itself. It is useful to classify carbon acids in terms of the activating group responsible for conferring the acidity. Table 4 contains pK_a data on a representative range of carbon acids; the powerful activating effect of both cyano and nitro groups is clearly evident. In general the acidities of carbon acids do not change greatly (+ 1-2 pK units) as the solvent is altered, a notable exception being nitromethane which is some 5-7 pK units less acidic in dimethyl sulphoxide than in water.

Table 4. Acidities of some carbon acids

Activating group	Compound	pK_a *
Cyano	Cyanoform	-5.1
	Malononitrile	11.2
	Methyl cyanide	~25
Nitro	Trinitromethane	0
	Dinitromethane	3.6
	Nitromethane	10.2
Sulphone	Dimethyl sulphone	28.5[a]
	Dibenzyl sulphone	22[a]
Keto	Acetylacetone	8.83
	Benzoyltrifluoroacetone	6.54
	Acetophenone	21.5[b]
	Methyl styryl ketone	21.6[b]
Hydrocarbon	Fluorene	19.4[c]
	Indene	19.9[c]
	Benzanthrene	21.2[c]
	1,1,3-Triphenylpropene	26.6[c]

* in H_2O at $25^{\circ}C$ unless specified otherwise

a in dimethyl sulphoxide

b in dimethyl sulphoxide - water mixtures

c in cyclohexylamine.

The strongest base that can exist in water is the hydroxide ion. The fact that it is so strongly hydrated - each hydroxide ion is thought to be strongly bonded to three water molecules - must however inhibit its ability as a catalyst. This can be seen very clearly when a dipolar aprotic solvent such as dimethyl sulphoxide is added to a 0.01M solution of tetramethylammonium hydroxide in water. The acidity function H_- which measures the ability of a basic solution to abstract a proton from a neutral acid, increases from a value of 12 in the purely aqueous conditions to 26 in 99.5 mole % Me_2SO, corresponding to an increase of 14 powers of ten in basicity.[36] This tremendous increase comes about in the main through desolvation of the hydroxide ion. Other dipolar aprotic solvents such as dimethylformamide and hexamethylphosphoramide act in a similar but less effective manner. Highly basic media involving alcohol alkoxide solutions, to which dimethyl sulphoxide has been added, can also be produced. Other non-aqueous media of high basicity include solutions of amide salts in liquid ammonia or some amine solvent such as cyclohexylamine.

Similar rate-equilibrium correlations are found for base-catalysed exchange reactions as for the acid-catalysed reactions. Thus for a given catalyst the rates of isotope exchange decrease with decreasing acidity, the degree of correlation being at its best when carbon acids having the same activating group and closely similar structures are considered. Significant deviations are observed in the case of aliphatic nitro compounds where isotope exchange is slower than for other carbon acids of similar pK_a's, and for β-diketones that are known to be extensively hydrated.

For the more acidic carbon acids (pK_a < 10) general
base catalysis is normally observed but with decreasing
acidity hydroxide ion catalysis becomes predominant and finally
exclusive. In studies in highly basic media it has been
found that the weaker the acid the greater the rate accelerations - these can be as high as 10^8 - 10^{11} with the resulting
benefit that carbon acids as weak as toluene can be labelled
by base-catalysed isotopic exchange procedures. Some of the
carbon acids that have been labelled by base catalysed
exchange, together with appropriate references, are given in
Table 5.

Table 5.   Examples of Compounds Labelled by Base Catalysed
Hydrogen Isotope Exchange

Experimental Conditions	Compound(s)	Position Labelled	Refs.
Neutral	β-Diketones e.g. acetylacetone	methylene	37
	β-Disulphones e.g. bisethyl-sulphonylmethane	methylene	8
	Cyclohexanones e.g. 2-acetyl-cyclohexanone	α to $>$C=O group	38
	Heterocyclics e.g. purine, caffeine	C-8	39,40
	Malononitrile	methylene	41
Alkaline (pH 7-12)	Acetophenone	methyl	17
	Diethyl malonate	methylene	42
	Phenylacetylene	acetylenic	15
	Nitromethane	methyl	43
	p-Nitrobenzyl cyanide	methylene	44
Highly basic media (H_12-30)	Fluorene	C-9	45
	Dimethyl sulphoxide	methyl	18
	Triphenylmethane	$>$C-H	46
	Toluene	ring and/or side chain	47

It is implicit in the above discussion that the carbon acid
is undergoing exchange as the neutral species; other forms
can also undergo reaction e.g. the salts of carboxylic acids[48]
can be deuteriated in the α-position merely by refluxing in
alkaline D_2O, the probable mechanism is one in which the
rate-determining step is cleavage of the C-H bond, leading
to the formation of an α-carbanion carboxylate intermediate:

$$Ar-\underset{\underset{H}{|}}{C}H-CO_2^- \ + \ OD^- \ \xrightarrow{\text{slow}} \ Ar-\bar{C}H-CO_2^- \ + \ HDO$$

$$\text{fast} \downarrow D_2O$$

$$ArCHD.CO_2^- \ + \ OD^-$$

A reaction between two negatively charged species is, on
electrostatic grounds alone, considerably less favourable than
one between a neutral species and an anion. Equally a reaction
between a positively charged carbon acid and an anion should
be favoured. The importance of electrostatic and heteroatom
effects can clearly be seen in isotope exchange studies involving heterocyclic compounds. Exchange from the C-2 position
in imidazole or the equivalent C-8 position in purines is
specifically catalysed by hydroxide ions; one can therefore
employ rate-pH profiles in order to provide details of the
reaction mechanism - these can also be used to obtain the
pK_a values for either ionisation or protonation. Thus adenine
itself gives a bell-shaped rate-pH profile (Fig.4) which
tells us that the anionic form (formed by ionisation of the
N-9-H) is unreactive and that reaction occurs between the
protonated form (protonation at N-7) and hydroxide ions.
Blocking the N-9 position, as in 9-alkyl purines or adenosine,
makes it impossible to form the anionic form and at high pH

the neutral form can now undergo exchange – the relative
reactivities of the two species differ by a factor of 10^8.
If the negative charge is developed at a site well removed
from the exchanging CH group there is no reason why the
anionic form can not undergo exchange – in the case of
theobromine for example all three forms, protonated, neutral
and anionic are reactive. The rate-pH profile for xanthine
is similar to that found for guanine where two mechanisms
are operative, hydroxide ion attack on the protonated and
neutral species. The results for xanthosine diverge from
those of xanthine at high pH, signifying that the dianionic
species, which is the predominant form present in solution,
must be undergoing exchange.

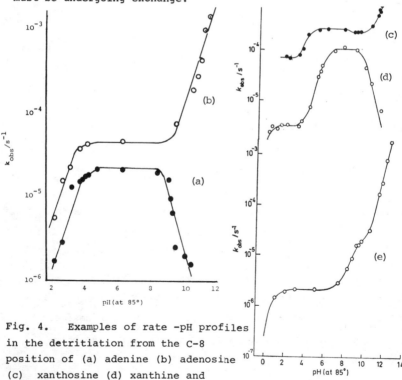

Fig. 4.   Examples of rate –pH profiles
in the detritiation from the C-8
position of (a) adenine (b) adenosine
(c) xanthosine (d) xanthine and
(e) theobromine

The characteristic shapes of the rate-pH profiles tell us a great deal about the optimum conditions for labelling these compounds as well as the storage conditions that should be employed. They can however provide no information when kinetically equivalent species are involved: e.g. in the case of adenosine 5′-monophosphate there is reason to believe that the zwitterion (H^+AMP^{2-}) as well as the species $AMPH^-$, formed through the equilibrium, are both involved in the overall exchange process[49].

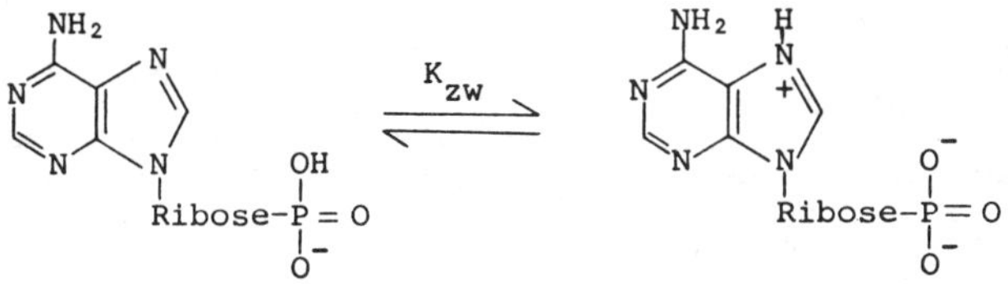

$$(AMPH^-) \quad \xrightleftharpoons{K_{zw}} \quad (H^+AMP^{2-})$$

In some cases carbon acids react with bases to give species other than carbanions. This is particularly true of aromatic nitro compounds, where several kinds of interaction have been recognised[50]. First, there is the charge-transfer complex (1) formed as a result of partial transfer of electronic charge from the base; complete electron transfer leads to the formation of a radical anion (2), readily recognised by its electron spin resonance spectrum. When the unshared electron pair of the base B^- forms a covalent bond to an aromatic carbon atom the benzenoid resonance is lost and a Meisenheimer complex (3) is formed. In the deuteriation (or tritiation) of nitroaromatics the experimental conditions should be carefully chosen so that nucleophilic displacement of the nitro group(s) does not take place in preference to exchange.

NO_2⌬NO_2 —B^- NO_2
$[$ NO_2⌬NO_2 $]^{\cdot -}$ NO_2
H B / NO_2⌬NO_2 / NO_2

(1)   (2)   (3)

(c) Metal-catalysed

Catalytic labelling of organic compounds can be
carried out under both heterogeneous and homogeneous conditions. The former require higher temperatures and
pressures; catalytic poisoning and the need to regenerate
the catalyst can also be complicating factors. In addition
reaction mechanism studies are much easier to carry out
on homogeneous systems. Nevertheless much more work has
been carried out under heterogeneous than under homogeneous
conditions. Group VIII transition metals have been the
most extensively studied catalysts - these include Pt,
Pd, Ni, Co, Fe, Ru, Ir, Rh and Os in either supported or
unsupported form. Isotopic water has usually been prefered
to enriched hydrogen gas as the source of the label mainly
because with the gas competing hydrogenation can occur
yielding by-products that can render purification difficult.
The operating temperatures have normally been in the range
$100-200^{\circ}C$. The catalysts, of which Pt is the most active,
are employed as the oxides or chlorides, activation being
achieved by either (a) hydrogen reduction, (b) borohydride
reduction or (c) self-activation by the substrate during
exchange.

The mechanisms of heterogeneous metal catalysed isotope
exchange can be divided into two categories - classical and
π-complex; each can be further subdivided into dissociative
and associative exchange mechanisms and are illustrated in
Table 6 using benzene as the organic substrate and pre-reduced
PtO_2 as the catalyst. The isotope label reacts with the metal

in the following manner:-

$$D_2 + 2Pt \rightleftharpoons 2Pt\overset{\displaystyle D}{|}$$

$$D_2O + 2Pt \rightleftharpoons \underset{\displaystyle Pt}{\overset{\displaystyle D}{|}} + \underset{\displaystyle Pt}{\overset{\displaystyle OD}{|}}$$

Table 6. Mechanisms of Heterogenous Metal-catalysed
Isotope Exchange [51]

Mechanism	Details	Comments

Dissociative / Classical / Associative mechanisms:

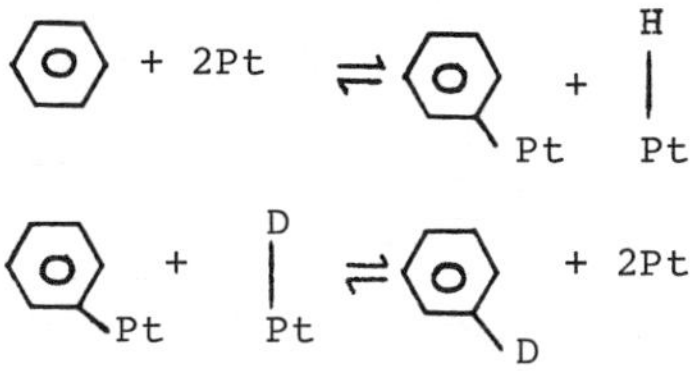

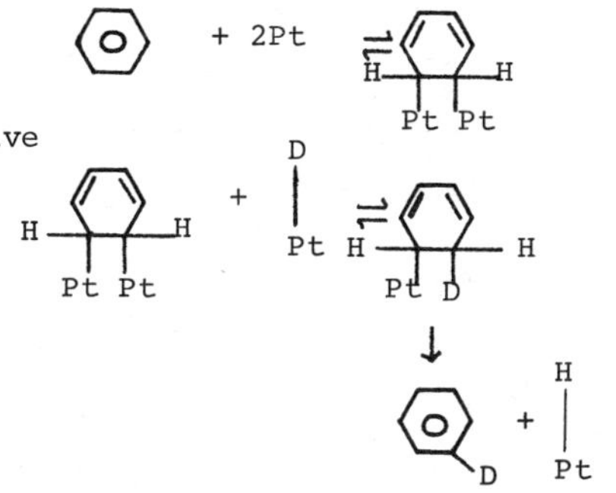

Comments: Compound is chemisorbed in a dissociative manner.

Comments: Compound is chemisorbed in an associative manner and exchange occurs via the half-hydrogenated state.

Dissociative / π-Complex / Associative mechanisms:

Comments: Molecule rotates through 90° from a horizontal π bonded complex to a vertical σ-bonded chemisorbed state

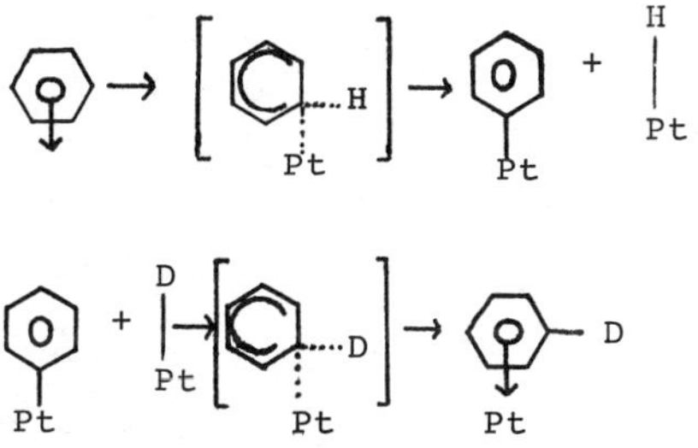

Comments: Mechanism is similar to a radical substitution reaction

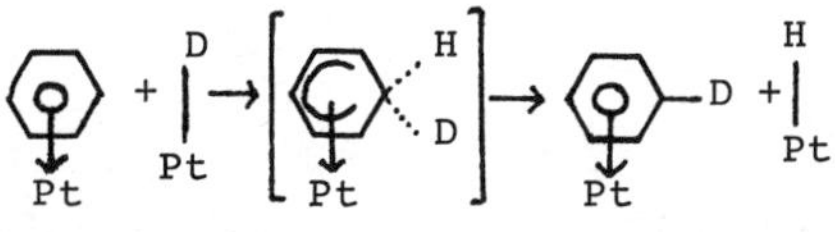

Heterogeneous Pt exchange has been successfully used to label a wide range of organic compounds - these include amino acids, polycyclic aromatic hydrocarbons, heterocyclics and steroids. Invariably good isotope incorporation is obtained although certain groups such as iodo and nitro tend to poison the catalyst.

The study of corresponding homogeneous hydrogen isotope exchange reactions started in 1967 with the observation by Garnett and Hodges[52] that the $PtCl_4^{2-}$ ion could induce such exchange in aromatic compounds. Shortly afterwards Shilov and coworkers[53] showed that alkanes could also be labelled and subsequently futher work has shown that the heterogeneous and homogeneous exchanges have many features in common. This has led to the suggestion that the homogeneous system could serve as a model for elucidating the mechanisms of a variety of heterogeneous catalytic processes.

The conditions for following homogeneous hydrogen isotope exchange used by Garnett and coworkers (Table 7) have been adopted by others. The presence of the acetic acid ensures homogeneity and the added hydrochloric acid reduces the tendency of the tetrachloroplatinate to disproportionate according to the equation

$$2PtCl_4^{2-} \rightleftharpoons Pt^O + Pt^{IV}Cl_6^{2-} + 2Cl^-$$

at the operating temperature (80-200°C). The precipitated Pt^O and the resulting Pt^{IV} do not promote exchange. In solutions of Pt(II) chloride the activity of the various platinum containing species[54] towards exchange is in the order $S_2PtCl_2 > SPtCl_3^- > PtCl_4^{2-}$ (S = solvent); ligands with the properties of soft bases (CN^-, CNS^-, PPh_3) decrease the rate of hydrogen exchange.

Under homogeneous conditions arene hydrocarbons are in general more ·reactive to hydrogen isotope exchange than alkanes: benzene, the least reactive arene is about twice as reactive as the most reactive alkane, cyclohexane.[55] Alkane reactivity decreases in the order primary > secondary > tertiary. In the alkyl benzenes exchange occurs both in the ring and the side-chain; when the latter is long preference for terminal methyl labelling is observed.[56] Arene and benzylic hydrogens are of the same order of reactivity; strong ortho-deactivation is observed in instances where steric hindrance is important.

For the other homogeneous catalysts that have been employed (Table 7) mineral acid stabilisation is not necessary and the exchange is carried out at slightly higher temperatures to compensate for the slower rates. Although these catalysts all have a common feature in that more than one hydrogen atom can be exchanged during a single residence of the substrate on the catalyst (the so-called multiple exchange factor), they also show different selectivities towards certain substrates: e.g. in n-butyl benzene the β and γ hydrogens are virtually unreactive towards $RhCl_3$ and the terminal methyl group of ethyl benzene is deuteriated to a lesser extent with a hexachloroiridate (III) catalyst then with $PtCl_4^{2-}$. Instances where heterogeneous exchange is difficult but homogeneous exchange occurs readily have also been reported:[52] acetophenone, nitrobenzene, bromobenzene and naphthalene, require a long time to exchange with a heterogeneous platinum catalyst whereas with tetrachloroplatinate they are easily labelled.

Table 7. Catalysts for Homogeneous Hydrogen Isotope Exchange

Catalyst	Procedure	Comments	Refs.
Na_2PtCl_4	The substrate (0.5ml) in a catalyst solution of CH_3COOH (2 mole), THO (1 mole), HCl ($2.6 \times 10^{-3}M$) and Na_2PtCl_4 ($2.6 \times 10^{-3}M$) is heated for 5h. at $120^{O}C$ in a sealed ampoule.	Alkanes, alkylbenzenes have been successfully labelled.	56
Na_3IrCl_6	The substrate is heated together with the catalyst (0.02M) in 25 mole % $MeCO_2D/D_2O$ for 160h at $130^{O}C$.	Toluene and xylenes are ring labelled. Longer chain alkyl benzenes are labelled in side-chain (mainly terminal methyl)	57
$RhCl_3$	The substrate is heated together with the catalyst (0.02M) in $MeCO_2D$: D_2O (1:1) for 96h at $130^{O}C$.	Alkyl benzenes are generally labelled. Exchange is slower than for Pt & Ir. Small poisoning effects by e.g. NO_2 group.	58

The heterogeneous metal catalysed exchange method using tritium gas which was developed by Evans and coworkers is discussed in Chapter 2. Buchman and Pri-bar[60] have adopted a similar approach using activated palladium, prepared by reduction of the oxide, as the catalyst. The method has been used to tritiate a number of tricyclic antidepressants to high specific activity.

Polyhydride catalysts of the type $TaH_3(\pi-C_6H_5)_2$, $IrH_5[P(CH_3]_2$ and $NbH_3(\pi-C_6H_5)_2$ in the presence of deuterium gas catalyse exchange in a number of aromatic hydrocarbons.[61] In contrast to Pt or $PtCl_4{}^{2-}$ only aryl C-H bonds exchange whilst little deuteriation occurs _ortho_ to bulky groups such as methyl; exchange seems to occur in a stepwise manner, again different to the platinum catalysts.

Lewis acids of the type $SbCl_5$, $AlCl_3$, BBr_3 and $EtAlCl_2$ have been widely used as hydrogen isotope exchange catalysts,[62-64] ethylaluminium dichloride being the most effective. For deuteriation studies C_6D_6 is used as the source of the label but in tritiation work the substrate-catalyst complex is hydrolysed with a small amount of high specific activity tritiated water. Ethylaluminium dichloride can induce rapid exchange at room temperature and is characterised by a high selectivity for aromatic protons and absence of steric effects. Both cyclic and non-cyclic alkanes can be labelled at elevated temperatures although there is some evidence of isomerisation. The preferred mechanism is as follows, with HCl acting as a co-catalyst:

$$HCl + EtAlCl_2 \longrightarrow H^+[EtAlCl_3]^-$$

The different characteristics of these catalysts now
lead to the possibility that they may be used in conjunction
with one another to provide specifically labelled compounds.
Thus the relatively large steric requirement of the Pt
catalysts means that with mesitylene only the methyl groups
are labelled whereas with $EtAlCl_2$ the label is entirely in the
ring. When bromobenzene is treated with BBr_3 the label is
mostly located at the <u>ortho</u> and <u>para</u> positions; treatment with
a Pt catalyst leaves only label at the <u>ortho</u> position.[63]
$[^2H_8]$ Toluene, prepared by heterogeneous Pt exchange, on
treatment with $SbCl_5$, leads to $[^2H_3]$ toluene.[65]

(d) Photochemical and Radiation-induced.

Although considerable work has been carried out on
photosubstitution reactions very little has been reported on
photochemical methods of inducing hydrogen isotope exchange –
the first review of the subject has only recently appeared.[66]
This may in part reflect the very wide choice of methods that
are already available; the danger of increasing the proportion
of side reactions may be another factor.

It has been known for some time that the basicities of
many aromatic molecules are increased in the excited state, and
the first example of photoinduced electrophilic aromatic substitution was reported in 1963 - this involved the deuteriation
of 9,10-dimethyl-anthracene.[67] Subsequent studies involving
toluene, anisole, various xylenes and nitrobenzene have been
reported; exchange occurs via the S_E2Ar^* mechanism, the counterpart of the acid-catalysed mechanism $(A-S_E2)$. Thus for the
photodedeuteriation of toluene we have:

(* denotes excited state)

Irradiation of a mixture of benzonitrile and tritiated
diethylamine (i.e. $(C_2H_5)_2NT$) in benzene with a high pressure
mercury arc lamp for up to 20h showed that tritium had been
incorporated in all ring positions, the order of reactivity
being para >> ortho > meta.[68] With propanethiol the rate of
tritium incorporation is a thousand fold faster than in acetic
acid, despite the latter's greater acidity. This result suggests
that the basicity of the excited state is not an important
factor in the mechanism but rather that complex formation occurs
between the singlet excited benzonitrile and the amine, followed
by electron transfer from the quencher to the quenchee to give
a radical cation and a radical anion respectively. Exchange
occurs from the radical cation:

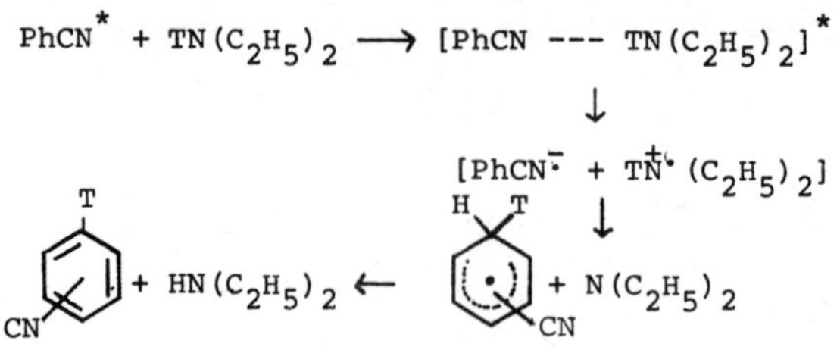

$$PhCN^* + TN(C_2H_5)_2 \longrightarrow [PhCN \text{ --- } TN(C_2H_5)_2]^*$$

$$\downarrow$$

$$[PhCN^{\overline{\cdot}} + T\overset{+\cdot}{N}(C_2H_5)_2]$$

The same group have reported[69] on the light induced tritiation
of methyl benzoate via a similar mechanism.

The large majority of the previous examples of isotopic
hydrogen exchange have been concerned with proton (H^+) transfers.
Exchange can however take place via hydrogen atom ($H\cdot$) transfer:

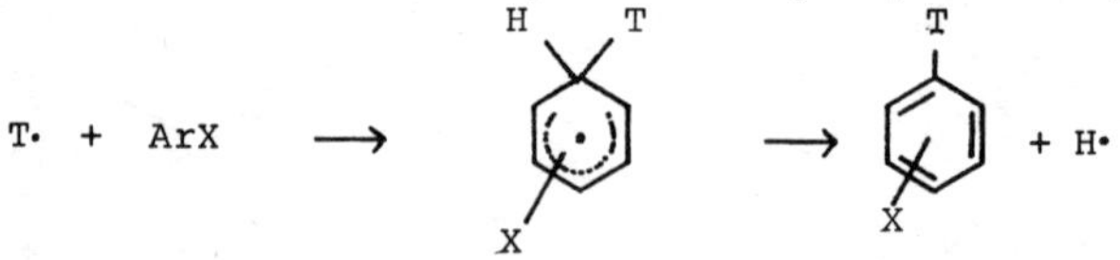

The hydrogen atom is the simplest possible neutral free radical
and its reactions have been studied in detail in the gas phase
in connection with research on combustion. In solution the
main emphasis has been on the role it plays in radiation biology.
Deuterium and tritium atoms can best be generated in solution
by photolysis of deuteriated and tritiated thiophenols and as
the reactions of the hydrogen atom in solution are now well
documented[70] this approach to the labelling of organic compounds
offers many opportunities.

The hydrogen atom is also one of the species produced in
the radiolysis of aqueous solutions, the others being the
solvated electron (e^-_{aq}), the hydroxyl radical ($OH\cdot$) and the
solvated proton ($H_3\overset{+}{O}$); radiolysis can be achieved by internal
(tritium β^-radiation) or external means e.g. ^{60}Co source. The
formation of highly reactive radical species opens up new
avenues of isotopic hydrogen exchange[71] although in the case of

the hydroxyl radical reaction with aromatic compounds only leads to hydroxylation or oxidation (see p. 393).

The mechanism of the hydrogen atom induced exchange is the same as for the corresponding photochemical reaction, namely hydrogen atom addition to form a cyclohexadienyl radical followed by hydrogen loss. Indeed it is a feature of radiation induced reactions that they usually involve one oxidation and one reduction step.

Isotope exchange induced by the solvated electron in aqueous solution can only take place at high pH because at low (<7) values the active species are rapidly converted into hydrogen atoms:-

$$e^-_{aq} + H_3\overset{+}{O} \longrightarrow H\cdot + H_2O$$

In the case of benzene the solvated electron reacts in the following manner,

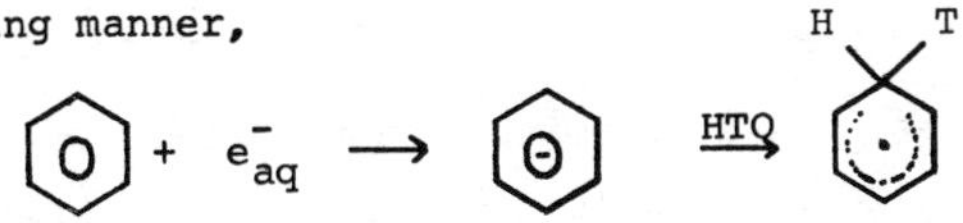

$$\text{C}_6\text{H}_6 + e^-_{aq} \longrightarrow [\text{C}_6\text{H}_6]^- \overset{HTO}{\Longrightarrow} \text{cyclohexadienyl radical (H, T)}$$

the cyclohexadienyl radical being subsequently oxidised. The limited experimental data seem to indicate that the solvated electron is not as effective as the hydrogen atom in inducing isotope exchange.

Saturated aliphatic compounds such as ethers and alcohols are unreactive towards solvated electrons. Nevertheless the radiation induced exchange reaction is retarded both by the addition of electron scavengers (e.g. Ag^+, Cu^{2+}, Ni^{2+}) and hydroxyl radical scavengers (N_3^-, Br^-). Such findings require the involvement of two primary radiolysis products and in the case of dioxane for example the following mechanism has been proposed:[72]

In principle these radiation induced exchange reactions
offer an alternative method of tritiating organic compounds
provided simultaneous degradation of the reactant is not a
problem. As long ago as 1957 Wilzbach[73] reported on the successful
tritiation of a number of organic compounds after they had been
exposed to tritium gas for different periods of time. Much
subsequent work has been devoted to (a) improving the degree of
incorporation and (b) minimising the extent of by-product formation. Garnett and coworkers[74,75] have shown that saturated and
unsaturated aliphatics as well as aromatic hydrocarbons, as
exemplified by cyclohexane, 1-hexene and benzene may be labelled
to specific activities in the μCi/mmole range when THO of
5Ci/mmole specific activity is used as the isotope source.
The radiochemical purity is very high (98%) and much better than
in the Wilzbach method where up to 70% of the incorporated
activity may appear as a mixture of addition and degradation
products, making radiochemical purification very difficult.
This work shows that radio-gas chromatography can provide a very
sensitive means of monitoring the appearance of by-products.
The possibilities of the method have been further explored in
the tritiation of sugars containing asymmetric centres but much
more work needs to be done in order to ascertain the upper limit
of attainable specific activity that is possible before degradation becomes a problem.

(e) Enzymic.

Although there are many examples of isotope exchange
reactions in biochemistry most of these involve the transfer
of another group as in hydrolysis, elimination and substitution reactions. Instances of hydrogen isotope exchange
as discussed in the present chapter are relatively few.
Undoubtedly the most important is the transamination-
catalysed isotope exchange of amino acids [76,77] which is a
convenient method for obtaining these compounds labelled
at the α and sometimes β positions. Several pyridoxal-
containing enzymes (the first five in Table 8) have been
used and the generally accepted mechanism once again
represents an example of carbon-hydrogen bond activation.
The initial step is the formation of a Schiff's base
(aldimine 4), followed by abstraction of the α-proton to
give a carbanion (5); exchange with the solvent (D_2O) and
addition of the deuteron at the C-4′ position of pyridoxal
gives rise to the ketimine (6). The imino function at C-2
facilitates abstraction of a β proton yielding the enamine-
carbanion intermediate (7). Reversal of steps 0-3 then
results in the formation of the labelled amino acid:-

Table 8. Enzyme-catalysed Hydrogen Isotope Exchange Reactions

Enzyme	Compounds Labelled	Refs.
Cystathionine γ-synthase	L-α-aminobutyrate	77
γ-Cystathionase	L-amino acids - alanine, serine, α-aminobutyrate, norvaline-all deuteriated in α (and sometimes β) positions. Homoserine exchange accompanied by elimination.	78
Glutamate-alanine transaminase	L-alanine, - glutamate, - aspartate	7
Glutamate pyruvate transaminase	L-alanine, glycine	76
E. coli aspartate aminotransferase	L-phenylalanine, - aspartate	79
Rabbit muscle aldolase	Dihydroxyacetone phosphate, D-fructose 1,6-bisphosphate D-fructose 1-phosphate	6

A ketimine intermediate is also involved when rabbit muscle aldolase catalyses exchange in ketones; this enzyme has a lysine ε-amino group (at the active site) which interacts with ketonic substrates. In the case of dihydroxyacetone phosphate further reaction occurs with the loss of one of the C-3 hydrogens to yield an enamine(8):

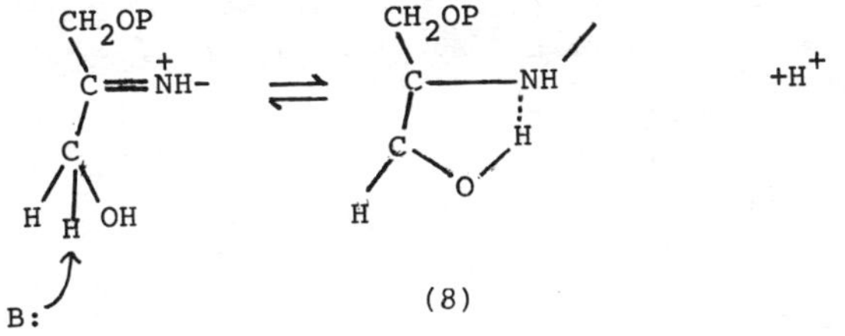

(8)

(f) Miscellaneous

Because of solubility difficulties it is sometimes necessary to carry out isotope exchange reactions under heterogeneous conditions. The difficulties that arise from the inability of the reactants to come together may be overcome by the addition of small quantities of an agent which transfers one reactant across the interface into the other phase. The phase transfer agent[80] is not consumed but performs the transport function repeatedly. Organic-soluble quaternary ammonium or phosphonium cations have been found to be excellent agents for the transport of anions from an aqueous phase to an organic phase and this kind of phase-transfer catalysis has been widely used in synthetic organic chemistry. Its application in isotopic hydrogen exchange studies can be seen by reference to the deuteriation of 2-octanone[81] in a 5% NaOD in D_2O solution. After 3h at $30^\circ C$ less than 5% incorporation of deuterium had taken place. Repeating the experiment in the presence of a quaternary ammonium salt led to complete equilibration of the 1- and 3-carbon hydrogens within 0.5h. The phase transfer can be depicted as follows:

$$NR_4^+ \; X^- \; + \; OD^- \; \rightleftharpoons \; NR_4^+ \; OD^- \; + \; X^- \qquad \text{aqueous phase}$$

$$NR_4^+ \; OD^- \; + \; SH \; \rightleftharpoons \; S^-NR_4^+ \; + \; HDO \qquad \text{organic phase}$$

$$S^-NR_4^+ \; + \; D_2O \; \rightleftharpoons \; SD \; + \; NR_4^+ \; OD^-$$

The deuteriation of a number of alkyl thiazoles in the presence[82]
of various quaternary salts can be sufficiently accelerated
so that the C-5(H) can be labelled as well as the more reactive
C-2(H). Isotopic hydrogen exchange under these conditions
becomes more attractive and the method is to be preferred to
the alternative procedures - reduction of the halogenated
derivatives by deuteriated acetic acid in the presence of
zinc dust or hydrolysis of the lithium derivative in the
presence of D_2O.

Electrochemical methods of synthesis find wide application
in organic chemistry.[83] Until recently it was customarily assumed
that the electrolytic reduction of alkyl halides proceeded
with no regeneration of the halide i.e. step 1 was strictly
irreversible

$$RX \; \underset{1}{\overset{+e}{\rightleftharpoons}} \; [RX^-] \; \longrightarrow \; R' \; + \; X^- \; \longrightarrow \; \longrightarrow \; RH$$

This can be subjected to examination by carrying out the
reduction in a deuterium containing solvent. For phenacyl
chloride in dimethylformamide containing 1% D_2O the results[84]
revealed considerable deuteriation of the methylene hydrogens
in competition with the reduction. The most likely explanation
is that the radical anion, on tautomerisation and equilibration
with D_2O, is re-oxidised via electron transfer to the electrode
or to another molecular species giving the exchanged starting
material:

$$Ph-C\underset{CH_2Cl}{\overset{O}{\diagup}} \xrightarrow{\ e\ } Ph-\overset{\cdot}{C}\underset{CH_2Cl}{\overset{\bar{O}}{\diagup}} \rightleftharpoons Ph-\overset{\cdot}{C}\underset{CHCl^-}{\overset{OH}{\diagup}}$$

$$\Big\updownarrow$$

$$Ph-C\underset{CHDCl}{\overset{O}{\diagup}} \xleftarrow{\ -e\ } Ph\overset{\cdot}{C}\underset{CHCl^-}{\overset{OD}{\diagup}}$$

Somewhat similar findings have been reported by Merz[85]
who studied the electrochemical reduction of 2,2-dichloro-
1,1-bis(p̲-tolyl)ethylene (9) under strictly aprotic conditions
(0.5M LiBr in dimethylformamide); the electrochemically produced
carbanion intermediate (10)

$$\overset{R}{\underset{R}{>}}C{=}C\overset{Cl}{\underset{Cl}{\diagup}} \qquad \overset{R}{\underset{R}{>}}C{=}\underline{C}\overset{Cl}{\diagup} \qquad R-C{\equiv}C-R$$

R = p̲-tolyl (9)   (10)   (11)

rearranged to give bis(p̲-tolyl)-acetylene (11). In the presence
of D_2O (0.2M) the only product was (12), the rearrangement being

$$\overset{R}{\underset{R}{>}}C{=}C\overset{Cl}{\underset{D}{\diagdown}}$$

(12)

totally suppressed. The deuterium incorporation was much
higher when LiBr was the supporting electrolyte than when
tetra-n-butyl-ammonium bromide was used, a result that was
rationalised in terms of the widely different solvating properties of the two cations. This work was also noteworthy in
that co-electrolysis of a mixture of bromobenzene and fluorene
at cathode potentials around -2.3V led to good incorporation
of deuterium in the 9 and 10 positions of fluorene. This
shows that deprotonation of weak carbon acids is possible
even in the presence of water - presumably the tetrabutylammon-

ium cation is able largely to exclude water from the double
layer region where protonation of highly reactive carbanions
seems to proceed. The production of electrogenerated bases
under such circumstances greatly increases the scope of isotopic
hydrogen exchange reactions.

Apart from the above example and also the radiation-
induced reactions there are relatively few cases of isotopic
hydrogen exchange proceeding through oxidation-reduction type
mechanisms. A further instance is highlighted by the work of
Regen[86] who found that when alcohols were heated at $200^{O}C$ in
the presence of the catalyst dichlorotris (triphenylphosphine)
ruthenium(II) $[RuCl_2(PC_6H_5)_3]$ deuterium bound to oxygen
exchanged with hydrogen exclusively at the α-carbon atom:

$$R_1R_2CHOD \rightleftharpoons R_1R_2CDOH$$

The most commonly used method for preparing such compounds
entails oxidation to either the corresponding aldehyde or
carboxylic acid followed by reduction with lithium aluminium
deuteride. Thus two separate reactions and an expensive
reagent are required in contrast to this simple, inexpensive
and rapid one-step method. Whilst all primary alcohols undergo
exchange those secondary alcohols investigated gave poor
results. Subsequent work has shown that tertiary alcohols
do not undergo exchange, thereby suggesting that the mechanism
is an example of a reversible hydrogen transfer in an oxidation
- reduction reaction:

$$R_1R_2CHOH + (Ru) \rightleftharpoons R_1R_2 \overset{H}{-}C-O-Ru + HCl$$

$$R_1R_2CDOH + (Ru) \leftarrow \overset{D^+}{\underset{}{}} R_1R_2C=O \ldots (RuH)$$

Subsequent tritiation studies[87] have shown that some label
is incorporated in the β position and this can be explained
by the fact that when the aldehyde (or ketone) is formed the
hydrogen atom β to the hydroxylic group becomes α to the
carbonyl and can easily undergo exchange in the presence
of the catalyst:

$$R_1CH_2-\overset{\overset{\displaystyle OH}{|}}{CH}-R \longrightarrow RCH_2-\overset{\overset{\displaystyle O}{||}}{C}-R$$

Further reduction of the aldehyde (or ketone) during the
course of the reaction gives some β-labelled alcohol. Such
an interpretation is supported by the fact that tritiation
of 2-heptanone, on heating in the presence of the catalyst
for 1h, occurs in the α position.

4. Factors Affecting Exchange

We have already seen how the acid-catalysed rates of
detritiation in aromatic systems vary, both as a function
of the substrate and also the acid. Similar correlations
also exist for the base-catalysed detritiation of carbon
acids. The Brönsted relationship, of which these are examples,
constitutes one of the most successful linear free energy
relationships. It can provide information concerning the
nature of the transition state through which each reaction
must pass, as well as making possible a prediction of the
rate of isotopic hydrogen exchange of a particular compound.
There are however instances when serious deviations may
arise. Such a situation may come about in different ways.
Firstly, the substrate may contain a group, situated near
to the exchanging site, which is itself capable of acting

as a catalyst. Thus in the case of o-isobutyrylbenzoic acid the carboxylate group is able to act as a catalyst in the following way:

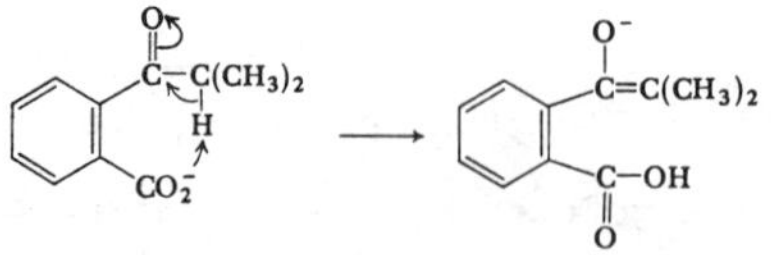

Several examples of <u>intramolecular catalysis</u> have been reported (Table 9) and although some of these refer to halogenation studies it follows from the mechanism of that reaction that similar effects would be observed in corresponding isotopic hydrogen exchange investigations. It is to be expected that intramolecular catalysis would be more important in a stereochemically fixed system than in a more flexible arrangement.

<u>Table 9.</u>   Examples of Intramolecular Catalysis in the
Ionisation of Carbon Acids

System	Comments	Ref.
1. Keto-acids of the form $CH_3CO(CH_2)_nCO_2H$.	The intramolecular rate is a maximum for n = 3 corresponding to a 6-membered cyclic transition state.	89
2. o-Isobutyryl- benzoic acid.	For this rigid system the intramolecular rate is higher than for the first system.	88
3. o-Carboxyaceto- phenone.	In self-buffered solutions the rate can be attributed solely to intramolecular proton transfer from the acetyl to the carboxylate group.	90
4. Derivatives of o-carboxyaceto- phenone.	The efficiency of the intramolecular process is considerably influenced by substituents, the major effect being steric in origin.	91
5. Detritiation of 2´, 6´- Dihydroxy- acetophenone and 2´-Hydroxyaceto- phenone		92

Another kind of catalysis that may affect the stability of the label is that in which the catalyst contains not only a group capable of inducing exchange but also another which

complexes with the substrate in such a way as to make the
complex much more reactive than the free substrate. The principle
behind this kind of **bi(or poly)functional catalysis**, is well
known in enzymology, and can be illustrated by reference to
the work of Hine and coworkers on isotopic hydrogen exchange
in aldehydes and ketones.[93] The latter react rapidly with primary
amine salts to give an iminium ion:-

$$\begin{array}{c}\diagup\\ \diagdown\end{array}\!C \!=\! O \;+\; RN\overset{+}{H}_3 \;\rightleftharpoons\; \begin{array}{c}\diagup\\ \diagdown\end{array}\!C \!=\! \overset{+}{N}\begin{array}{c}^H\\ _R\end{array} \;+\; H_2O$$

If therefore the salts contain a basic group (B) exchange via
the following pathway becomes possible:-

In the iminium ion the neutral oxygen atom has been replaced
by a positively charged nitrogen atom. This more strongly
electron-withdrawing centre makes the hydrogen atoms in the
iminium ion more acidic than those in the original carbonyl
compound. The effectiveness of bi-functional catalysis can be
estimated from the Brönsted equation that holds for mono-
functional catalysis.

Complex formation is also the starting point for **metal-
ion catalysis.** Although there have been many reports, e.g. of
ester hydrolysis and hydration reactions, where metal ions

lead to quite startling rate accelerations, few investigations
involving isotopic hydrogen exchange reactions have been
carried out. In addition to being able to stabilise the transition state through complex formation metal ions can act as
Lewis acids, thereby fulfilling a role analogous to that of
a proton in specific acid catalysis; they can also increase
electrostatic attraction (or minimise repulsions) between a
charged base and the substrate.

In the deuteriation[94] of acetonyl phosphonate (13) magnesium
ions are found to accelerate the rate greatly, probably as a
result of complexation between the metal ion and substrate:-

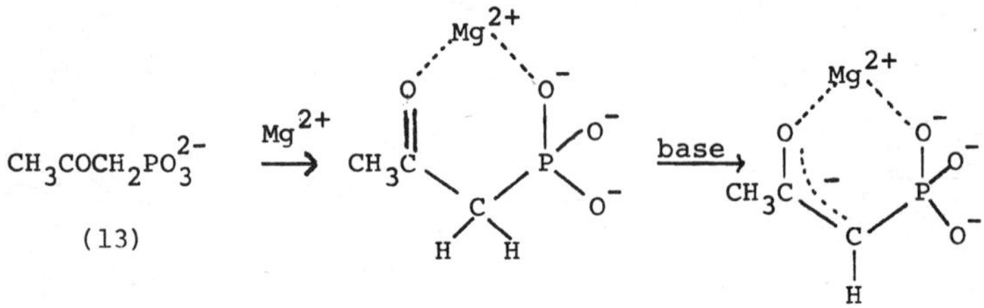

$$CH_3COCH_2PO_3^{2-} \xrightarrow{Mg^{2+}} \quad \xrightarrow{base}$$

(13)

Similar effects have been noted in some halogenation studies
(Table 10).

Heterocyclic compounds frequently contain more than one
site for complex formation and those that have the imidazole
or purine ring system where exchange from the C-2(H) or equivalent C-8(H) occurs are very suitable compounds for studying
the possible effects of metal ions on rates of isotopic hydrogen
exchange. In addition these compounds usually undergo exchange
either as the protonated (14) or neutral form (15) so that
a direct comparison of the reactivity of the metal complexed
form (16) can be made. If the site of complex formation is

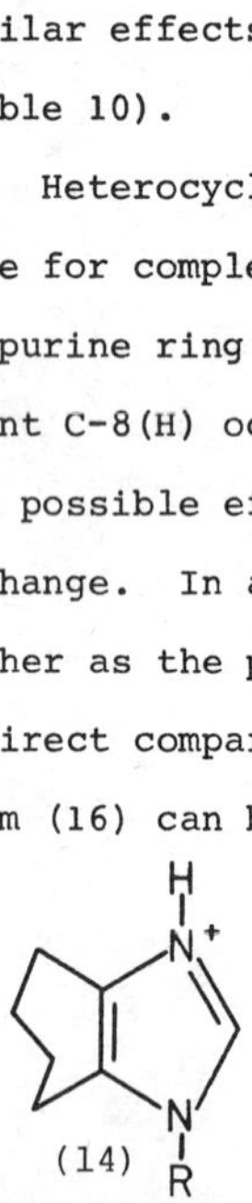

(14) 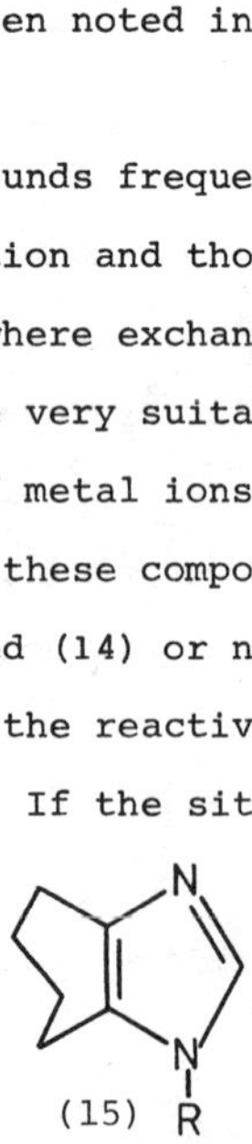   (15) 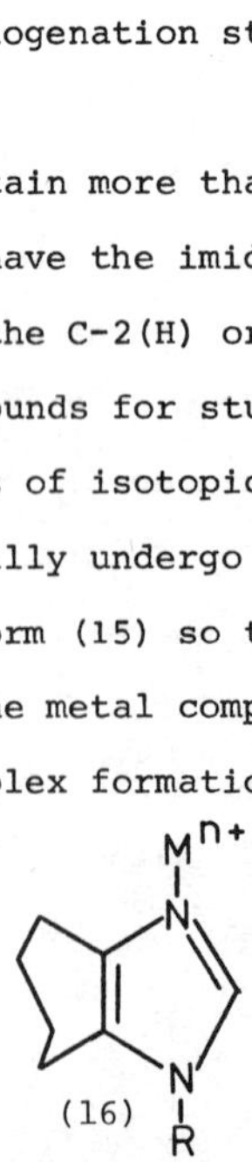   (16)

Table 10. Examples of Metal-Ion Catalysis in the
Ionisation of Carbon Acids

System	Comments	Ref.
1. Halogenation of ethyl acetoacetate and ethyl 2-oxocyclopentane carboxylate in the presence of Cu^{2+}	Effects are small presumably because of weak complex formation.	96, 97
2. Deuteriation of acetonyl phosphonate in the presence of Mg^{2+}	Rate increases of up to 2×10^3 are observed.	94
3. Effect of Mg^{2+} on base-catalysed enolisation of methyl acetonyl phosphonate.	Acceleration for charged bases is about twice that for uncharged bases so that electrostatic effects are probably important but not exclusive.	95
4. Acetate-catalysed iodination of 2-acetyl pyridine	Complex formation with Zn^{2+}, Ni^{2+} and Cu^{2+} leads to large rate enhancements, the catalytic constants being between 5×10^3 and 2×10^5 larger that that of the uncomplexed substrate.	98
5. Dedeuteriation of 1-methyl $[5-^2H]$-tetrazole in the presence of Cu^{2+} and Zn^{2+}	σ complex formation led to large rate enhancement	99
6. Detritiation from $[8-^3H$-1-methyl-inosine and 1-methylguanosine.	The metal complexed forms are between 10^4-10^6 more reactive than the neutral forms but less reactive than the protonated forms.	100

immediately adjacent to that undergoing exchange one would
expect the reactivity of the metal complexed form to be close
to that of the protonated form and this has been found to be
the case for 1-methylinosine and 1-methylguanosine.

The reactivity of the 2-methylene hydrogens of amino acids
is greatly increased as a result of chelation and the principle
can be used to label a range of amino acids. Williams and
Busch[101] were the first to demonstrate the magnitude and specificity
of the effect through their [1]H n.m.r. study of the deuteriation
of Co(III)-chelated glycine and alanine. The lability of
protons at the C-2 position of certain amino acids and peptides
has also been demonstrated for a series of glycine-like Co(III)
chelates.[102] The methylene protons of malonate will also exchange[103]
when the malonate is chelated to Co(III) and deuteriated
aspartic acid has also been prepared using the same principle.[104]
All these observations serve as clear evidence of the electron
withdrawing action of the central metal ion upon the chelate
ring to enhance the lability, as well as the acidity, of the
2-methylene hydrogens.

The reactive intermediates most frequently encountered
in isotopic hydrogen exchange reactions are carbonium ions
and carbanions and the conditions under which these can undergo
rearrangements are fairly well documented.[105,106] Thus for
carbanions examples of 1,3 and 1,5 proton transfers, geometric
isomerism, cyclisation reactions, homoconjugation and sigma-
tropic rearrangements have been reported,[106] nearly always
under conditions of high basicity. There is however one
other kind of rearrangement which because it occurs under
relatively mild conditions is partly responsible for the
biochemist's preference for [14]C (or ^{13}C) labelled compounds.

In attempts to develop an assay for the enzyme phenylalanine
hydroxylase based on the release of tritium from $[4-^3H]$-phenyl-
alanine it was noticed that in the conversion to tyrosine
approximately 95% of the tritium migrated to the 3-position,
with 5% being lost to the medium:

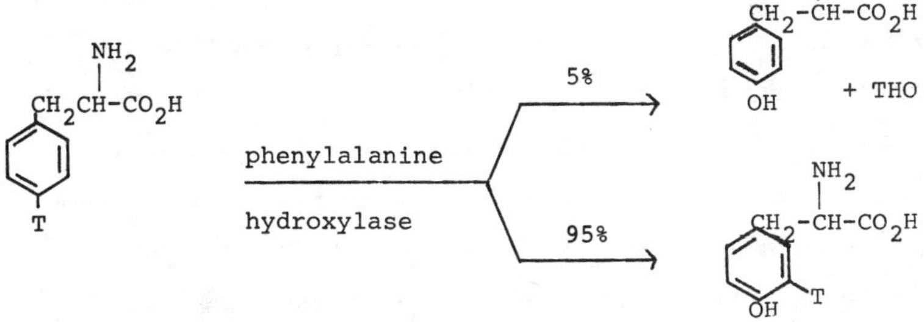

This particular rearrangement was the first example of what
became known as "the NIH shift"; more than a hundred examples
have since been reported.[108,109] Aromatic hydroxylation can
be induced by enzymes such as cytochrome P450, by hydroxyl
radicals (produced via either the Fenton reagent, ultra-violet
radiation of hydrogen peroxide, γ-radiation of water) or by
transition metal-O_2 systems. Although the mechanism has been
the subject of several studies very little in the way of kinetic
investigations has been carried out.

In enzymatic hydroxylation arene oxides are known to be
active intermediates; the subsequent isomerisation to the
phenol usually involves the keto form:

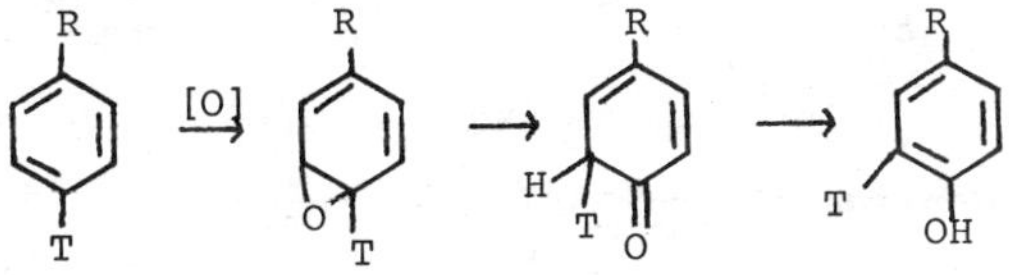

More recently arene oxides have been implicated in the carcin-
ogenicity of certain aromatic compounds; the discovery of the
"NIH shift" has therefore had important implications in both
chemistry and biochemistry.

5. Conclusion.

For most applications of deuterium and tritium labelled
compounds it is necessary to know (a) the degree of isotope
incorporation, (b) the specificity or pattern of labelling and
(c) the stability of the label. All three factors are dependent
on the mechanism of the hydrogen isotope exchange procedure.
It follows therefore that a better understanding of the mechanistic details of various exchange procedures should ensure
that when these compounds are used the dangers associated with
possible loss of label are properly appreciated. The development of ^{2}H and ^{3}H n.m.r. spectroscopy, both with the power
to determine the pattern of labelling and degree of isotope
incorporation, will greatly assist in this respect. An
important corollary is that the attraction of generally
labelled compounds which are so easily prepared by one-step
catalytic procedures is greatly increased. Such studies[110]
have the added benefit of providing details concerning the
selectivity of different catalysts. Fig. 5 gives such examples:

[G-^{3}H] quinoline labelled using heterogeneous platinum is
very uniformly labelled whereas because of steric hindrance
most of the label in [G-^{3}H] lutidine is in the methyl groups.
Labelling of propylbenzene using Raney nickel leads to specific
incorporation into the benzylic position whereas homogeneous
$PtCl_4{}^{2-}$ catalysis of acetophenone exchange leads to general
labelling, with considerable ortho-deactivation.

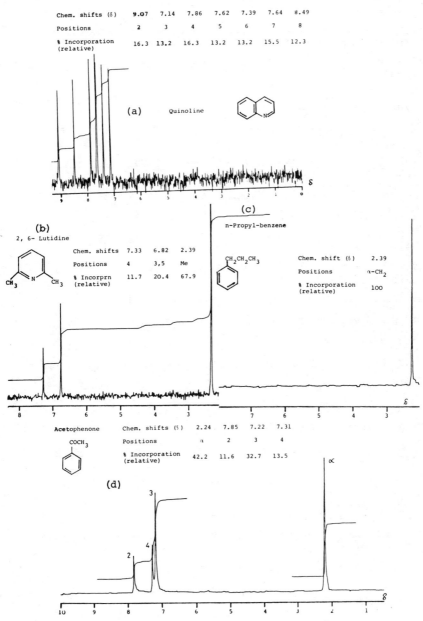

Fig. 5 ^{3}H n.m.r. spectra (^{1}H decoupled) of (a) [G-^{3}H]-quinoline
(b) [G-^{3}H]-lutidine, (c) [α-CH$_2$-^{3}H]propylbenzene and
(d) [G-^{3}H]-acetophenone

References

[1] A.F. Thomas, "Deuterium Labelling in Organic Chemistry",
Appleton-Century-Crofts, New York, 1971.

[2] E.A. Evans, "Tritium and its Compounds". 2nd edtn. Butterworths.
London 1974.

[3] "Isotopes in Organic Chemistry" (ed. E. Buncel and C.C. Lee),
Elsevier Amsterdam, vols 1-4.

[4] J.R. Jones, "The Ionisation of Carbon Acids", Academic Press,
London, 1973, Chapter 1.

[5] H.H. Mantsch, H. Saito and I.C.P. Smith, Progr. in N.M.R.
Spectroscopy, 1973, 11, 211.

[6] G. Lowe and R.F. Pratt, Eur. J. Chem., 1976, 66, 95.

[7] A.J.L. Cooper, J. Biol. Chem., 1976, 251, 1088.

[8] B.G. Cox, F.G. Riddell and D.A.R. Williams, J. Chem. Soc. B,
1970, 859.

[9] A. Streitwieser, jr., D.E. Van Sickle and W.C. Langworthy,
J. Amer. Chem. Soc., 1962, 84, 244, 251.

[10] T. Oshima and N. Tamiya, J. Biochem., 1961, 78, 116.

[11] E. Casadevall and P. Metzger, Tet. Lett., 1970, 48, 4199.

[12] A.J. Kresge and Y. Chiang, J. Amer. Chem. Soc., 1967, 89, 4411.

[13] G.J.Thomas Jr. and J. Livramento, Biochem., 1975, 14, 5210.

[14] D.H. Clague and C. Masters, J.C.S. Dalton, 1975, 858.

[15] E.A. Halevi and F.A. Long, J. Amer. Chem. Soc., 1961, 83, 2809.

[16] A.J. Kresge and Y. Chiang, J. Amer. Chem. Soc., 1961, 83, 2877.

[17] J.R. Jones, R.E. Marks and S.C. Subba Rao, Trans. Farad. Soc.,
1967, 63, 111.

[18] R. Stewart and J.R. Jones, J. Amer. Chem. Soc., 1967, 89, 5069.

[19] J.A. Elvidge, E.A. Evans, J.R. Jones, C. O'Brien and J.C.
Turner, J.C.S. Perkin 11, 1973, 432.

[20] A. Kankaanperä, L. Oinonen and P. Salomaa, Acta. Chem. Scand.,
1977, 31A, 551.

[21] J.R. Jones and S.E. Taylor, to be published.

[22] J.A. Elvidge, J.R. Jones, C. O'Brien and E.A. Evans, J.C.S.
Chem. Comm., 1971, 394.

[23] M. Lajunen and H. Pilbacka, Acta. Chem. Scand., 1976 30A, 391.

[24] K. Perring, Ph.D. Thesis. University of Surrey. 1979.

[25] V.M.A. Chambers, E.A. Evans, J.A. Elvidge and J.R. Jones, Tritium nuclear magnetic resonance (tnmr) spectroscopy. Review no.19. The Radiochemical Centre, Amersham 1978.

[26] C.K. Ingold, C.G. Raisin and C.L. Wilson, J. Chem. Soc., 1936, 915, 1637.

[27] A.J. Kresge and Y. Chiang, J. Amer. Chem. Soc., 1959, 81, 5509.

[28] A.J. Kresge and S.G. Mylonakis, Y. Sato and V.P. Vitullo, J. Amer. Chem. Soc., 1971, 93, 6181.

[29] F. Hibbert, Comprehensive Chemical Kinetics (ed. C.H. Bamford and C.F.H. Tipper), Elsevier, Amsterdam. 1977, 8, 97.

[30] R.Taylor, Comprehensive Chemical Kinetics (ed. C.H. Bamford and C.F.H. Tipper), Elsevier, Amsterdam. 1972, 13, 1.

[31] G.A. Olah and R.H. Schlosberg, J.Amer. Chem. Soc., 1968, 90, 2726.

[32] G.A. Olah, J. Shen and R.H. Schlosberg, J. Amer. Chem. Soc., 1970, 92, 3832.

[33] G.E. Calf, B.D. Fisher and J.L. Garnett, Chem. Comm., 1966, 731.

[34] J.C. Stephens and L.C. Leitch, J. Labelled Cmpds., 1967, 3 65.

[35] N.W. Werstiuk and T. Kadai, Proc. of the 1st. Int. Conf. on Stable Isotopes in Chemistry, Biology and Medicine, May 9-11 1973. Oak Ridge. USAEC p.13-19.

[36] D. Dolman and R. Stewart, Can. J. Chem., 1967, 45, 911.

[37] D.J. Barnes and R.P. Bell, Proc. Roy. Soc., 1970 318A, 421.

[39] J.A. Elvidge, J.R. Jones, C. O'Brien, E.A. Evans and H.C. Sheppard, J.C.S. Perkin II, 1973, 1889.

[40] J.R. Jones and S.E. Taylor, J.C.S. Perkin II, in press 1979.

[41] F. Hibbert and F.A. Long, J. Amer. Chem. Soc., 1971 93, 2836.

[42] J.A. Elvidge, D.K. Jaiswal, J.R. Jones and R. Thomas, J.C.S. Perkin II, 1976, 353.

[43] R.P. Bell and D.M. Goodall, Proc. Roy. Soc., 1973 294A, 273.

[44] F. Hibbert and F.A. Long, J. Amer. Chem. Soc., 1972, 94, 2647

[45] D.W. Earls, J.R. Jones, T.G. Rumney and A.F. Cockerill, J.C.S. Perkin II, 1974, 1806.

[46] J.R. Jones, J. Labelled Cmpds., 1968, 4, 197.

[47] T.S. Chen, J. Wolinska-Mocydlarz and L.C. Leitch, J. Labelled Cmpds., 1971, 6, 285.

[48] J.G. Atkinson, J.J. Csakvary, G.T. Herbert and R.S. Stuart, J. Amer. Chem. Soc., 1968, 90, 498.

[49] J.R. Jones and S.E. Taylor, J.C.S. Perkin II, (in the press).

[50] Ref. 4. ch.2.

[51] G.E. Calf and J.L. Garnett, Adv. in Het. Chem., 1973, 15, 137.

[52] J.L. Garnett and R.J. Hodges, J.C.S. Chem. Comm., 1967, 1001.

[53] N.F. Gol'dshleger, M.B. Tyabin, A.E. Shilov and A.A. Shteinmann, Zhur. Fiz. Khim, 1973, 43, 749.

[54] R.J. Hodges, D.E. Webster and P.B. Wells, J.C.S. Dalton, 1972, 2571.

[55] R.J. Hodges, D.E. Webster and P.B. Wells, J.C.S. (A), 1971, 3230.

[56] J.L. Garnett and R.S. Kenyon, J.C.S. Chem. Comm., 1971, 1227.

[57] J.L. Garnett, M.A. Long, A.B. McLaren and K.B. Peterson, J.C.S. Chem. Comm., 1973, 749.

[58] M.R. Blake, J.L. Garnett, I.K. Gregor, W. Hannan, K. Hoa and M.A. Long, J.C.S. Chem. Comm., 1975, 930.

[59] E.A. Evans, H.C. Sheppard, J.C. Turner and D.C. Warrell, J. Labelled Cmpds., 1974, 10, 569.

[60] O. Buchman and I. Pri-bar, J. Labelled Cmpds and Radiopharmaceuticals, 1978, 14, 263.

[61] G.W. Parshall, Acc. Chem. Res., 1975, 8, 113.

[62] M.A. Long, J.L. Garnett and R.F.W. Vining, J.C.S. Perkin II, 1975, 1298.

[63] M.A. Long, J.L. Garnett and J.C. West, Tet. Lett., 1978, 4171.

[64] C.Mantescu and A.T. Balaban, Can. J. Chem., 1963, 41, 2120.

[65] J.L. Garnett, M.A. Long, R.F.W. Vining and T. Mole, J.C.S. Chem. Comm., 1972,1172.

[66] W.J. Spillane, Isotopes in Organic Chemistry (ed. E. Buncel and C.C. Lee) 1978, 4, 51.

[67] J.P. Colpa, C. Maclean and E.L. Mackor, Tetrahedron, Supplement 2, 1963 19, 65.

[68] M. Yoshida, H. Kaneko, A. Kitamura, T. Ito, K. Oohashi, N. Morikawa, H. Sakuragi and K. Tokumaru, Bull. Chem. Soc.

Japan, 1976, 49, 1697.

[69] A. Kitamura, H. Kaneko, N. Morikawa, K. Oohashi, H. Sakuragi K. Tokumaru, M. Umeda and M. Yoshida, Bull. Chem. Soc. Japan, 1977, 50, 2195.

[70] W.A. Pryor and J.P. Stanley, Intra-Science Chem. Repts., 1970, 4, 99.

[71] V. Gold and M.E. McAdam, Accounts Chem. Res., 1978, 11, 36.

[72] V. Gold and M.E. McAdam, Proc. Roy. Soc., 1975, 346A, 427.

[73] K.E. Wilzbach, J. Amer. Chem. Soc., 1957, 79, 1013.

[74] D.H.T. Fong, J.L. Garnett and M.A. Long, J. Labelled Cmpds., 1972, 8, 695.

[75] D.H.T. Fong, C.L. Bodkin, M.A. Long and J.L. Garnett, Aust. J. Chem., 1975, 28, 1981.

[76] M. Babu and R.B. Johnston, Biochemistry, 1976, 15, 5671.

[77] S. Guggenheim and M.Flavin, J. Biol. Chem., 1969, 244, 6217.

[78] W. Washtien, A.J.L. Cooper and R.H. Abeles, Biochemistry, 1977, 16, 460.

[79] E. Gout, S. Chesne, C.G. Bequin and J. Pelmont, J. Biochem., 1978, 171, 719.

[80] E.V. Dehmlow, Angew Chem. Int.Edtn. 1974, 13, 170.

[81] C.M. Starks, J. Amer. Chem. Soc. 1971, 93, 195.

[82] W.J. Spillane, H. J.-M. Dou and J. Metzger, Tet. Lett., 1976, 2269.

[83] A.J. Fry, "Synthetic Organic Electrochemistry," Harper and Row, New York 1972.

[84] A.F. Diaz, W.P. Chengand M. Ochoa, J. Amer. Chem. Soc., 1977, 99, 6319.

[85] A. Merz and G. Thumm, Tet. Lett., 1978, 679.

[86] S.L. Regen, J. Org. Chem., 1974, 39, 260.

[87] M. Saieed, Ph.D. Thesis, University of Surrey, 1979.

[88] E. Harper and M.L. Bender, J. Amer. Chem. Soc., 1965, 87, 5625.

[89] R.P. Bell and M.A.D. Fluendy, Trans. Farad. Soc., 1963, 59, 1623.

[90] R.P. Bell, B.G. Cox and J.B. Henshall, J.C.S. Perkin II, 1972, 1232.

[91] R.P. Bell, D.W. Earls and J.B. Henshall, J.C.S. Perkin II, 1976, 39.

[92] R.P. Bell and D.W. Earls, J.C.S. Perkin II, 1976, 45.

[93] J. Hine. Acc. Chem. Res., 1978, 11, 1.

[94] R. Kluger and P. Wasserstein, J. Amer. Chem. Soc., 1973, 95, 1071.

[95] R. Kluger and A. Wayda, Can. J. Chem., 1975, 53, 2354.

[96] K.J. Pedersen, Acta Chem. Scand., 1948, 2, 252.

[97] K.J. Pedersen, Acta Chem. Scand., 1948, 2, 385.

[98] B.G. Cox, J. Amer. Chem. Soc., 1974, 96, 6823.

[99] H. Kohn, S.J. Benkovic and R.A. Olofson, J. Amer. Chem. Soc., 1972, 94, 5759.

[100] J.R. Jones and S.E. Taylor, J.C.S. Perkin II, in press.

[101] D.H. Williams and D.H. Busch, J. Amer. Chem. Soc., 1965, 87, 4644.

[102] J.B. Terrill and C.N. Reilley, Inorg. Chem., 1966, 5, 1988.

[103] P.F. Coleman, J.I. Legg and J. Steele, Inorg. Chem., 1970, 9, 937.

[104] W.E. Keyes and J.I. Legg, J. Amer. Chem. Soc., 1973, 95, 3431.

[105] N.C. Deno, Isotopes in Organic Chemistry (eds. E. Buncel and C.C. Lee), Elsevier, Amsterdam, 1975, 1, 1.

[106] D.H. Hunter, Isotopes in Organic Chemistry (eds. E. Buncel and C.C. Lee), Elsevier, Amsterdam, 1975, 1, 137.

[107] G. Guroff, C.A. Reifsnyder, and J.W. Daly, Biochem. Biophys. Res. Comm., 1966, 24, 720.

[108] T. Metsura, Tetrahedron, 1977, 33, 2869.

[109] D.M. Jerina, Chemtech., 1973, 3, 120.

[110] unpublished results, University of Surrey.